Environmentalism and Global International Society

Environmentalism and Global International Society reveals how environmental values and ideas have transformed the normative structure of international relations. Falkner argues that environmental stewardship has become a universally accepted fundamental norm, or primary institution, of global international society. He traces the history of environmentalism's rise from a loose set of ideas originating in the nineteenth century to a globally applicable norm in the twentieth century, which has come to redefine international legitimacy and states' global responsibilities. He shows how this deep norm change came about as a result of the interplay between state and non-state actors, and how the new environmental norm has interacted with the existing primary institutions of global international society, most notably sovereignty and territoriality, diplomacy, international law, and the market. This book shifts the attention from the presentist focus in the study of global environmental politics to the *longue durée* of global norm change in the greening of international relations.

ROBERT FALKNER is an associate professor of International Relations and the Research Director of the Grantham Research Institute on Climate Change and the Environment, at the London School of Economics and Political Science (LSE). He has published widely on global environmental politics and international political economy, including *The Handbook of Global Climate and Environment Policy* (2016) and *Business Power and Conflict in International Environmental Politics* (2008).

Cambridge Studies in International Relations: 156

Environmentalism and Global International Society

Cambridge Studies in International Relations is a joint initiative of Cambridge University Press and the British International Studies Association (BISA). The series aims to publish the best new scholarship in international studies, irrespective of subject matter, methodological approach or theoretical perspective. The series seeks to bring the latest theoretical work in International Relations to bear on the most important problems and issues in global politics.

156 *Robert Falkner*
Environmentalism and Global International Society

155 *David Traven*
Law and Sentiment in International Politics
Ethics, Emotions, and the Evolution of the Laws of War

154 *Allison Carnegie* and *Austin Carson*
Secrets in Global Governance
Disclosure Dilemmas and the Challenge of International Cooperation

153 *Lora Anne Viola*
The Closure of the International System
How Institutions Create Political Equalities and Hierarchies

Environmentalism and Global International Society

Robert Falkner
London School of Economics and Political Science

CAMBRIDGE
UNIVERSITY PRESS

University Printing House, Cambridge CB2 8BS, United Kingdom

One Liberty Plaza, 20th Floor, New York, NY 10006, USA

477 Williamstown Road, Port Melbourne, VIC 3207, Australia

314–321, 3rd Floor, Plot 3, Splendor Forum, Jasola District Centre, New Delhi – 110025, India

79 Anson Road, #06–04/06, Singapore 079906

Cambridge University Press is part of the University of Cambridge.

It furthers the University's mission by disseminating knowledge in the pursuit of education, learning, and research at the highest international levels of excellence.

www.cambridge.org
Information on this title: www.cambridge.org/9781108833011
DOI: 10.1017/9781108966696

First published 2021

A catalogue record for this publication is available from the British Library.

ISBN 978-1-108-83301-1 Hardback

To Kishwer and Sophia

Contents

Acknowledgements

Most academic work is a collective effort, so it is only right to start with an acknowledgement of the support and encouragement that I received from colleagues and friends during the writing of this book.

Andrew Hurrell, my doctoral supervisor at Oxford University, first introduced me to English School (ES) theory in the 1990s, and the conversations I had with him back then still resonate with me. Little did he (or I, for that matter) anticipate at the time that I might end up writing an ES book on the rise of environmentalism – after all, my doctoral research was focused on the international political economy of the global environment. But such is the power of ideas, good ideas, that they linger in your mind and spark new thoughts at unexpected moments. I am grateful for all the advice, inspiration and support that Andy has provided throughout all these years.

My second introduction to the English School came many years later, after I had joined the Department of International Relations at the London School of Economics and Political Science (LSE). Working alongside Barry Buzan, including as associate editor (with Barry and Kim Hutchings) of the *European Journal of International Relations*, I came to appreciate Barry's critical role in the 'reconvening' of the English School. His 2004 book *From International to World Society?* sparked in me a renewed interest in ES theory and influenced my thinking about long-term normative change in international relations. In recent years, I have had the fortune of working with Barry on several projects in which we explore the fluctuating fortunes of certain fundamental norms in international society. This collaboration has been enlightening and fun in equal measure. I have learnt a great deal from Barry (including mind mapping) and suspect that I shall never be able to repay my intellectual debt to him.

There are many colleagues and friends who I have had the benefit of working with over the years and who have helped me develop the ideas that went into this book project, through joint research projects or publications, or simply in conversations in and around seminar rooms.

I would like to thank, in particular, Ken Abbott, Helmut Anheier, Alina Averchenkova, Steven Bernstein, Frank Biermann, Sander Chan, Jeff Chwieroth, Jennifer Clapp, Mick Cox, Peter Dauvergne, Robyn Eckersley, Sam Fankhauser, Fergus Green, Aarti Gupta, Tom Hale, Veerle Heyvaert, Kathy Hochstetler, Matt Hoffmann, Bob Keohane, Mathias Koenig-Archibugi, Markus Lederer, Michael Mason, James Morrison, Pete Newell, Joana Setzer, Henry Shue, Hannes Stephan, Hidemi Suganami, Stacy VanDeveer, Leslie Vinjamuri, David Vogel, John Vogler, Andrew Walter, Peter Wilson and Steve Woolcock. I am also grateful to Barry Buzan for commenting on a first full draft of the book. I fear I may have left out some but hope they will forgive me.

I would like to thank the following institutions for giving me an opportunity to present some of the arguments from this book project: Cardiff University, the Grantham Research Institute on Climate Change and the Environment and the International Relations department at LSE, the Munk School of Global Affairs and Public Policy at Toronto University, Oxford University, Reading University and the 2017 Annual Convention of the International Studies Association in Baltimore.

I owe two institutions a special debt for their support: the International Relations department at LSE, both for material support and a period of sabbatical leave to get me started on this project, and the Grantham Research Institute on Climate Change and the Environment, which I joined in 2017 as its research director and where I was able to complete the book.

I am grateful to the wonderful cohort of PhD students that I have had the fortune of supervising at LSE: Christopher Wright, Carola Kantz, Jonas Meckling, Nico Jaspers, Richard Campanaro, Michael Bloomfield, Kai Monheim, Robyn Klingler-Vidra, Philip Schleifer and Marian Feist, and now Carlotta Clivio, Eunjeong Park and Katharina Kuhn. PhD supervision should be a two-way street, and time and again I've come away deeply enriched by the experience of working with so many gifted young scholars.

Special thanks go to two anonymous reviewers and the editors of the Cambridge Studies in International Relations, Evelyn Goh, Christian Reus-Smit and Nicholas Wheeler, for their helpful and constructive comments on the proposal and manuscript. Thanks are also due to John Haslam and his team at Cambridge University Press for steering me so ably through the publication process.

Since this is a book about expanding humanity's moral horizon to the non-human environment, I hope I might be forgiven for thanking at least some of the animals and landscapes, wild and cultivated, that have provided me with inspiration and solace during the writing process:

Bavaria's Wörthsee, the olive groves of Umbria, Brompton Cemetery, Fuji the rescue cat, and the orangutans of Sabah's Danum Valley. It is one of the peculiarities of human psychology that we often only know the true value of something when it's lost, and the ecological costs of this psychological disposition are all too evident from our destructive relationship with the natural environment. For all the material prosperity that humanity aspires to, let us not forget that we will always need Thoreau's 'tonic of wildness', the sense of being part of a larger community of living beings and inanimate objects. We owe this community a duty of care.

Finally, my greatest debt is to my family, Kishwer and Sophia, for their unwavering support, patience and above all love. I dedicate the book to both of them.

Abbreviations

ASEAN	Association of Southeast Asian Nations
ASPCA	American Society for the Prevention of Cruelty to Animals
CAN	Climate Action Network
CBD	Convention on Biological Diversity
CBDR	common but differentiated responsibilities
CBDR-RC	common but differentiated responsibilities and respective capabilities
CERES	Coalition for Environmentally Responsible Economies
CITES	Convention on International Trade in Endangered Species
CLRTAP	Convention on Long-range Transboundary Air Pollution
COP	Conference of the Parties
CSD	Commission on Sustainable Development
ECA	export credit agency
ECOSOC	United Nations Economic and Social Council
EEC	European Economic Community
EPA	Environmental Protection Agency
ES	English School
EU	European Union
FAO	Food and Agriculture Organization
FCO	Foreign and Commonwealth Office
FSC	Forest Stewardship Council
GATT	General Agreement on Tariffs and Trade
GCF	Green Climate Fund
GDP	gross domestic product
GDR	German Democratic Republic
GEF	Global Environment Facility
GEP	global environmental politics
GHG	greenhouse gas
GIS	global international society

GM	genetically modified
GMO	genetically modified organism
GONGO	governmental and non-governmental organisation
GPM	great power management
GRI	Global Reporting Initiative
ICBP	International Council for Bird Protection
ICC	International Criminal Court
ICRW	International Convention for the Regulation of Whaling
IFC	International Finance Corporation
IIED	International Institute for Environment and Development
INGO	international non-governmental organisation
IOPN	International Office for the Protection of Nature
IPCC	Intergovernmental Panel on Climate Change
IR	International Relations
ISO	International Organization for Standardization
ITCPN	International Technical Conference on the Protection of Nature
IUCN	International Union for Conservation of Nature and Natural Resources
IUPN	International Union for the Protection of Nature
IWC	International Whaling Commission
MARPOL	International Convention for the Prevention of Pollution from Ships
MDB	multilateral development bank
MEA	multilateral environmental agreement
MNC	multinational corporation
MPA	marine protected area
NAAEC	North American Agreement on Environmental Cooperation
NAFTA	North American Free Trade Agreement
NASA	National Aeronautics and Space Administration
NATO	North Atlantic Treaty Organization
NEPA	National Environmental Policy Act
NGFS	Network of Central Banks and Supervisors for Greening the Financial System
NGO	non-governmental organisation
ODS	ozone-depleting substances
OECD	Organisation for Economic Co-operation and Development
OEEC	Organisation for European Economic Co-operation

OSCE	Organization for Security and Co-operation in Europe
PCBs	polychlorinated biphenyls
POPs	persistent organic pollutants
PPP	public–private partnership
PPP	polluter pays principle
PRI	Principles for Responsible Investment Initiative
RSPB	Royal Society for the Protection of Birds
RSPCA	Royal Society for the Prevention of Cruelty to Animals
RTA	regional trade agreement
SAICM	Strategic Approach to International Chemicals Management
SBSTTA	Subsidiary Body on Scientific, Technical and Technological Advice
SDGs	Sustainable Development Goals
SPNR	Society for the Promotion of Nature Reserves
SPWFE	Society for the Preservation of the Wild Fauna of the Empire
TCFD	Task Force on Climate-related Financial Disclosures
UN	United Nations
UNCCD	United Nations Convention to Combat Desertification
UNCED	United Nations Conference on Environment and Development
UNCHE	United Nations Conference on the Human Environment
UNCSD	United Nations Conference on Sustainable Development
UNECE	United Nations Economic Commission for Europe
UNEP	United Nations Environmental Programme
UNESCO	United Nations Educational, Scientific and Cultural Organization
UNFCCC	United Nations Framework Convention on Climate Change
UNSC	United Nations Security Council
UNSCCUR	United Nations Scientific Conference on the Conservation and Utilization of Resources
WCS	World Conservation Strategy
WSSD	World Summit on Sustainable Development
WTO	World Trade Organization
WWF	World Wildlife Fund (until 1986), World Wide Fund for Nature (thereafter)

1 Introduction
The Greening of Global International Society

This book explores a profound transformation in international relations: the adoption of environmental stewardship as a fundamental international norm. At the first United Nations (UN) environment conference in 1972, international society declared 'the protection and improvement of the human environment' (Stockholm Declaration) to be the duty of all governments. This was the first time that states collectively accepted a normative commitment to protect the environment, not only within their territories but also at the global level. In subsequent decades, international society followed up this commitment with hundreds of international environmental treaties and created several international organisations dedicated to supporting global environmental protection. This flurry of environmental diplomacy in the last third of the twentieth century contrasts with the preceding half a century of failed efforts to establish environmental protection as an international policy priority. Today, hardly a day passes in the diplomatic calendar without some international forum discussing environmental threats or negotiating global response measures. It is no exaggeration to say that environmental matters have become omnipresent in international relations. If the nineteenth century was the age of nationalism, and the twentieth century the age of democracy, then the twenty-first century may well turn out to be the 'age of ecology' (Radkau, 2011).

In this book I seek to show that the dramatic expansion of global environmental politics (GEP) since the 1970s is not simply a case of yet another collective action problem being added to the list of global policy issues. It signals both a profound shift in the role and identity of states across the world and a significant step in the normative evolution of *global* international society (GIS). Environmentalism has become a fundamental international principle – or primary institution in English School (ES) parlance – that suggests the beginning of a transformation in international legitimacy. Framing the emergence of GEP in this way has important consequences for our understanding of the relationship between environmentalism and international relations. It opens up a new perspective on how environmental ideas have reshaped the normative order of international society; whether environmentalism has strengthened, weakened or modified existing primary institutions (e.g. sovereignty, territoriality, diplomacy); but also how environmentalism in turn has been influenced and changed by its engagement with GIS.

From an International Relations (IR) perspective, the story of the rise of environmental stewardship is an encouraging one. When viewed within the context of the slow pace of societal development in the international system, understood as the redefining of the moral purpose of the state (Reus-Smit, 1999) and a shift in the foundations of international legitimacy (Clark, 2005), it can be argued that the greening of international society amounts to a significant and comparatively rapid transformative change. The rise of global environmentalism is a distinctive case of international normative discontinuity that has few equals in the twentieth century. This is all the more remarkable as the adoption of environmentalism was not a response to some systemic deficiency in the society of states. States did not, initially at least, accept a duty of global environmental care in order to stabilise the international balance of power or prevent international order from collapsing. Global environmentalism, understood as a social movement that initiated green normative change in international society, arose out of the normative maelstrom of domestic politics, first in the most industrialised economies and then in other parts of the world. Its roots can be traced back to the first environmental organisations of the nineteenth century, though it only developed international political salience after the Second World War, against the backdrop of the looming legitimacy crisis of the nation-state as the guardian of society's well-being and prosperity. As I shall demonstrate subsequently, the international norm of environmental stewardship emerged as a new social purpose, first in domestic politics and then in international relations, growing out of a world societal demand to tame the ecological excesses of global industrialism.

This progressive account of the rise of environmental stewardship looks less persuasive, however, when viewed in the context of the worsening global environmental crisis. The 'great acceleration' (McNeill and Engelke, 2016) of humanity's detrimental impact on the planet since the mid-twentieth century has severely tested the problem-solving capacity of both states and international society. Despite having created numerous multilateral environmental agreements and introduced environmental mandates into other parts of the global governance system, GIS has not managed to curtail, let alone reverse, some of the worst forms of environmental degradation: from the global climate crisis (United Nations Environment Programme, 2018) to mass biological extinction (Ceballos, Ehrlich and Dirzo, 2017), and from the continuing destruction of tropical forests (Food and Agriculture Organization, 2018) to the dumping of plastic waste in the oceans (Dauvergne, 2018), the society of states seems to be at a loss when it comes to addressing some of the most pressing global environmental problems. Indeed, when measured against the scale of the planetary ecological challenge that humanity faces, it would seem that international society is the wrong place to look for a global environmental rescue.

How, then, can we square these seemingly contradictory observations about environmentalism's progress in the international realm? There can be little doubt that, despite some isolated success stories, GIS has repeatedly fallen short of environmental expectations. If the international normative structure has started to be 'greened', as I argue in this book, then this hasn't gone nearly far enough. Indeed, it remains an open question whether the international states system can develop an effective and timely response to the climate change threat, species extinction, resource depletion or biodiversity loss. As I shall discuss in this book, many environmentalists place their hopes, not in international society, but in world society, that is the myriad of non-state actors that have become engaged in the search for global environmental solutions. Yet, given the persistence of the states system and fierce international political contestation around environmental issues, it is clear that international cooperation among states will have to be part of any global environmental response. The environmental crisis may call into question the legitimacy of the states system, but no alternative to the current world political system is currently available. There are many ways in which the current international approach to environmental action can be improved, by strengthening international organisations, boosting environmental aid, enhancing international fairness and justice, increasing institutional transparency and accountability, and improving non-state actor participation in international processes. Yet, most such advances in global environmental governance require the acquiescence or actual support of international society, and powerful states can easily hold up progress towards greater environmental sustainability. It matters, therefore, how GIS is constituted, and how its normative structure can be adapted. The analysis in this book seeks to enhance our understanding of how international environmental norm change has come about in the past, in the hope that this might improve our ability to accelerate it in the future.

This book has been in the making for many years. I first applied ES theory to the rise of global environmentalism in a paper for a special issue of *International Affairs* published twenty years after the Rio 'Earth Summit' (Falkner, 2012). Working subsequently with Barry Buzan on a project to trace the emergence of new primary institutions, in environmental politics (Falkner and Buzan, 2019) and in global political economy, we developed an analytical framework for empirically tracking the rise of new fundamental norms and assessing their interactions with the existing international order. This book builds on and extends the framework developed by Falkner and Buzan (2019), providing a fuller historical account of the rise of environmental stewardship and introducing further nuances into the story of environmentalism's interaction with GIS.

International Transformations and the English School

The study of change and continuity is at the heart of the IR discipline. Change is, of course, ubiquitous in international politics as much as in all social life. However, profound transformations in the normative structure of international society are rare by comparison (Holsti, 2004: chapter 1). Ever since the emergence of the Westphalian order in the seventeenth century, the modern society of states has been characterised by remarkable continuity in some of its key constitutional elements: the principles of sovereignty and territoriality, which still define contemporary statehood; the rules of diplomacy, which continue to regulate the behaviour of states towards each other; and the operation of the balance of power, which gives order to interstate relations when power is unequally distributed or the power distribution shifts. As Bull and Watson (1984) and others have shown, these fundamental institutions of the Westphalian international order originated in Europe and were gradually globalised, particularly through the creation of colonial empires and then decolonisation. As Europe's international society became a truly global international society, many of its normative foundations remained largely intact.

This is not to say that the international normative structure doesn't change. Far from it, key principles have either changed their meaning, declined in importance (e.g. war; see Buzan, 2004: 196) or disappeared altogether (e.g. dynastic succession; see Buzan, 2004: 246), and new principles have emerged that have had a lasting impact on international legitimacy (e.g. nationalism; see Mayall, 1990). Yet other new norms have struggled to develop the kind of momentum that would make them a candidate for primary institution status in GIS (e.g. human rights; see Buzan and Schouenborg, 2018: 94). Change at the level of international society's constitutional structure is thus possible, though it is likely to be a slow-paced, drawn-out and contested process.

Many theories of IR struggle with the notion of deep-seated normative change. Realists assume that anarchy, the predominant structural feature in international relations, has remained unaltered over centuries, and that within an anarchic system the main changes occur at the level of the distribution of power capabilities of states. Powers rise and fall, and international order oscillates between multipolarity, bipolarity and unipolarity, but the underlying logic of international behaviour remains the same. The (potential) emergence of environmentalism as a normative principle that expands the core purpose of the state and affects what it means to be a legitimate member of international society does not feature in this theoretical perspective. States may well pursue environmental objectives if these support their national interest, but environmentalism as such is not expected to alter the structure of the international system. The only conceivable way in which environmentalism could become an imperative for the system as a whole is the emergence of a global ecological catastrophe that poses a systemic risk to the survival of a sovereignty-based international order. Much like the threat of a major asteroid strike from space, an environmental apocalypse that threatens the survival of both states and the international system would necessitate a collective response that pushes aside all concern for relative gains and power imbalances.

Liberal and constructivist IR theories do a much better job at accounting for the greater variety of social purposes that can be absorbed into the normative structure of international society. Liberals and constructivists place greater emphasis on the role of ideas and norms, on the possibility for states and state leaders to learn their way out of calamitous international anarchy and on the role of domestic politics within states to change the parameters of behaviour among states. Both liberals and constructivists are able to adopt an evolutionary view of international society in which the arrival of new norms gives rise to the possibility of a restructuring of its normative order. Liberals remain wedded to a rationalist outlook, which treats ideas and norms as intervening factors that do not affect the identity of the state, thereby leaving the nature of the international system largely unchanged by the arrival of new ideas. It is constructivists that have gone furthest in developing a deeper understanding of how normative change can bring about a change in the identity of states themselves, and therefore an evolution in the international structure of international relations.

For a study of the causes, drivers and impacts of long-term and deep-seated norm change in international relations, which is at the heart of my project, liberalism and constructivism offer useful starting points. However, it is the ES that provides a unique, and in my view ideal, vantage point from which to analyse the arrival of a new norm and its interaction with other established norms of international society. With its focus on the societal dimensions of international relations and the patterns of long-term historical change, it brings into play the normative dimensions of international life and is particularly sensitive to the malleability and historical situatedness of fundamental norms. Its conceptual triad of international system, international society and world society offers a comprehensive framework for examining dynamics of social action within, but also between, these pillars of the global political system. Although the ES used to be associated almost exclusively with the concept of international society and the question of why and how states form a society, at least since Buzan's (2004) reworking of ES theory it has become clear that it is better seen as a master theory that seeks to work out how the three pillars are related, how they interact and how they influence each other (Dunne, 2010: 733–4).

The ES distinction between primary and secondary institutions helpfully distinguishes different levels of international change (Buzan, 2004; see also Holsti, 2004). Within the IR subfield of GEP, analysts usually focus on the creation of secondary institutions, that is international environmental treaties, public or private governance mechanisms and international organisations, and usually only within specific environmental issue areas (such as climate change, biodiversity and chemicals). Rarely does the GEP literature examine the broader historical and normative pattern of institutional development in the environmental field as such. By distinguishing between environmentalism as a primary institution and specific environmental regimes or regime complexes as secondary institutions, ES theory opens up a perspective on how deeply embedded the environmental norm has become in international relations, how it informs the creation of specific secondary institutions and how it relates to other primary institutions that make up international society's constitutional order.

ES theory is also more explicitly normative in its orientation than either liberal or constructivist IR theory. To be precise, normative theory in the ES tradition comes in two forms: a philosophical or ethical tradition that seeks to determine 'the right or the good or the proper form of action' and a sociological or anthropological form that discerns 'the norms or practices of a particular society' in a descriptive sense (Mayall, 2009: 210). The ES tradition has combined both types of normative reasoning, which allows it to develop a much richer discussion of societal development in international relations. Its conceptual dyad of pluralism and solidarism has been used as a set of markers for different normative positions on the desirability of different types of international order. It can also be employed as part of an analytical–empirical account of international societal evolution, as a set of criteria for measuring and evaluating normative development in international relations. As such, it opens up a fruitful perspective on the greening of international society, which both provides analytical categories for describing empirical developments in GEP and helps to bring to the surface some of the implicit normativity in GEP debates.

It was once not uncommon to dismiss the ES as marginal or irrelevant to the study of GEP (Paterson, 2005). This perception was fuelled not least by the ES's early neglect of environmental issues and Bull's influential but reductionist treatment of the rise of environmentalism as a fundamentally anti-statist force that, if successful, would render the states system obsolete. The situation has changed, however, in more recent years, with a growing number of ES scholars engaging with the rise of GEP and reflecting on its significance as a source of normative development in international relations (see Chapter 2). GEP has now emerged as a major empirical case for ES debates on pluralist versus solidarist trends in international society and for the respective roles of states and non-state actors and the intensifying interplay between world society and international society. As I hope to show in this book, the ES can benefit from closer analysis of the rise of environmentalism, as an important test case for progressive societal development. At the same time, scholars of GEP would benefit from greater engagement with the ES's theoretical tradition, which allows them to distinguish between different levels of environmental normative development and place the rise of GEP in a larger historical context. In short, it is time to move beyond what until recently was a case of mutual neglect between ES and GEP scholarship in IR. This book seeks to set the scene for a closer and mutually beneficial engagement between ES and GEP scholarship.

The Argument and Structure of the Book

The argument that I develop in this book can be summarised as follows. Environmentalism has emerged as a fundamental norm of GIS. The origins of the gradual greening of international society can be traced back to the early twentieth century, but it was not until the 1970s that environmental stewardship was accepted by the society of states as a primary institution, as an integral part of its normative structure. Environmental protection was first declared to be the duty of all states at the first UN environment conference in Stockholm in 1972. The new environmental norm overcame initial contestation and reached universal recognition by the time of the 1992 Rio Earth Summit. Much like other fundamental norms and political ideologies, environmental stewardship contains an inner normative core that is stable over time and a set of peripheral concepts and principles that are adjustable and relate the core norm to specific historical contexts. At its core is an ecological consciousness and concern, that is an awareness of the threat that human activities pose to the health of the planet and an ethic of care for the natural environment that translates into an environmental responsibility for states and international society. Beyond this inner core, environmentalists differ over questions of ethical motivation (whether environmental concern is based on anthropocentric or ecocentric values) and political strategy (whether effective environmental protection requires radical changes to the international system or mere adaptive reforms). The precise meaning of environmental stewardship and how it is to be applied in international relations remain contested, and in this sense the international environmental norm is malleable and open to change. The gradual greening of international society is thus a story of how states came to accept a fundamental commitment to global environmental protection and how they struggled over its precise meaning and relationship with other fundamental norms of GIS.

The rise of environmental stewardship is an important case of international norm transfer from world society to international society. Environmental ideas first arose in domestic and transnational societal debates in the nineteenth century and slowly morphed into a loosely connected global environmental movement. From the early 1900s onwards, environmental campaigners and scientists lobbied governments on numerous occasions to take up the task of global environmental protection. Several attempts to create an international environmental agenda were launched in the first half of the twentieth century, but these largely failed. Only with the rise of the modern environmental movement in the 1960s did sufficient political momentum build behind the internationalisation of environmental protection. The norm transfer from world to international society was not a straightforward process, it was negotiated and contested, and it required leadership by powerful states that championed global environmental protection. Its success is all the more remarkable as environmental stewardship did not arise out of a systemic need to maintain international order. Global environmentalism reflects a new social purpose that international society came to adopt, in response to norm entrepreneurship originating outside the state-centric realm.

Once adopted as a fundamental norm, environmental stewardship underwent a process of social consolidation and globalisation. This involved ongoing contestation and resistance, as well as normative accommodation and change. Over time, however, international society's commitment to environmental protection hardened, even if the implementation of international environmental rules has been patchy and their environmental effectiveness remains uncertain. As the commitment to environmental stewardship deepened, the interplay between the environmental norm and other primary institutions of GIS has intensified. Environmental stewardship has provided a good fit with some fundamental norms, leading to an expansion of diplomacy and international law, while it has had little or no impact on other primary institutions, such as the balance of power, war, great power management and nationalism. The international environmental norm has posed a challenge to some primary institutions that stand in the way of a more internationally centralised response to environmental problems, most notably sovereignty, territoriality, the market and developmentalism. As such, the environmental norm has undergone considerable normative accommodation, which has enabled it to establish itself as part of the international normative structure. It has also worked against strict interpretations of certain pluralist institutions, such as sovereignty and territoriality, contributing to a certain reinterpretation, though not a fundamental transformation, of the meaning of modern sovereign statehood.

The book is divided into three main parts. The *first part* sets out the theoretical and conceptual contexts within which my analysis is situated. Chapter 2 introduces the main tenets of the ES. It discusses how ES theory has engaged with environmental issues in the past, moving from a near total neglect of environmentalism in its early stages to the development of an ever more comprehensive framework for studying environmentalism as a source of international normative change. The chapter concludes with a summary of the analytical framework that underpins this book.

Thereafter, Chapter 3 reviews the origins and evolution of environmentalism as a set of ideas, as a political ideology in its own right and as a social and political movement that has reshaped politics in the twentieth century and beyond. The first part introduces the main variants of environmentalism that have emerged out of the broad tradition of environmental thinking in the nineteenth century. The second part considers how environmentalism, an ideology not originally concerned with international relations, came to be applied to questions of international order. Using the ES's conceptual dyads of pluralism and solidarism, and international society and world society, four ideal types of a global green order are identified: *Green Westphalia* and *global environmental governance* representing the pluralist and solidarist ends of the spectrum within state-centric international society, and *eco-localism* and *eco-globalism* as the corresponding concepts in world society.

The *second part* of the book takes a historical perspective on the emergence of environmentalism as a fundamental norm of international society. Chapter 4 traces the origins of the environmental movement and its gradually expanding impact on national politics in the nineteenth century. It reviews the largely failed efforts to establish an international agenda for environmental protection, from the pre-First World War years to the League of Nations, highlighting the limitations of an environmental movement that still lacked broader mass support across the industrialised world.

Chapter 5 examines the historical context in which international society came to adopt environmental stewardship as a primary institution. It opens with a review of the post-1945 situation, when international society took the first tentative steps towards a greater environmental role. It then charts the emergence of the modern environmental movement from the 1960s onwards, and specifically its impact on national environmental policy in leading industrialised countries. The chapter concludes with an analysis of the 1972 Stockholm conference, the constitutional moment in the greening of international society. Stockholm was the moment when environmental stewardship was adopted as a fundamental international norm, but it also witnessed deep divisions, mainly between developed and developing countries, over how the norm was to be interpreted.

Chapter 6 considers how, once formally adopted, the new environmental norm evolved into a truly global primary institution, applicable not only in the Global North but also in the countries of the Global South. The chapter traces the consolidation of environmental stewardship as a primary institution, through the creation of a large number of secondary institutions and through the institutionalisation of environmental diplomacy.

Chapter 7 completes the set of historical chapters by following the evolution of GEP into the contemporary era, identifying processes of both normative consolidation and contestation. It explores the interaction between international and world society in strengthening the applicability of environmental stewardship in new forms of transnational environmental governance involving non-state actors. And it examines continuing areas of contestation over how environmental stewardship is to be interpreted and how the environmental norm relates to other primary institutions.

The *third part* of the book abandons the chronological structure of part two and adopts an analytical focus on the underlying drivers of the greening of international society. Engaging the conceptual toolkit of the ES, Chapter 8 considers the impetus in GEP for a strengthening of the solidarist elements within the state-centric order, from cosmopolitan norms in international law to shifts in the meaning of primary institutions such as sovereignty and territoriality.

Chapter 9 examines the extent to which pluralist constraints continue to curtail normative development in GEP, but also considers the potential for a pluralist logic of coexistence to contribute to a strengthening of international environmental cooperation. The focus here is on the securitisation of the global environment, particularly as existential environmental threats emerge as a challenge to international order and stability, and the potential role that great power responsibility and management might play as part of GEP.

Chapter 10 focuses on non-state actors in GEP and the interaction between world society and international society more generally. It examines the contribution of world society actors as norm entrepreneurs and as drivers of the greening of international society; it focuses on their involvement in the diplomatic process of creating the secondary institutions that govern the environmental policy field; and it considers the reality and potential for environmental action through world society-based transnational environmental governance. The chapter pays close attention to the pluralist and solidarist foundations of world society and also explores the potential for the current interplay between world society and international society to intensify, leading to greater convergence and integration between the two (so-called integrated world society).

Chapter 11 summarises the main arguments of the book and reviews the significance of the findings for ongoing theoretical and empirical debates in ES and GEP scholarship. It charts an agenda for future research, on the normative transformations in international society, on norm transfer and interaction between world and international society, and on the potential for a strengthening of environmentalism as a fundamental norm in international relations.

Part I

Theory

2 English School Theory and Global Environmental Politics

The English School (ES) of International Relations (IR) provides the main theoretical and conceptual framework for the analysis in this book. Given its focus on the societal dimensions of international relations and long-term historical change, ES theory is particularly well suited for investigating the drivers and impacts of the rise of global environmentalism in global international society. This chapter proceeds in three stages: The first section introduces the conceptual landscape of the ES. Given space constraints, I can only provide a brief overview of those core tenets that are of particular relevance to the analysis in this book. Fortunately, several full-length accounts of the ES tradition exist that interested readers will find useful (Dunne, 1998; Linklater and Suganami, 2006; Buzan, 2014). The second section reviews how various ES authors have interpreted the rise of global environmental politics, moving from an initial neglect of environmental issues among the founders of the ES to a more comprehensive engagement with global environmentalism in recent years. Building on this, the third section outlines the analytical framework that will inform the empirical analysis in the remainder of the book. Readers familiar with the ES tradition may wish to skip the first section of this chapter and go straight to the second section, which focuses on the link between ES and environmental scholarship.

The English School, International Society and Normative Change

The origins of the ES can be traced back to the late 1950s, when an eclectic group of scholars and practitioners of international diplomacy began to hold regular meetings under the auspices of the British Committee on the Theory of International Politics (Dunne, 1998: chapter 5). The purpose of these meetings was to develop a new approach to IR that conceived of interstate relations as social relations, in contrast to the mechanistic notion of an international system that dominated North American IR in the post-war era. The members of the British Committee came from diverse disciplinary backgrounds – international relations, history, philosophy and theology – but were united by the core conceptual idea of an international society, with rules, norms and social practices governing the behaviour of states. Ironically, it was only in the 1980s that this group came to be referred to as a coherent school of thought, at a time when its distinctive theoretical approach was already seen to be on the wane (Jones, 1981). By the 1990s, however, the situation had reversed and the ES re-emerged as a major theoretical tradition in IR, helped not least by the rise of other sociological perspectives and the growing interest in the role of ideas and norms more generally (Checkel, 1998; Wendt, 1999). Having been 'reconvened' in the late 1990s (Buzan, 1999), the ES today counts among its members a growing number of scholars from around the world, including also in many non-Western countries (Zhang, 2003). Whether 'English School' is still an appropriate name for what has increasingly become a global approach to IR theorising is a legitimate question, but alternative names ('classical approach'; 'international society tradition') have so far failed to dislodge the ES label.

Societal Dimensions

The idea of an international society is the ES's core conceptual contribution to IR (Buzan, 2004: 1). In Bull's famous words, a society of states 'exists when a group of states, conscious of certain common interests and common values, form a society in the sense that they conceive themselves to be bound by a common set of rules in their relations with one another, and share in the working of common institutions' (Bull, 1977: 13). In contrast to the mechanistic view that characterises rationalist approaches, the ES and other social theories of IR draw our attention to the norms and rules that make up the social context in which states operate. The ES thus brings into play not just power and wealth as the central attributes of actors, but also the 'normative vocabulary of human conduct' (Jackson, 1992: 271), such as rights and customs, membership and recognition, and equality and reciprocity. International society and its members are highly interdependent in that the members, that is states, shape international society and are also shaped by it (Buzan, 2004: 8). This notion of the mutual constitution of structures and agents, a common feature of both ES and constructivist theory (Hurd, 2008), has important consequences for the study of long-term international change. It leads us to consider not just how international actors bring about change in the normative structure of international society (as is common in the IR literature on norms and norm entrepreneurs), but also how new norms in turn alter the behaviour and identity of those actors.

It was once common to identify the ES exclusively with the 'international society' concept and contrast it with other IR theories that made 'international system' (e.g. realism) or 'world society' (e.g. liberalism, critical theory) their conceptual core. To some extent, this is indeed how early ES theorists situated themselves in the theoretical landscape, as can be seen in Wight's (1991) famous 'three traditions' framework, consisting of the 'three R's': realism, rationalism and revolutionism (see also Cutler, 1991). This categorisation no longer makes sense, however. In fact, as Buzan (2004) has shown, the ES today offers a more comprehensive theoretical lens through which we can study the entire range of international orders, from international system to international society and world society. Rather than treating them as mutually exclusive, it asks how they are interrelated and how they interact. The system–society distinction, once the defining difference between Realism's mechanistic system thinking and the ES's sociological perspective, has itself become increasingly questionable (Buzan, 2014: 171–2). An anarchical international system cannot do without some basic norms, such as sovereignty (Onuf, 2002: 228), while all international societies are underpinned by a basic system-type interaction quality (James, 1993). Likewise, recent ES theorising has begun to explore the social dynamics that weave international and world society together, exploring processes of norm transfer from one to the other (Clark, 2007). More recent ES contributions therefore go beyond Wight's rigid distinction of three domains, opening up the possibility of investigating the relative strength of system versus society dynamics in interstate relations, the interplay among the three domains and areas of overlap and potential merging (Buzan, 2004: 257–61).

To be sure, the ES has much in common with other social theories of IR and forms part of a wider sociological shift in the discipline. However, it possesses certain characteristics that enable it to make a distinctive contribution to the study of long-term normative change: First, it places the study of international order and change in a comparative historical framework. By tracing international society's evolution from its European origins to international expansion and beyond, it creates a framework for comparing different international societies across time and space. Second, the ES combines this historical framework with a distinctly normative perspective. It establishes a set of criteria that allow scholars to evaluate societal development from a normative point of view, thereby connecting empirical analysis and normative theorising within one theoretical tradition in IR.

Historical and Normative Dimensions

The historical study of the evolution of international society has been central to the ES from the beginning, and it remains at the heart of much of its empirical–analytical project today. Many of the core ES texts draw on historical cases to develop their theoretical position: Bull (1977) explores the emergence of a Christian international society in early modern Europe and its transformation into a global international society as the basis for his pessimistic outlook on its future. Bull and Watson (1984) bring together a large number of case studies of individual nations joining or challenging international society, from the Ottoman Empire to China, Japan and India, which provides the backdrop for their own framework of the expansion of international society. As Buzan (2014: chapter 5) points out, Bull and Watson's expansion story has animated a great deal of subsequent ES scholarship: on how the Westphalian states system established distinctive fundamental institutions (sovereignty, war, balance of power, diplomacy, international law, great power management); how the expansion of colonial empires and the later process of decolonisation led to the globalisation of these institutions; and how this process helped to integrate developing countries into an increasingly global international society still characterised by Western attributes. Over time, the ES focus on the expansion of Europe's Westphalian order has gradually given way to a broader research agenda on the historically bounded characteristics of different international societies – across time, by going back in history beyond the roots of Westphalia (Watson, 1992; Suzuki, Zhang and Quirk, 2014); and across space, leading to the study of societal formations from the global level down to the regional and subregional (Diez and Whitman, 2002; Buzan and Gonzalez-Pelaez, 2009; Buzan and Zhang, 2014). This historical project is very much alive today, with a new focus on the globalisation of international society challenging the earlier expansion story (Reus-Smit and Dunne, 2017) and the question of the strength and weakness of GIS today receiving greater attention (Buzan and Schouenborg, 2018).

The ES is also distinctive in its explicit normative orientation, which goes beyond a purely historical account of how the constitutional structure of international society has evolved. In this regard, the classical ES follows in a long tradition of philosophical reflection on the good life and how to structure a good society and state (Williams, 2011). It has created a unique conceptual vocabulary to assess the legitimacy of international order and the desirability of the particular values and institutions that it is built on. Centred on the pluralism–solidarism distinction (Bain, 2014), this vocabulary allows the ES to debate the right balance between particularist and universal values, and between the need to protect the existing international order and the desire to change it in order to promote global justice. It also provides normative criteria for assessing particular instances of deep-seated international change that (at least partially) challenge the existing order, be it the rise of nationalism (Mayall, 1990) or human rights (Vincent, 1986). ES theorists don't necessarily agree on how to assess the consequences of such normative change. Indeed, the pluralist–solidarist debate is often portrayed as being about such fundamental normative differences: Pluralists are usually fearful of the destabilising effects of rapid and radical norm change and defend an international order based on national sovereignty, non-intervention and the protection of value diversity. In contrast, solidarists are more likely to embrace historical change that makes it possible to advance cosmopolitan values, even if that comes at the cost of undermining the Westphalian roots of international society. The ES thus may be rife with internal disagreements about the desirability of international change, but it manages to bring the often hidden normative biases in IR research to the surface and makes them the subject of focused and structured debate.

Pluralism versus Solidarism

The pluralism–solidarism distinction serves two purposes. On the one hand, it defines the two principal positions in the ES's long-standing normative debate. As discussed earlier, this debate is about the right balance between the preservation of order and the pursuit of justice in international relations, and the trade-offs that exist between pluralist and solidarist futures. On the other hand, the pluralism–solidarism debate also operates at the empirical–analytical level, setting out markers of societal development in international relations (for an overview, see Hurrell, 2007: chapters 2 and 3; Williams, 2005; and Buzan, 2004: 45–62). When used as an empirical–analytical tool, pluralism and solidarism define different stages in the evolution of international society: pluralist international society operates with a minimal set of norms focused on preserving the status quo, while solidarist international society is based on more comprehensive and demanding norms that go beyond minimalist coexistence to promote the common good. The ES thus provides analytical categories for a comparative perspective on societal development in international relations, distinguishing between 'thin' and 'thick' international societies (Buzan, 2004: 59) and comparing them across time and space. This perspective also yields a framework for assessing the emergence of new fundamental norms (e.g. human rights, environmentalism), their interaction with other existing fundamental norms and thus their overall impact on international order.

It is important to note in this context that pluralism and solidarism, although conceptually distinct and separate, coexist in reality. In contrast to the prevalent normative framing of a 'great conversation' in the ES, which has tended to view pluralism and solidarism as mutually exclusive states of affairs, the empirical–analytical perspective leads us to treat them as two different elements of international order that usually coexist and overlap. Simply put, 'world order is and always has been both pluralist and solidarist' (De Almeida, 2006: 68). The parallel existence of pluralist and solidarist elements can be seen in the mixture of primary institutions that define different types of international societies, with some (e.g. territoriality, balance of power, war) more closely identified with the former, and others (e.g. international law, democracy) representing the latter. In the same way, we can conceive of newly emerging norms as promoting pluralist (e.g. nationalism in the nineteenth century) or solidarist (human rights in the twentieth century) principles, interacting with the existing mix of primary institutions and pushing international order in one direction or another.

Primary and Secondary Institutions

The pluralist–solidarist divide reflects the ES's predominant interest in understanding the processes of long-term international change, particularly in the context of the expansion story of European international society. The question of international change – its direction and scope, chances of success and underlying causes – is of course central to most major IR theories and debates. However, as Holsti (2004: 6) points out, IR lacks a consensus on 'not only *what* has changed but also on *how we can distinguish* minor change from fundamental change'. In this regard, the ES's major contribution is to direct our attention to reconfigurations in the underlying normative structure of international society as manifestations of fundamental change. The first generation of ES scholars may not have been explicit enough in defining this focus, but Holsti's (2004) and Buzan's (2004) reworking of ES theory has helped to clarify the ES's theory of international change as one of deep-seated institutional change. The institutions that both Holsti and Buzan have in mind are the fundamental norms that define the basic order of international society, rather than the treaties or organisations that states create to deal with issue-specific cooperation problems. Such fundamental change can take many different forms. At one end of the spectrum, it concerns the large-scale and long-term transition of global order, be it from international system to international society, or from international society to world society. At the other end, it can involve the emergence, decline or disappearance of particular institutional features within an otherwise static international society. Watson's (1992) historical interest in the expansion and evolution of international society falls into the former category. Vincent's (1986) discussion of human rights and Mayall's (1990) study of the rise of nationalism tend to fall into the latter. Clark's (2005) exploration of how international legitimacy and the conditions for legitimate membership in international society have evolved can be seen to encompass both types of international change.

The distinction between fundamental and issue-specific institutions, arguably the ES's central conceptual innovation in the theory of international change (Knudsen, 2018: 25), is well established even if early ES theorists (Bull, 1977; Wight, 1977) were not explicit and careful enough in their terminology. In *The Anarchical Society*, Bull (1977: 67) speaks of 'fundamental' or 'constitutional' institutions (e.g. sovereignty, diplomacy, balance of power, war) that make up the deep social structure in interstate relations. Liberal IR theorists similarly speak of 'fundamental practices' (Keohane, 1988: 385), which are to be distinguished from international regimes that seek to manage international cooperation in issue-specific areas only. Both Holsti (2004) and Buzan (2004) have since clarified the language around fundamental and issue-specific institutions, which have become known as 'primary' and 'secondary' institutions in ES parlance. Buzan (2014: 16–17) defines primary institutions as

> deep and relatively durable social practices in the sense of being evolved more than designed. These practices must not only be shared amongst the members of international society, but also be seen amongst them as legitimate behaviour. Primary institutions are thus about the shared identity of the members of international society. They are constitutive of both states and international society in that they define not only the basic character of states but also their patterns of legitimate behaviour in relation to each other, and the criteria for membership of international society.

Secondary institutions, by contrast, refer to intergovernmental arrangements that serve specific purposes in given issue areas. They include international treaties and organisations (or 'regimes' in liberal IR) and are distinctive in that they are purposefully created rather than evolved.

Dimensions of International Change

This distinction is important for the understanding of international order and change in this book. If we want to assess the rise of environmentalism as a (potentially) fundamental change in international society, we need to focus on its impact on primary institutions as the main markers of such change. Of course, any new norm is also likely to bring about change in secondary institutions. In fact, given the rapid expansion of secondary institutions since they first appeared in the late nineteenth century (Buzan, 2014: 17), we are likely to find more instances of change at the level of secondary rather than primary institutions. However, while it may be easier to alter existing or introduce new secondary institutions, change at this level tends to be less consequential for the long-term evolution of international society. In contrast, international change at the level of primary institutions tends to be less frequent but more consequential when it occurs. For example, creating a new international organisation to monitor states' compliance with an environmental treaty may be a significant achievement of international diplomacy, but is of limited impact on states' behaviour if the underlying norm of sovereignty, which restricts the ability of any international body to enforce compliance, remains unchanged. In contrast, a comprehensive reinterpretation of states' duties towards each other, and the conditions for legitimate membership in international society, can have profound consequences for the level of compliance with treaty obligations that can be achieved in international society.

The distinction between primary and secondary institutions is central to the ES theory of fundamental international change, but it is worth emphasising that the two institutional levels are closely connected and influence each other. Primary institutions express the fundamental norms that constitute international order, and as such set the normative context within which states create and alter international regimes and organisations. In this sense, secondary institutions reflect the underlying normative principles of international society. At the same time, secondary institutions are not wholly derivative of primary institutions. They reproduce international society's underlying order, but changes in secondary institutions can also initiate or contribute to deep-seated normative change. As sites of social reproduction, secondary institutions are also the focal points for normative contestation around fundamental norms. As international regimes and organisations emerge, evolve, decay or die, such changes at the level of secondary institutions leave their traces in the underlying normative structure of international society. Primary institutions are thus the key markers of major international change, but it is in the evolving relationship between primary and secondary institutions that processes of change usually play themselves out.

If newly emerging norms are to become enduring features of international society's constitutional structure, then we should expect them to become primary institutions in their own right. But norm change involves tensions and trade-offs, between existing and emerging norms, and between those that seek to bring about change and those that are keen to prevent it. Successful normative change can be expected to bring about some degree of resolution to those conflicts, but as is the case with all social systems international society too is characterised by ongoing tensions between its different normative components. The rise of nationalism illustrates the contested nature of all such fundamental norm change: The idea of the nation as the ultimate source of sovereign state authority (popular sovereignty) came to the fore in the nineteenth century out of a series of clashes with competing norms of political legitimacy (dynasticism, religion, empire). By the twentieth century, it had become firmly established and led to the creation of new states as part of a broader push for self-determination and decolonisation. In doing so, it successfully marginalised the long-established principle of dynastic succession, redefined sovereign statehood as a core principle of Westphalian order and entered into a long-standing and still unresolved clash with the liberal norm of open market-based economic exchange (Mayall, 1990). In similar fashion, we can expect an emerging norm of environmentalism to sit uneasily alongside some primary institutions (e.g. sovereignty, territoriality, market), potentially challenging them or leading to their reinterpretation. Enduring normative tensions are thus part and parcel of any process of fundamental international change, and we can interpret the presence and intensity of such tensions as one indicator of the strength or weakness of international society at any given point in time (Buzan and Schouenborg, 2018: 202).

The ES also opens a perspective on the question of what ordering or regulative work primary institutions perform in international society. Bull spoke of 'fundamental' or 'constitutional' institutions that define the normative order of international society, which have since become known as primary institutions. But whereas Bull failed to differentiate between different types of primary institutions, others have suggested a sub-categorisation that specifies their different functions. Holsti (2004: 24–25) speaks of 'foundational' institutions that define the identity of states (e.g. sovereignty) and 'procedural' institutions that regulate their interactions (e.g. diplomacy). This echoes Ruggie's (1998: 22) distinction between 'constitutive' and 'regulative' rules. Reus-Smit (1999) pushes in the direction of constitutive rules and adds a more foundational level of international order by introducing the moral purpose of the state as underpinning all primary institutions. Allan (2018) has recently gone one step further and introduced the notion of 'cosmological ideas' that constitute discourses on state purpose. In contrast, Schouenborg

(2011) stepped back from the effort to introduce a deeper hierarchy of institutions and instead proposes a more pragmatic, functional–institutional, approach. He distinguishes five functional categories (legitimacy and membership; regulating conflicts; trade; authoritative communication; international organisation) that allow us to categorise primary institutions according to the ordering and regulative work they perform.

World Society

The conventional complaint about the ES has been that its state-centric perspective has led it to ignore the growing importance of non-state actors currently operating in world politics(Paterson, 2005). There is some truth to this, certainly as far as the first ES generation is concerned. However, a better interpretation would be that while the ES has always considered world society to be an essential part of its conceptual triad, the very notion of world society has suffered from theoretical neglect and conflicting conceptual uses. As Little (2000: 411) notes, the idea of world society is 'the most problematic feature' of the ES's three master concepts. Clark (2007: 6) refers to it as 'notoriously slippery', especially when compared to the better-established concepts of international system and international society. More recently, however, the world society concept has gone through a kind of renaissance (Pella, 2013). Several representatives of the English School's second generation have rethought the world society concept as part of the ES's intellectual project, proposing new ways to turn it into the focus of a productive research agenda (Buzan, 2004, 2018; Clark, 2007).

The traditional ES focus on world society has tended to reduce the concept to a mere cipher for the actual or imagined community of humankind, with individuals seen as the main counterpoints to states as (potential) units of a global political community. Representing the human level in international relations, world society either signifies the unrealised alternative to a state-centric international society (Bull, 1977) or represents the endpoint of a historical transformation towards the vision of cosmopolitan world community, as in Wight's (1991) 'Revolutionism'. In this perspective, the world society concept has strong normative overtones, representing a solidarist vision for realising cosmopolitan values, while its empirical reality either remains unexplored or is assumed to be insignificant.

More recently, a number of ES authors have reworked the concept of world society to resolve some of these ambiguities and limitations. Clark's work (2007) is significant as he moves the discussion from normative deliberation (à la Revolutionism) to the empirical level. Through analysis of historical turning points after major conflicts, when the international order was renegotiated during peace conferences, he demonstrates the high level of interaction and mutual influence between states and non-state actors. Clark thus shows how world society actors, defined as individuals and non-governmental organisations (NGOs), have shaped the evolution of the normative order of international society, though his approach conforms with the classical notion of world society championing a cosmopolitan international order, based on values such as racial equality, human rights and democracy (Clark, 2007). Ralph (2007) similarly traces the cosmopolitan influence of world society on the society of states, in his case via the solidarist institution of the International Criminal Court.

Buzan goes one step further by removing the world society concept from its use in the English School's normative debate and establishing it as part of his structural version of ES theory. In *From International to World Society?* (2004), Buzan disaggregates world society into 'transnational society' (populated by non-state collective actors) and 'interhuman society' (comprising individuals) alongside interstate society, a terminology that has not been widely adopted in ES circles. More recently, he has proposed an alternative categorisation of three versions and uses of world society: *normative world society*, which comes close to Bull's normative notion of the 'great society of humankind' and Buzan's interhuman societies that are based on large-scale collective identities; *political world society*, which captures the role played by the many organised non-state actors that operate outside international society but seek to influence it; and *integrated world society*, which serves as an umbrella term to include all major domains, 'creating an ideal-type for a prospective future' (Buzan, 2018: 130). Normative and political world society have an identifiable empirical presence and historical roots, while integrated world society remains a normative aspiration or teleological endpoint, and is thus less relevant from an empirical–analytical perspective (Buzan, 2018: 131).

The ES debate on world society has made considerable progress in clarifying the concept and establishing it to its rightful place as a master concept alongside international society and international system. World society is far from being a settled concept, however, and debate continues about the relationship between international and world society, the social institutions that lend structure to world society and the link between normative and empirical uses of the concept (Stivachtis and McKeil, 2018). For the context of this study, it suffices to note the shift towards a more structural understanding of world society in ES theory and beyond, which helps set up an analytical focus on the various roles that world society actors are playing in processes of fundamental change in international society. As will be discussed subsequently, the rise of environmentalism and its impact on the norms underpinning the society of states requires us to put the social enmeshment between world and international society at the centre of the analysis.

Methodological Questions

Methodological questions have never featured prominently in classical ES theorising, as some critics have lamented (Finnemore, 2001; Jones, 1981); they have either been ignored or marginalised, or subsumed in some general commitment to non-positivist eclecticism (Navari, 2014: 205–6). As the ES has consolidated into an increasingly coherent research tradition more recently, it has generated growing interest in establishing its methodological foundations (see the contributions to Navari, 2009). There is thus greater methodological reflexivity to be found in ES writing today (Buzan, 2014: 22–3), even if it remains the case that its distinctive perspective on international relations is 'united not by particular methods' but by a 'concern with certain central questions' (Hurrell, 2001: 489). It should be noted that, in epistemological terms, most ES work sides with the *Verstehen* tradition in the social sciences (Linklater and Suganami, 2006: 101), which has consequences for the methodological choices ES authors make. Its central premise of the existence of a social realm that is constituted by its own rules and intersubjective meanings leads to an empirical focus on understanding the meaning that actors attach to social interactions. Unsurprisingly, therefore, the ES tends to eschew the imperative of developing scientific–causal models to explain actors' behaviour, as is central to the *Erklären* tradition in the social sciences (Hollis and Smith, 1991). This is not to suggest that ES authors ignore causal analysis (Linklater and Suganami, 2006: 101), but to highlight that they tend to operate with different models of causality. In social systems, normative structures influence agents and vice versa, and it is the two-way logic of mutual constitution rather than the linear logic of cause–effect that characterises ES research (Buzan, 2014: 22).

Methodological pluralism, whether by design or default, continues to define the ES. Scholars interested in comparing the structural features of different international societies across space and time work with the comparative historical method (Buzan and Little, 2000; Little, 2009). Those that focus on the diplomatic practices and norms that characterise particular international societies will need to develop 'insider accounts of social reality' (Wilson, Peter, 2012: 580). They will want to establish and scrutinise the assumptions, understandings and motives of those actors at the heart of diplomacy, 'statespeople' as Robert Jackson (2009) calls them, whether they focus on text, speech or law (Navari, 2014: 213). And scholars concerned with developing the normative dimensions of ES theory are engaged in a merging of the sociology of norms with explicit normative theory, or practical reasoning about 'the scope of moral action in a world of sovereign states' (Reus-Smit, 2009: 72). ES theory thus engages and relies on established methodologies in other disciplines and fields of study, from history and law to sociology and philosophy.

As will be discussed subsequently, my objective in this book is to develop a theoretically grounded account of the emergence of a new fundamental norm in international society – environmental stewardship – and to explore its interaction with the existing international normative structure. As such, this places my project on the analytical–empirical side of the ES tradition, although I will seek to engage its normative tradition when examining the social consolidation and evolution of environmentalism in international relations. My approach in this book is not dissimilar from constructivists' efforts to establish the existence and significance of international social structures (cf. Checkel, 1998: 334): to document the emergence of a new fundamental norm; to establish the close connection between the new norm and other institutional features of international society; to identify a correlation between the new norm and shifts in states' behaviour and identity; and to examine the interaction and mutual shaping between the emerging and existing international norms.

Having introduced the core tenets of ES theory and reviewed its distinctive contribution to the study of fundamental international change, the next section provides an overview of how ES authors have dealt with the rise of environmental politics in international relations, from initial neglect towards reluctant engagement and now a more fully developed research programme on environmentalism as a driver of international normative change.

English School Perspectives on Global Environmental Politics

Initial Neglect of Environmental Issues

It has taken the ES longer than some other IR traditions to develop a sustained focus on global environmental politics. For much of its early history, ES authors tended to ignore international environmental issues.

This was certainly the case with Charles Manning, Herbert Butterfield, Martin Wight and Adam Watson, and to a lesser extent also Hedley Bull. The situation began to change in the 1990s, especially after the 1992 Rio Earth Summit, when the first ES authors started to take note of the rise of global environmentalism, though well into the 2000s we find scholars bemoaning the English School's neglect of environmental themes. Bellamy (2005: v) states in the introduction to *International Society and Its Critics* that there had been 'virtually no engagement with [...] environmentalism', and Linklater and Suganami (2006: 2) similarly observe that global environmental politics is one of the '[p]ast areas of neglect on the part of the English School' (see also Buzan, 2004: 186). Unsurprisingly, therefore, Dunne's (1998) history of the ES, which traces its evolution from the British Committee to Bull's and Vincent's later works, does not mention environmental issues at all.

The reasons for this neglect of environmentalism are not difficult to fathom. As far as the first generation of ES theorists is concerned, most of them made their main intellectual contributions at a time when environmental protection was not yet established on the international agenda. Wight had worked out his mapping of the three 'Rs' of international theory by the 1950s; Manning's key text *The Nature of International Society* was published in 1962; and the first collection of papers produced by the British Committee appeared only a few years later (Butterfield and Wight, 1966). As will be discussed in later chapters, environmentalism only began to leave a lasting mark on international relations from the late 1960s. Until after the 1972 UN Conference on the Human Environment (UNCHE), few if any IR scholars paid serious attention to various states' efforts to deal with transboundary environmental harms. The intellectual foundations of the ES had thus already been laid by the time that environmental stewardship was emerging as a primary institution of international society.

Could it be that the early ES's neglect of environmentalism has deeper, intellectual, reasons? That a theoretical tradition concerned mainly with a state-centric agenda of international order and security was blind to the beginnings of a profound normative change that was initiated and influenced by non-state actors? Undoubtedly, the origins of the ES are closely tied up with the intellectual climate of the Cold War in the 1950s, when most IR scholars were preoccupied with great power conflict and the threat of all-out nuclear war. Much like post-war Realists, the early ES theorists took questions of how to maintain order in an international society dominated by a few great powers to be the central focus of the discipline. To be sure, the British Committee was made up of an eclectic group of scholars and diplomats, with backgrounds in history, theology, philosophy and law, but their thematic concerns all revolved around what can be described as the 'high politics' of the Cold War era: international anarchy, the balance of power, diplomatic practice, the use of force and disarmament. By contrast, environmentalism, still a predominantly domestic issue in the post-1945 era, did not fit easily into this intellectual terrain. Even when ES authors started to widen their empirical focus from the 1970s onwards to include novel normative developments based on cosmopolitan values, it was human rights and not environmentalism that was at the centre of ES interest. A younger generation of ES authors (Vincent, 1986; Dunne and Wheeler, 1999; Wheeler, 2003) largely built their case for a more solidarist international order on the expanding scope for humanitarian intervention. As Buzan (2004, 43) notes, this exclusive focus on human rights 'blocked off any other considerations of what might constitute either solidarist international society or cosmopolitan world society', from the rise of the market norm to global environmentalism.

The Environmental Challenge and Bull's Pluralist Response

Hedley Bull's *The Anarchical Society* is the first major ES text to address environmentalism's arrival on the international stage. Published in 1977, five years after the first UN environmental conference in Stockholm, the book contains a brief but influential passage on environmentalism that was to have considerable impact on how ES theory framed discussions of environmentalism in IR – and how the ES was perceived by scholars of GEP. Bull is not interested in the rise of global environmental issues per se – there is no mention of the Stockholm conference, the newly created UN Environmental Programme (UNEP) or the environmental treaties that had been negotiated after Stockholm. Instead, Bull engages with environmentalists' arguments in order to consider – and ultimately dismiss – the then widely held view in environmental circles that the states system was increasingly incapable of dealing with the urgent global challenges that humanity faced. Bull was writing against the background of 1970s environmental discourses that viewed state-centric politics with scepticism and proposed alternative world orders along cosmopolitan (Falk, 1971) or authoritarian (Ophuls, 1977) lines. Richard Falk's *This Endangered Planet* was a particularly influential text at the time, and Bull uses it as a foil to advance his own argument in defence of state-centric international order.

Bull makes three interconnected points in support of his state-centric and pluralist perspective (see Falkner, 2017b, for a more extensive discussion of Bull's views): *First*, because international society is pluralist in nature, in terms of the diverse societal values and interests that various states represent, it is unlikely to create, let alone sustain, deep international cooperation in the pursuit of common objectives, including environmental ones. *Second*, the existence and stability of state-centric international society is in itself a precondition for finding effective environmental solutions: 'Without such a basis of minimum order it is scarcely possible that common issues of the environment can be faced at all' (Bull, 1977: 294). *Third*, Bull concedes that in the long run state-centric solutions may not be sufficient to deal with environmental dangers. This is the closest Bull comes to a solidarist position on global environmentalism, but he quickly returns to his pluralist theme: It is only through the states system that 'a greater sense of human solidarity in relation to environmental threats may emerge' (Bull, 1977: 294). In sum, pluralist international society is a necessary, if not sufficient, condition for tackling global environmental problems, and environmentalists' demands for deep interstate cooperation, let alone a central international authority, to rein in global environmental destruction are bound to go unfulfilled.

Bull's engagement with the rise of global environmentalism is fairly limited. Rather than opening up an ES perspective on environmentalism as a transnational movement and source of normative change, it ultimately serves to close down the English School's engagement with GEP. Bull neither discusses the empirical record of international environmental cooperation in the 1970s nor does he consider the emerging role of environmental NGOs as non-state actors. Having dismissed Keohane and Nye's (1971) transnationalist research agenda (Bull, 1977: 277–81), he sees no need to explore the transformative agency and normative power of the environmental movement. Bull's primary interest is in countering any call for the creation of a larger political system – be it a world state or a global political community – that would advance universal values and serve common interests such as environmental protection. *The Anarchical Society* provided a blueprint for later pluralist writings on GEP, but by failing to engage with the full breadth of the environmental challenge it narrowed rather than broadened the ES perspective: by framing environmentalism as an essentially anti-statist political force and dismissing it as a politically naive project, Bull foreclosed a more serious engagement by ES scholars with the rapidly expanding field of GEP research; and by reinforcing the then widespread perception of the ES as narrowly state-centric, Bull made it far too easy for GEP scholars to dismiss the ES as an irrelevant theoretical tradition.

Environmentalism and the ES's Second Generation

Of the second generation of ES theorists who followed in the footsteps of Bull, Hurrell was the first to develop a more sustained interest in environmental politics. Writing in the early 1990s, against the backdrop of the 1992 Rio Earth Summit and a peak of global environmental concern, Hurrell produced pioneering IR scholarship that helped connect the ES with environmental scholarship in IR (Hurrell, 1994, 1995; Hurrell and Kingsbury, 1992a). Hurrell echoes Bull's state-centric premise when declaring that 'the difficulties of inter-state co-operation must still constitute the starting-point for any study of the prospects for global environmental management' (Hurrell and Kingsbury, 1992b: 5). Critically, however, he goes beyond Bull's narrow pluralist agenda when exploring the legitimacy crisis that the Westphalian states system faces as it struggles to address global environmental problems. For Hurrell, the ecological crisis points to three profound deficiencies of a state-centric international system: the fragmentation of political authority into a large number of sovereign nation-states is unlikely to produce the kind of globally coordinated response that the ecological crisis requires; a growing number of nation-states are proving themselves to be incapable of providing adequate environmental governance solutions at the domestic level; and the ecological crisis is eroding the normative appeal of the nation-state and its role as the primary focus of human loyalties (Hurrell, 1994, 2007: 216–17).

Hurrell thus takes Bull's scepticism as the starting point for his own analysis but develops the outlines of a solidarist critique of state-centric pluralism. While noting that 'there is little chance of escaping from the centrality of the state' (2007: 235), he nevertheless accepts that there are 'very good reasons as to why so much discussion of the ecological challenge is couched in terms of a move "beyond sovereignty"' (Hurrell, 2007: 224). He warns that conflicts over values and perceptions of justice make it difficult to realise the common interest in GEP (2007: 231) but acknowledges that the rise of environmentalism presents a serious challenge to 'the practical viability and the moral acceptability of state-based pluralist international order' (2007: 222). Hurrell is keenly aware of the limitations of cosmopolitan solidarism in international politics but points to the proliferation of non-state actors as a sign of the emergence of complex new forms of global governance beyond the states system (Hurrell, 2007, 224–8). In a sense, Hurrell performs a delicate balancing act between the pluralist and solidarist instincts of the ES when considering the evolving field of GEP. The potentially transformative developments that are occurring within GEP need to be acknowledged even though 'the scale of changes that have actually taken place' should not be overstated (2007: 236).

At around the time that Hurrell was exploring international society's legitimacy challenges against the backdrop of a global environmental crisis, Reus-Smit and Jackson began to assess environmentalism's transformative potential for the international normative order. Reus-Smit develops the idea of a green moral purpose of the state that is emerging in the late twentieth century, although he concedes that the results of this 'ideological revolution [...] remain unclear' (1996: 119). Jackson (1996) similarly detects a normative shift in international society towards stronger environmental values. He introduces the very concept of environmental stewardship, or trusteeship, which he further elaborates in his book *The Global Covenant* (2000: 175–8), and interprets it as a new international norm that has 'come into view which invokes the responsibilities of national leaders for the health of the planet: responsibility for the global commons' (Jackson, 2000: 176). Jackson agrees with Bull that a pluralist international society need not stand in the way of a significant degree of international cooperation on environmental matters. National sovereignty does not stop states from 'collaborative action to deal with joint or common problems' (Jackson, 1996: 176), and the expansion of international law and treaty-making in the late twentieth century testifies to that.

What marks out Jackson's argument is that he not only conceives of environmentalism as an emergent primary institution (although he does not discuss it in those terms) but also identifies the new environmental norm as a responsibility that applies to state leaders. Whereas Reus-Smit locates environmentalism in the deep constitutional structure of international society, Jackson ties it to the moral landscape that state leaders operate in when they justify their actions. Jackson's work, in particular, serves to build a bridge from Bull's narrow pluralist position to a normatively more ambitious solidarist perspective that reimagines the state as 'a servant of the common good of the earth' (Jackson, 2000: 176). The care for the planet points to an environmental ethic that reaches well beyond human-centric responsibilities as defined in traditional IR discourses. Jackson still harbours doubt about the strength of this moral revolution (2000: 177–8), but makes it clear that even a pluralist international society is not static in normative terms.

Linklater (2011) takes a further step towards a solidarist interpretation of environmentalism in his work on the prevention of transboundary harm. Echoing the distinction between pluralism and solidarism in ES debates, he distinguishes between international and cosmopolitan harm conventions, with the former contributing to order and constraining force in interstate relations and the latter aiming for the protection of individuals and social groups against a wide range of harms. In that environmentalism has given rise to a 'new moral and political vocabulary', it pushes international society to be 'concerned not with "national interests" or international order but with the well-being of the species and the fate of future generations' (Linklater, 2011: 37). Environmentalism thus enlarges the moral landscape for international relations by including non-human species, and it has also begun to embed in international institutions and law 'a cosmopolitan version of the harm principle that includes duties to avoid damaging the commons' (Linklater, 2011: 40; see also Elliott, 2006). In this way, environmentalism becomes a progressive force in international normative development, from the narrow horizon of pluralist order towards a more encompassing notion of solidarist ecological justice.

An Emerging ES Research Programme on Global Environmentalism

In recent years, the coverage of global environmental issues within the ES has expanded significantly, to a point where GEP is now becoming more widely studied as a substantial empirical test case for claims about normative development in international relations. Falkner (2012), Buzan (2014: 161–3) and Falkner and Buzan (2019) trace the rise of global environmentalism as a primary institution in international society and identify its potential to elevate state-centric solidarism to greater prominence in international relations. Palmujoki (2013) explores the growing institutional fragmentation of global climate governance and its implications for international societal development. Rather than implying a pluralisation of GEP, Palmujoki (2013: 196) identifies a 'move from international society to world society and post-national perspectives'. In his interpretation, the diversification of climate action – horizontally from the climate regime to other international forums and vertically from states to the non-governmental sphere – suggests a strengthening and broadening of climate protection as a fundamental international norm. Ahrens (2017) also explores the potential for further solidarisation in the international climate regime. In her view, the growing involvement of non-state actors in climate governance and the European Union's (EU) role in creating the 2015 Paris Agreement point to a 'solidarising moment' (Ahrens, 2017: 16) in GEP. Falkner (2017a) similarly examines the possibility of the pluralist logic of coexistence gradually giving way to more solidarist forms of international climate cooperation, but concludes that the Paris Agreement's flexible and voluntary commitment structure points to a reversal to more pluralist forms of interstate cooperation.

Another focus in recent ES research on GEP has been the interaction between environmentalism and other fundamental norms of international society. Falkner and Buzan (2019) trace the growth of environmental stewardship into a primary institution on a global scale and offer a preliminary discussion of how its emergence is interacting with existing primary institutions, such as diplomacy, law, sovereignty and great power management. Clark (2011) explores what role hegemony, as a legitimate form of power asymmetry, can play in solving the climate change conundrum, and what legitimacy challenges any hegemon faces when seeking to organise an international response to such a challenging global environmental problem. Based on a study of China's shifting engagement with international climate politics, Kopra (2018, 2019) examines the changing relationship between emerging notions of climate responsibility and the established norm of great power responsibility (see also Zhang, 2016: 814–15). Malhotra (2017) explores the interaction between environmental stewardship and international law, and particularly the failure to establish the concept of ecocide crime as part of solidarist legal development.

As we can see from this brief discussion of recent scholarship, the ES has moved well beyond its initial neglect of global environmental politics to a fuller consideration of the implications of environmentalism for the international order. There is now convergence among ES authors around the idea that environmental stewardship has emerged as a fundamental international norm. Debate continues, however, on just how strongly it has become embedded in the international constitutional order and what effect it is having on the direction of international societal development. ES scholars have deployed the conceptual dyad of pluralism and solidarism as a measure of normative development in international relations. Much like nationalism and human rights, the rise of global environmental politics now serves as a major site of empirical analysis to test claims about the sources and direction of international norm change, and specifically about the potential for progressive or transformative changes in the international order. The ES has also opened a new perspective on the interaction between international society and world society, focusing on processes of norm transfer and mutual normative shaping, which follows the transnationalist research agenda but deals with larger questions of societal development, including the overlap between and potential merging of international and world society.

The Greening of Global International Society: A Framework of Analysis

The question of international change – its causes, form and direction – is at the heart of IR. It is present in attempts to delineate major epochs of world history, whether based on shifts in the distribution of power (Waltz, 1979) or changes in the material and ideational foundations of modernity (Buzan and Lawson, 2015). And it informs the major theoretical debates that are commonly taken to have defined the evolution of IR as a discipline. International change, if understood in such terms, is not about the endless but often minor and inconsequential changes that characterise social life at all levels, but the more fundamental and long-lasting shifts that transform the nature of international relations (Holsti, 2004). In this book, I adopt a similar focus on the long-term and deep-seated impact that the rise of environmentalism has had on international society. Global environmental politics is of course replete with changes of all kinds: international environmental conferences and institutions have proliferated, as have political declarations, action plans, pledges and commitments, and aid programmes. But whether this adds up to more than just the sum of individual actions or is merely part of a growing cacophony of environmental diplomacy is not immediately obvious. My interest is in understanding the rise of global environmentalism as the source of (potentially) fundamental change, located at the level of the normative structure of international relations. This book is about whether GIS has been 'greened', in the sense of environmentalism having become a fundamental international norm, comparable to the emergence of nationalism in the nineteenth century or human rights in the twentieth century.

Primary Institutions as Indicators of Change

Identifying and measuring transformative change requires some form of metric or indicator of international societal development. Holsti (2004: 18) makes a convincing case for international institutions as the 'essential markers of change in the domain of international politics'. What Holsti and others have in mind are 'fundamental institutions' (Reus-Smit, 1997) or 'fundamental practices' (Keohane, 1988), that is primary institutions in ES parlance. Primary institutions contain the essential characteristics of international society, the enduring rules of international interaction that give it recognisable presence and stability. Changes at the level of primary institutions thus become a useful marker for the kind of deep-seated change that we are interested in when thinking about the long-term evolutionary patterns in international relations. One complication in this context is that no fixed list of primary institutions exists. As has been noted by others, the empirical process of identifying primary institutions is open to selection bias (Wilson, Peter, 2012). Indeed, ES authors have come up with different lists of primary institutions, from the narrow classical set as defined by Bull (1977) to Buzan's (2004: 187) expanded list of master and derivative primary institutions. The ambiguity that is pervasive in the discussion of primary institutions is not without reason, however. Because human societies are constantly creating or changing fundamental norms in response to new and complex challenges of social organisation, accounts of any society's contemporary normative structure will vary across time and space. The question of a specific international society's normative structure is an empirical one, and we are left with 'definition plus empirical observation' (Falkner and Buzan, 2019: 135) as the only viable strategy for establishing whether a new primary institution has been added to the normative mix or an existing one has become obsolete.

My empirical strategy for establishing the rise of environmentalism, its emergence as a fundamental norm, and its adoption, globalisation and contestation as part of international society's normative structure is based on two central steps. As set out by Falkner and Buzan (2019: 136), when identifying an emergent primary institution we first expect to see 'a clearly defined value or principle applicable across international society (whether global or regional)'. This value or principle will be historically situated, it is likely to evolve in response to changing social and historical circumstances and will thus not be fixed in time. However, despite the endless normative permutations that we can observe, it needs to have a sufficiently clearly defined normative core in order to become a socially relevant institution in international relations. Major political ideologies, such as liberalism and nationalism, are of this quality. Understood as 'combinations of political concepts organized in a particular way' (Freeden, 1996: 75), ideologies provide a coherent world view that connects political thought with political action. They invariably contain an inner core of values that is essential to defining their identity, with other more tangential values grouped around that core. As Freeden (1996: 77) states, ideologies are 'characterized by a morphology that displays core, adjacent and peripheral concepts', and it is peripheral concepts that play the important role of relating core beliefs and values to specific historical and geographical contexts. They thus act as a bridge between what is essential to an ideological belief system and what is of a more temporary and contingent nature. They put abstract principles into political operation and define their specific relevance in given historical contexts (Freeden, 1996: 79–80).

The second step in my empirical strategy involves identifying evidence of social impact in international relations. For a new norm to qualify as an (emergent) primary institution, we 'expect to observe a significant degree of social consolidation' (Falkner and Buzan, 2019: 136). Such consolidation can occur through two principal mechanisms: the creation of secondary institutions, which reproduce the emerging norm and thereby strengthen, develop and entrench it; and observable patterns of change in the behaviour and identity of states, which is in accordance with the new norm. The growth of secondary institutions around new international norms has been a central plank of the study of international change in IR for some time. In fact, much of the GEP literature has followed liberal institutionalist IR theory and applied the regime lens when investigating the rise of environmental politics. Both the conditions for the successful creation of environmental regimes and their effectiveness as regulative governance instruments have been the focus of a vast literature (for an overview, see Mitchell, 2010: chapters 5 and 6; Andresen, 2013). In this regard, this study can build on the existing literature in GEP, but it is important to note that my focus is more narrowly on how growth in environmental regimes reinforces the legitimacy of environmentalism as a primary institution, and how changes and contestation at the level of environmental regimes affects the evolution of the underlying environmental norm. Similarly, the rise of environmentally oriented behaviour among states has been studied before (for an overview, see Barkdull and Harris, 2002; Falkner, 2013), as has the question of a changing identity and moral purpose of the state (Eckersley, 2004a; Bäckstrand and Kronsell, 2015). Again, I build on this research but sharpen the focus on how such developments speak to the strengthening of the emergent primary institution of environmentalism.

With regard to the different ordering and regulative functions that primary institutions perform in international society, I adopt a pragmatic approach to the sub-categorisation of primary institutions. We can generally distinguish between constitutive and regulative institutions, with the former defining the nature of international society and the identity of its members and the latter guiding their interactions. The extent to which environmental stewardship has affected the international normative structure can be assessed by its impact on constitutive versus regulative rules, with the former denoting a deeper level of international change. However, questions or constitutive and regulative effect are often difficult to disentangle, and keeping up a strict differentiation between the two types of primary institutions is problematic, particularly when applied to a variety of international societies across space and time (Schouenborg, 2011). I return to this question in the concluding chapter.

Spatial Dimensions of Normative Change

ES theory opens up an important perspective on the spatial dimensions of any fundamental change in international relations. Using its comparative historical approach to societal development provides a useful framework for exploring the geographic reach of normative innovation, be it at the level of individual primary institutions or entire societal structures. As part of its core expansion story, the ES originally focused on the process through which Europe's Westphalian international order came to be extended to the global level. Starting out as a regionally defined social structure that emerged out of Europe's tumultuous seventeenth century, international society was slowly but steadily globalised through a mixture of colonisation, socialisation, competition and reform (Bull and Watson, 1984; Reus-Smit and Dunne, 2017; Buzan and Schouenborg, 2018). The expansion of Europe's international society was an uneven and unequal process. It involved different modes of interaction, between European empires that considered each other as equals and between 'civilised' European states and 'uncivilised' or 'semi-civilised' non-European states that needed to be brought up to the 'standard of civilisation' for full membership in international society (Keene, 2002; Suzuki, 2009). Over the course of the eighteenth and nineteenth centuries, Europe's colonial empires expanded dramatically and coexisting regional international societies were replaced by an emerging global international society (Buzan and Schouenborg, 2018), even though societal development has continued to vary across different regions.

Individual fundamental norms, too, can be found to have gone through various degrees of globalisation or regionalisation. Whereas nationalism has followed a path of full globalisation from the nineteenth to the twentieth century (Mayall, 1990), other fundamental norms (human rights, democracy) have been championed only by some states while being resisted by others, and are therefore only applicable in certain regional contexts. The normative landscape of GIS is thus uneven, characterised by considerable regional differentiation (Acharya, 2009; Buzan and Schouenborg, 2018). When exploring novel normative developments, we need to consider their spatial reach as much as their overall strength. This also raises the question of whether emerging norms, such as environmentalism, are culturally bounded, whether they have certain national or regional roots, whether they have outgrown those roots and whether they have become applicable across a sub-global group of countries/regions (e.g. the West) or have reached the point of full globalisation.

Norm Conflict, Contestation, Decay and Renewal

The ES framework also opens up a perspective on the interaction between emerging norms and the existing set of primary institutions. In order to become an established international norm, environmentalism needs to find its place in the normative structure of international society, with consequences for other elements of this order. Several types of impacts and interactions are possible. First, a new norm may offer a smooth fit with existing norms, either reinforcing them or leaving them unaffected. Nationalism's rise in the nineteenth century, for example, fits well with the existing norms of territoriality and sovereignty. Both ended up being strengthened as the foundations of Westphalian order, though now based on notions of popular sovereignty. Second, a new norm may clash with certain existing norms, putting pressure on them to change or leading to their eventual decline. In the case of nationalism, we find that once established it sat uneasily with the long-standing primary institutions of colonialism and dynastic succession. Nationalism led to the fairly rapid demise of the dynastic principle as the basis for legitimate sovereign statehood, but it took well into the second half of the twentieth century for the principle of colonial rule to lose all legitimacy against the by then universally accepted principle of self-determination. Third, tensions between norms can persist for some time, not every normative conflict leads to the demise of one norm or the other, and new norms themselves may change in response to pressure from existing primary institutions. Normative order is subject to a constant process of change, with individual norms coming under pressure to adapt or being pushed aside as others emerge and strengthen. Over the last three centuries, some of GIS's primary institutions have survived (e.g. territoriality, sovereignty, diplomacy), others have declined in importance (e.g. war) or disappeared altogether (e.g. dynastic succession), while yet others have emerged only partially and remain contested (e.g. human rights, democracy).

Much of the IR literature on fundamental norms has focused on their structuring and shaping power (Checkel, 2001), which tends to treat norms as stable and fixed entities (Hoffmann, 2010: 5). The ES framework and the more recent norms literature (e.g. Wiener, 2008) suggest, however, that we need to view norms as constantly malleable and open to change, even when institutionalised in international society. Norms arise and come to be established in society through a process of mutual constitution, and as such are open to contestation. Of course, norms need to be stable, up to a point, in order to have a discernible impact on social relations. But norms are complex constructs, containing more than one normative prescription, and need to be applied in historically and socially contingent contexts. Their meaning can change over time, and different actors interpret norms differently or live by different norms. As Sandholtz (2008: 101) put it, 'normative structures … cannot stand still.' The same can be said about primary institutions in international relations, which are always open to dynamics of change. As states reproduce them in their social interactions, the meaning of primary institutions may shift, particularly when normative ambiguities exist or actors seek to move common understandings in new directions (Sandholtz, 2008; Bailey, 2008). Norm complexity and malleability can lead to normative change but it is also a source of norm robustness (Welsh, 2019). Contestation over emerging or existing norms and shifts in the meaning of norms are thus part and parcel of the study of deep-seated normative change.

The ES framework, and in particular its distinction between pluralism and solidarism, opens up a useful perspective on the direction and strength of normative development. Traditionally, this distinction has been taken to depict two different normative positions within the ES tradition – about what 'ought' to happen in international relations, and specifically how to balance the conflicting demands for international order and justice (Bain, 2014; Linklater and Suganami, 2006: 59–68). More recently, Buzan's (2004) reworking of the ES as a structural theory has brought to the fore the analytical use of the pluralist–solidarist distinction to investigate the wider state of normative development in international society. In this view, pluralism and solidarism represent two different endpoints of normative development, two distinctive states of what 'is', rather than 'ought'. We can use the ES conceptual dyad to describe two distinctive interaction logics in international relations and categorise existing or emerging primary institutions as belonging to either the pluralist logic of coexistence or solidarist logic of cooperation. Thus, war, balance of power and great power management are primarily concerned with maintaining a minimal order between states, while human rights, market and development aspire to achieve a greater degree of global cooperation and value convergence (Falkner and Buzan, 2019: 136).

This perspective allows us to detect changes in the normative focus of primary institutions, from a pluralist to a solidarist logic or vice versa. Applied to environmentalism, we can examine whether the rise of environmental stewardship has promoted more solidarist approaches to international policy-making, and whether this solidarisation of GEP is pointing in a state-centric or cosmopolitan direction, or a complex mixture of both (Buzan, 2004: 114–20; Hurrell, 2007: 224–8). Alternatively, we may also identify cases where environmental stewardship is based on a pluralist logic of international interaction (Falkner, 2017a), either because environmentalism has failed to overcome the deep value diversity that helps to entrench pluralist order or because the pluralist logic is sufficient to generate an effective collective response to existential ecological threats (Buzan, 2004: 233). In this way, the pluralist–solidarist distinction becomes a measure of overall normative development, of establishing how deeply environmentalism has become embedded in the normative structure of international society and the extent to which it has itself strengthened or weakened certain normative features of international society.

World Society and International Norm Change

The ES framework outlined earlier is focused on norm change within the society of states. While this might suggest a narrowly conceived state-centric perspective, it is important to note that the framework lends itself for an investigation of how international and world society are interrelated and how world society actors play into the greening of international society. As discussed earlier, the ES has left behind its state-centric origins and is now a fertile ground for research and debate on the role that non-state actors play in a globalising world order (e.g. Buzan, 2004, 2018; Stivachtis and McKeil, 2018). Some questions remain about the continuing ambiguity in the way ES authors define world society (Stivachtis and McKeil, 2018: 2), but recent theoretical advances yield well-defined pathways for empirical research. Within the context of this study, we can identify two principal ways in which world society is implicated in the normative development of international society: one, as a source of new norms that are inserted into international society, usually by non-state actors that operate outside but interact with the states system; and two, as the endpoint of a global process of merging international society and world society into a more comprehensive globalised world order.

The first dimension sees world society actors as norm entrepreneurs seeking to shape international society's normative structure. This dimension matches Buzan's (2018: 129) notion of a *political* world society, which 'comprises all the non-state social structures visible within humankind as a whole that have both significantly autonomous actor quality, and the capacity and interest to engage with the society of states to influence its normative values and institutions'. In contrast to the *normative* interpretation of world society as the 'great society of humankind', which has dominated ES thinking in the past, the empirically grounded notion of political world society helps connect the ES with the transnationalist research agenda that demonstrates actual non-state actor impact on international outcomes (Risse-Kappen, 1995). But while transnationalism tends to focus on specific outcomes brought about by non-state actors, usually in the context of creating secondary institutions, the ES perspective directs our attention to primary institutions, the social structure of international relations. As Clark (2007) and Pella (2013) demonstrate, norm transfer from world to international society has helped to redefine the foundations of international legitimacy at crucial turning points in the evolution of international society. In campaigns to ban the slave trade, establish racial equality and promote human rights, world society actors have mobilised universal cosmopolitan sentiment, with humankind serving as the moral referent. However, if we seek to understand the full spectrum of possible world society interventions in international society, we need to keep an open mind about world society's underlying normative project. If we accept that political world society is inevitably pluralist (Williams, 2005), then any study of world society's normative influence needs to capture both solidarist and pluralist outcomes. Similarly, the study of global environmentalism needs to be open to pluralist and solidarist variants of creating a green global order (see Chapter 3).

The second dimension concerns the (potential) merging of international society with world society and the creation of a larger, more comprehensive, global order made up of states and non-state actors. This perspective brings into play a third meaning of world society that is implied in ES debates but has only recently been made explicit by Buzan (2004, 2018). It alludes to what Buzan (2018: 130–1) calls *integrated* world society, an overarching social structure for humanity that includes all kinds of actors and organisations, all the way from individuals and social groups to states and intergovernmental organisations. It is an abstract concept, which represents the end point of a global societal development that integrates the political structures of international society and world society in its political form. Due to its hypothetical nature as a normative aspiration or teleological prediction, the concept of integrated world society is of only limited value for an empirical–analytical project (Buzan, 2018: 131), but it identifies an important dimension of societal development that is relevant to the study of emergent international norms. The interaction between world society and international society is not restricted to isolated interventions by non-state actors that produce discrete outcomes in interstate relations. In reality, states and world society actors are increasingly engaged in sustained and intricate forms of interaction that can lead to lasting patterns of cooperation and even co-dependence, as can be seen in the increasingly complex structures of global governance (Hurrell, 2007: 95–117). The analytical focus in the study of environmentalism's impact on international society thus needs to include the possibility of a broader societal development that sees elements of world society and international society merging into a larger global structure. Theories of global governance have sought to capture some of these dynamics (Weiss 2016; Acharya, 2016), though the precise social mechanisms that are behind this merging remain unclear.

Conclusions

This chapter has presented an analytical framework for studying the long-term processes of normative change in international relations, which manifest themselves in the emergence of new fundamental norms. Grounded in ES theory, which provides a structural theory of societal development in international relations, it has introduced the core concepts that guide the empirical analysis in subsequent chapters: the distinction between primary and secondary institutions, with the former serving as markers of deep-seated international change; the pluralism–solidarism divide, which offers a framework for assessing the direction of normative change, from thin to thick layers of institutionalised cooperation; and the respective roles played by states and non-state actors in the process of norm transfer from world society to international society, and vice versa.

Environmental issues may have played only a marginal role in the early history of the ES tradition, but recent ES authors have developed a richer seam of scholarship on global environmentalism as a potentially transformative force in international relations. As I have argued in this chapter, the ES's focus on historically situated changes in the international normative structure provides a unique framework for studying the impact that new norms, such as environmental stewardship, have on existing fundamental norms. By the same token, emergent norms interact with existing ones and are shaped by them in turn. The study of the rise of global environmentalism is also about the study of how engagement with international relations has shaped environmental ideas and values themselves.

Having set out the theoretical context in which this study is located, the next chapter introduces environmentalism as a political ideology and movement, both its normative core and the variants of environmentalism that have evolved over time. It identifies the principal ways in which environmentalism can be mapped onto the international realm, as pluralist or solidarist projects of greening international society or world society.

3 The Idea of Environmentalism

Environmentalism[1] became a global force in the second half of the twentieth century and has left a lasting mark on domestic and international politics around the world. If, as I argue in this book, the norm of environmental stewardship has joined the ranks of primary institutions in international society, then its impact on the international normative structure is, potentially at least, comparable to that of other new fundamental norms (e.g. nationalism, market). In order for a new norm to qualify as a primary institution, it has to be based on a value or principle that is applicable across international society, whether at a regional or global level (on the other criteria for primary institutions, see Chapter 2). The main task of this chapter is to delineate environmentalism's normative core and to identify the main ways in which it can be mapped onto the international realm.

The chapter is divided into two parts. The first part introduces the diverse roots of environmental thinking and identifies the normative core around which modern environmentalism is built. It traces the evolution of different strands of environmentalism and outlines the main debates that have shaped the evolution of environmental thinking and activism since the nineteenth century. The second part of this chapter outlines the different ways in which different environmental ideas can be applied to the international realm. Employing the ES's conceptual dyads of

[1] The term 'the environment' only came to be used in its contemporary sense in the 1960s, as a focus for government policy and as a comprehensive label for the natural world, encompassing ecological systems and species at all levels from the local to the planetary (Warde, Robin and Sörlin, 2018: chapter 1). The term 'environmentalism' similarly emerged around this time to describe the political movement and ideology that sought to promote environmental protection. Before the term came into popular use, several other concepts denoted a concern for the natural environment: the protection, preservation or conservation of nature, the commons, landscapes, natural habitats and natural monuments, or animal and plant species. Because they all fed into the broad tradition of environmental thinking and campaigning as it is understood today, I shall use the term 'environmentalism' for the entire history of environmental thought and practice from its origins in the nineteenth century until today.

pluralism/solidarism and international/world society, it identifies four ideal types of how a green global order can be created: 'Green Westphalia' and 'global environmental governance', representing the pluralist and solidarist variants of a green international society; and 'eco-localism' and 'eco-globalism' as the pluralist and solidarist versions of a green world society.

The Origins and Evolution of the Environmental Idea

Environmentalism is a broad church that combines a large variety of ethical and political positions (Schlosberg, 1999). Much like other political ideologies, the environmentalist tradition is centred on a normative core, in this case the belief that the natural environment should be protected. However, it includes a wide range of diverging and often conflicting ideas about how this objective can be achieved and how environmental values relate to other ethical and political concerns. Ideologies are not monolithic and rigid intellectual constructs. As Freeden (1996: 77) explains, they are 'characterized by a morphology that displays core, adjacent and peripheral concepts'. While core beliefs define an ideology's political orientation, giving it historical permanence and intellectual continuity, adjacent and peripheral concepts introduce an element of flexibility, making the ideological core applicable to specific historical and geographical contexts. Adjacent and peripheral concepts are important as they translate abstract principles into political reality and allow an ideology to adapt to changing political circumstances (Freeden, 1996: 79–80). To understand environmentalism's emergence as a fundamental norm in international society, we therefore need to keep in view both its stable intellectual core and the varying peripheral concepts that have given rise to a great number of different environmentalist strands and traditions over the last two centuries.

Roots of Environmental Thinking

Although environmentalism, understood as a political ideology and movement, is a distinctly modern phenomenon that first came to prominence in the nineteenth century, environmental thinking in a wider sense has much deeper historical roots. The natural environment has inexorably shaped human existence, and throughout human history different societies have come up with different ways to define their relationship to nature. For the best part of the history of *Homo sapiens*, natural forces exercised an overwhelming influence on societal development, and humans needed to learn to adapt to nature in order to survive and prosper. Lacking the means to control nature, early humans viewed their natural environment with a sense of fear and awe. Over time, they learned to harness the power of fire as a tool of environmental management, and the domestication of plants and animals gave rise to the first agricultural societies (Morris, 2011: 99–100). If humanity was to develop some degree of mastery over nature, however, it needed to develop a better understanding of its mysterious processes, and with expanding knowledge about the workings of nature came a greater appreciation for it. Throughout human history, fear of nature's destructive power was thus closely entwined with respect, and even admiration, for its complexity and beauty.

It was the world's major religions that first developed explicit ethical frameworks to include notions of individual or collective responsibility for nature protection (Stoll, 2015). Buddhism foreshadowed modern ecological thinking with its emphasis on the interconnectedness of all life forms. By cultivating the values of modesty and compassion, it also bequeathed an ethic of care and restraint to modern environmentalism (Schumacher, 1973). Christianity, although strongly associated with the instrumentalist view that humans ought to subjugate nature for their own benefit (White, 1967), became a major intellectual source for the idea of ecological stewardship, based on the Christian notion of humans as tenants rather than freehold owners of the earth (Sarre, 1995: 117). Hinduism, Islam and Judaism likewise contributed to the rise of distinctive ethical stances that demand respect for the environment as God's creation. If nature is sacred, as most major religions assume, then humans are commanded to use natural resources in a respectful and sustainable manner (Grim and Tucker, 2014).

The first explicit attempts at nature protection can be traced back many centuries, even if they do not necessarily reflect explicitly environmental motives. In medieval times, powerful rulers and wealthy landowners occasionally set aside large swathes of land, usually forests, to preserve strategic natural resources or protect hunting grounds. From the mid-sixteenth century onwards, the Spanish crown expanded its control over forested areas in an effort to safeguard long-term timber supplies for its imperial fleet (Wing, 2012). France's Grand Ordinance of the Waters and Forests of 1516 was designed to protect royal hunting areas, while a similar ordinance issued by Jean-Baptiste Colbert in 1669 sought to halt the rapid deforestation that threatened the French Navy's continued supply of timber (Grove, 1995: 59). From the mid-eighteenth century onwards, widespread deforestation in Germany's mining districts led to the first systematic efforts to rejuvenate forests. Forest authorities relied on natural and artificial regeneration, focusing mainly on the planting of pine and spruce, a practice that had started in some German regions as early as the sixteenth century (Johann, 2006: 3–4). Such measures foreshadowed later conservationist practices but were usually rooted in ideas about resource management rather than environmental protection.

The Industrial Revolution and Modernity

The decisive turning point in the evolution of environmental thinking came with the onset of the modern industrial age in the late eighteenth and early nineteenth centuries. It was the economic forces unleashed by the Industrial Revolution that provided the critical impetus for the emergence of a distinctive environmentalist tradition, first in literature, art and nature writing, and later in scientific and political discourses (Radkau, 2011: 38). A series of technological innovations based on extracting energy from fossil fuels, most notably the steam engine and the ability to convert mechanical energy into electricity, shifted the balance of power between nature and humanity decidedly in favour of the latter, thereby enabling human societies to escape the 'Malthusian trap' that had previously kept population growth and economic expansion within tight ecological limits (Clark, G., 2007). As industrialism spread from Britain to Continental Europe and then North America and other parts of the world, it left ever more visible scars on the surface of the planet. Black smoke billowing from factories and toxic waste flowing into rivers and lakes caused growing alarm about the destructive powers that had been unleashed by the Industrial Revolution. For the first time in history, humanity's desire to harness nature's power had resulted in serious and possibly irrevocable harm to the planet's health. A different conception of humanity's place in the natural world soon began to emerge, one that emphasised concern for, as much as mastery over, the natural environment. The global transformation of the nineteenth century brought about by the Industrial Revolution is thus rightly seen as the beginning of contemporary environmental concern (Buzan and Lawson, 2015: 96). Modernity and the industrial age, widely praised for setting humanity on a course of hitherto unimagined gains in prosperity, created its own antidote in the form of the environmentalist counter-reaction.

Historically speaking, environmentalism is a product of the modern age, but its relationship with modernity is an ambiguous one. To a large extent, environmentalism was, first and foremost, a reaction to, and in some cases a rejection of, modern industrial society. Late eighteenth- and nineteenth-century environmental thinkers drew heavily on anti-Enlightenment sentiment, questioned the idea of achieving societal progress through technological innovation, and denounced industrialisation's destructive impact on nature. Romantic poets, philosophers and artists were among the first to express a new environmental sensibility that resisted modern ideas of progress and rationalism (Safranski, 2009: 53–4). Among the philosophers of the Romantic era, the so-called primitivists saw in nature the original source of human identity and happiness. From Jean-Jacques Rousseau to Lord Byron, Romantic writers extolled the virtues of a return to nature and the pleasure that could be found in wilderness untouched by the human hand (Nash, 2001: 49–50). William Wordsworth famously gave expression to this anti-modern stance in his poem 'The Excursion' of 1814, which bemoaned 'the darker side' of industrial change and the 'outrage done to nature' by manufacturing towns (quoted in Clayre, 1977: 175–6). Romanticism promoted the notion that nature possessed an intrinsic value, as a source of aesthetic and religious inspiration, and that the wild forces of nature, usually feared for their destructive potential, in fact possessed an elusive and subliminal quality that enriched the human existence. In this view, the conservation of nature was less about managing the ecological excesses of the industrial system than about preserving nature's inherent harmony and beauty. Romantic environmentalism's main objective was to restore the natural order of life. Politically speaking, it was as conservative as it was conservationist.

Anti-modernist sentiment created a gulf between some forms of environmental thinking and the progressive ideologies (Buzan and Lawson, 2015: chapter 4) that came to shape politics in the nineteenth century. Many environmentalists have been critical of liberalism's emphasis on individual liberty and called for constraints on the free operation of the market. They have challenged nationalism's belief in the legitimacy of nationally defined political systems and instead demand global cooperation across borders to effectively deal with transboundary environmental problems. And they have questioned socialism's reliance on the full development of capitalist industrialism as a necessary step towards a socialist economy. In this sense, major strands of early environmentalism came to challenge the progressivist narratives that defined the great transformation of the nineteenth century. This anti-modern or 'post-modern' distrust of science, technology and Enlightenment thinking carries on to inform countercultural environmental thinking today (Pepper, 1996: 5).

Somewhat paradoxically, however, environmentalism gained in influence in global politics precisely because it was able to align itself with various progressive causes, especially after the Second World War. Over time, anti-modernist sentiment lost out to the scientific (e.g. Charles

Darwin, Alexander von Humboldt) and progressive (e.g. William Morris, John Ruskin) roots of nineteenth-century environmentalism, paving the way for a much broader political movement with an increasingly rationalist outlook. This process already started during the second half of the nineteenth century, with utilitarian thinkers promoting the sustainable use of natural resources based on scientific principles. The scientific conservation movement added a distinctly functionalist and human-centred logic to nature protection that sat uneasily alongside the spirituality of Romantic conservationism (Hays, 1959). By the early twentieth century, ecology had emerged as a branch of modern science, gaining recognition in society and politics and boosting environmentalism's influence in governmental bureaucracies. And during the 1960s and 1970s, environmentalism became part of the new progressive social movements in industrialised countries, undergoing a profound transformation from an elite project to a mass movement. Environmental protection now counted alongside civil rights, women's rights, nuclear disarmament and social justice as one of the core elements of progressive politics.

By the time most states had come to accept a duty to protect the global environment, a process that played itself out over the course of the twentieth century and reached its nadir in the 1970s (see chapter 5), environmental ideas had gone mainstream in democratic politics and environmental protection had become part of the daily bureaucratic routines of the modern nation-state. Environmental ideas could now be found on all sides of the political spectrum, leading to socialist (Pepper, 1993; O'Connor, 1994), liberal (Wissenburg, 2013) and conservative (Scruton, 2012) variants while feeding its own ideological tradition (Freeden, 1996; Newell, 2020). It was this blending of environmental thought with a wide range of political ideologies and movements – some anti-modern or post-modern, some modernist and progressive – that eventually helped propel environmentalism to its prominent place in global politics.

Colonialism's Environmental Legacy

Colonialism provided an important context within which environmental concern emerged and spread worldwide. Europe's colonial empires became a major source of ecological knowledge; they acted as sites for developing innovative conservation practices and also became a focus for the first transnational environmental campaigns. By the time international society accepted environmental stewardship as a fundamental norm, most colonial empires had been dismantled, but the colonial experience cast a long and troubling shadow on the growth of global environmentalism well after the completion of decolonisation (see Chapter 6).

Colonialism's contribution to the growth of ecological knowledge has been documented by several environmental historians (Grove, 1995; Anker, 2002; Barton, 2002). From the eighteenth century onwards, naturalists and scientists routinely accompanied colonial explorers on their journeys to distant parts of the world, in order to discover and explore unknown animal and plant species and bring them back to Europe for further scientific study and commercial exploitation. These expeditions also provided scientists with unique insights into the global interdependencies that connect different ecosystems. Alexander von Humboldt, the German polymath and explorer who extensively travelled in Latin America between 1799 and 1804, meticulously documented the continent's flora and fauna as well as geological formations and climatic conditions. He was the first to derive from his observations a holistic concept of nature, a notion of the planet as a single living entity, which foreshadowed twentieth-century ecological concepts such as James Lovelock's Gaia (Wulf, 2015: 7). The British geologist and biologist Charles Darwin similarly visited South America to gather extensive ecological data when he travelled as a passenger on HMS *Beagle* during its second voyage from 1831 to 1836. Darwin was able to experience first-hand how human influence could cause ecological turbulence on a large scale. While most colonists were unaware of the consequences of their actions, Darwin could see clearly how the arrival of European settlers and their animals had disturbed the ecological balance in the grasslands of the Rio de la Plata region in South America, causing alien weeds to spread and native species to disappear (Worster, 1994: 123; Crosby, 2004: 160–1). Humboldt, Darwin and other nineteenth-century naturalists were thus able to put together the first pieces of evidence for what later came to be known as the Columbian Exchange (Crosby, 2003), the large-scale transfer of plants, animals and human populations, alongside communicable diseases, among Europe, Africa and the Americas.

By the eighteenth century, colonial administrators themselves had begun to support the scientific study of the territories they controlled, not least because of the peculiar ecological challenges they faced in often unfamiliar natural environments. Over time, colonial experts and scientists developed international networks to exchange ecological information (Grove and Damodaran, 2006: 4346–7), thereby reinforcing 'the growth of a sense of a global environmental crisis' (Grove, 1995: 485). It is no exaggeration to argue that it was in the context of Europe's colonial expansion between 1660 and 1860 that it became possible for the first time to imagine the *global* scale of ecological interdependencies (Grove, 1995; Anker, 2002).

While this expanding body of ecological knowledge did not immediately translate into environmental action on a global scale, it did lead to the first distinctive environmental management approaches in colonial territories, long before they were more widely pursued in Europe and North America. In this sense, 'Europe's tropical colonies played a central role in the rise of modern nature conservation more generally' (Ross, 2017: 240). Deforestation and soil erosion were one of the most urgent environmental problems that colonial rulers had to deal with. This was particularly the case in small island colonies, such as Mauritius and St Helena, which served as defensive posts and suppliers of natural resources along transcontinental shipping routes. The ecological damage caused by excessive timber and agricultural production led some colonial administrators to experiment with new approaches to forest and soil conservation. Colonial botanical gardens were founded as sources of ecological learning and knowledge exchange, and state-led programmes of soil and forest conservation increasingly drew on local ecological expertise as much as on existing agricultural and forestry policies in Europe. By the middle of the nineteenth century, forest protection measures had become a firmly established part of English and French colonial administration in Africa and Asia (Grove, 1995; Barton, 2002).

Colonial administrations also had to contend with the growing demands of European environmentalists for the protection of wildlife in Africa and Asia. After a surge in game hunting at the end of the nineteenth century, several European colonial powers introduced restrictions on the hunting of large mammals in Africa and began to establish nature reserves. Among the first such conservation measures was the Cape Act for the Preservation of Game, introduced in 1886 and extended to the British South African Territories in 1891 (Prendergast and Adams, 2003: 252). The protection of wild animals also became the focus of the first transnational lobbying efforts by European environmentalists. In 1903, a small group of British conservationists and hunters founded the Society for the Preservation of the Fauna of the Empire, which lobbied Britain's Colonial Office and put pressure on colonial governments to extend and enforce conservation measures (Prendergast and Adams, 2003). Similar campaigns were also initiated in the United States as it began to build an informal empire. After the United States annexed the Philippines in 1899, animal welfare campaigners worked closely with American administrators to eradicate acts of animal cruelty, such as cockfighting and horsetail docking, as part of a wider campaign to eliminate 'uncivilised' forms of behaviour among locals (Davis, 2013). Despite opposing US militarism and colonialism, environmentalists nevertheless hoped that America would at least spread an 'empire of kindness' (Davis, 2013: 195).

Colonialism thus played an important role in the history of global environmental protection, as a source of ecological knowledge, as a testing ground for novel conservation practices and as the subject for transnational environmental campaigning. The flow of environmental insights and the evolution of conservation policy was a complex process, involving 'a multifaceted exchange of ideas, norms, and practices between metropoles and colonies, as well as between rival empires' (Ross, 2017: 240). Because colonies were part of the domestic fabric of governance within colonial empires, and not members of international society in their own right, their influence over the growth of global environmentalism flowed indirectly, via European colonial powers. It was only later, after decolonisation, that the South's own contributions to environmental thinking and practice came to the fore in international debates. Colonialism thus served as an early incubator for environmentalism, though it also bequeathed a toxic legacy of environmental paternalism that was to complicate the globalisation of environmental stewardship in the late twentieth century (see chapter 6).

National and Regional Varieties of Environmentalism

Nineteenth-century environmentalism has many different national roots, and it was only in the twentieth century that a more globally integrated environmental movement began to emerge. The first organised forms of environmentalism arose in a number of countries, most notably in North America and Europe, and gradually spread to other parts of the world. In the beginning, most environmental efforts were local struggles, rooted in a 'sense of place' (Armiero and Sedrez, 2014: 2), with people seeking to protect their special bond to local ecosystems, species or areas of natural beauty. What connected them was a shared sense of environmental concern, even if the referent of this concern varied considerably across societies. As the transnational exchange of environmental ideas and scientific knowledge began to speed up in the second half of the nineteenth century, organised environmentalism underwent a certain degree of ideational convergence. Still, national characteristics continued to shape environmental thinking and action well into the twentieth century. Ultimately, environmentalism, as much as its core concepts of 'nature' and 'sustainability', is a social construction, an idea that is heavily influenced by cultural and historical contexts (Hurrell, 2007: 234; Warde, Robin and Sörlin, 2018).

In North America, representations of nature as wilderness became the central leitmotif for the nascent environmental movement and had a lasting impact on how the American public came to think about nature preservation (Nash, 2001). During America's westward expansion in the nineteenth century, the pioneers saw untamed nature as a constant threat to their survival. Once the quest to push the 'American Frontier' westward had come to an end, wilderness started to take on a different meaning. The subsequent economic development and integration of the United States brought with it the prospect of the remaining wild areas gradually disappearing. What had previously been regarded as the necessary march of human civilisation now threatened to cause irreparable damage to the natural heritage of North America and eliminate the spiritual value that settlers had come to recognise in it. A combination of Romantic sensibility for wilderness and nationalist pride in the natural beauty of America's landscape thus led to a re-valuation of wilderness as a national treasure that needed to be preserved, not transformed.

By contrast, notions of wilderness protection played a far less prominent role in European environmentalism, not least because few areas in Europe were still left untouched by human civilisation by the time industrialisation took off. In Germany, nature protection (*Naturschutz*) came to be associated more closely with the idea of preserving the country's homeland (*Heimatschutz*), especially its cultivated landscapes and scenic landmarks (Lekan, 2004: chapter 1; Rollins, 1997: 70). The notion of *Heimatschutz* was based on the idea that the appearance of the land both reflected German culture and also actively shaped the country's national character (Lekan, 2004: 22). With national sentiment on the rise after German unification in 1871, late nineteenth-century environmentalists were able to tap into the growing sense of national pride when arguing for the preservation of areas of natural beauty. The protection of the country's mythical forests (Wilson, Jeffrey, 2012), which repeatedly became a national policy concern in the nineteenth and twentieth centuries, is but one example of how German environmentalism and nationalism reinforced each other and created a peculiar mix of environmental values. A similar blending of nationalist pride and newly awoken interest in natural landmarks occurred in late nineteenth-century Sweden (Sundin, 2005).

In Britain, too, public affection for the countryside and architectural landmarks, rather than lofty ideas about wilderness, were at the heart of the first nation-wide environmental campaigns. The National Trust (founded in 1895) and the Council for the Preservation of Rural

England (founded in 1926) played an important role in mobilising popular and political support for the protection of parks, country houses and rural environments (Bunce, 1994). In France, Romantic ideas about the preservation of areas of natural beauty surfaced in artistic and literary circles and fed into a national parks movement, which borrowed ideas from North American examples (Pincetl, 1993: 84–5). More so than in other European countries, however, a long-standing tradition of rational forest and water management going back to the late medieval period and relying on state intervention dominated environmental approaches, as can be seen in the work of the Société Nationale de Protection de la Nature (1854) (Bess, 1995: 832).

Outside Europe, environmental ideas circulated widely in the English-speaking world and developed considerable resonance in nineteenth-century Australia and New Zealand (Wynn, 1979; Beattie, 2011). In countries of the Global South, environmental ideas and practices of sustainable agricultural management have had a long history that predates the arrival of European colonists, though these had comparatively little influence on the development of environmentalism as a political movement. As discussed earlier, colonial rule provided an important context for the creation of environmental awareness, knowledge and conservation practice. Where Southern societies developed their own environmental tradition, they often did so in response to colonial rule and the devastating ecological consequences it had on local forests and soils. In Brazil, for example, environmental debates started as early as the late eighteenth century and were mainly focused on the ecological damage caused by colonists (Padua, 2000). In African colonial societies, nature conservation policies were introduced by colonial rulers but often ended up serving as disciplinary instruments of social control or economic exclusion (Davis, 2000; Neumann, 2001). Unsurprisingly, therefore, Southern traditions of environmental thinking prioritised questions of ecological sovereignty and social justice, and the creation of sustainable livelihoods (Guha and Martinez-Alier, 1997; Martinez-Alier, 2002). As will be discussed in Chapter 5, the emerging 'environmentalism of the poor' was to provide a radical alternative to Northern environmental approaches, though it was only with decolonisation after the Second World War that it gained wider traction in international society, in response to efforts by Northern countries to set the international environmental agenda.

Competing Versions of Environmentalism

The history of global environmentalism is marked by a series of often-intense debates between competing visions for a greener future. These debates concern, at a fundamental level, the question of how humans should relate to their natural environment: does nature possess an intrinsic value that humans need to respect and protect, or should environmental protection be determined mainly by the benefits it creates for human society? Does environmentalism involve the expansion of our ethical horizon to include the non-human world, or is it sufficient to ground it in utilitarian notions of human interest? In other words, should environmentalism be based on ecocentric or anthropocentric norms? Environmentalists are also divided by questions of practical reasoning and political strategy: should environmental objectives be pursued through reform and accommodation within the existing political–economic order, or as part of a strategy of radical change? Can environmental ideas sit next to and work with the established norms of state-centric international society, or does successful environmental action point towards a new form of post-national, post-Westphalian, politics? These debates have characterised the environmental movement's engagement with international diplomacy throughout the twentieth century, and they continue to do so today.

In the early phase of environmental thinking, the main fault line was between what Worster (1994) calls 'arcadian' and 'imperial' versions of ecology. The arcadian vision of ecology draws on the anti-modern instincts that animated the nature writers and Romantic artists of the late eighteenth and early nineteenth centuries. The English clergyman Gilbert White, whose *Natural History of Selbourne* (1789) became a posthumous publishing sensation and inspired generations of nature writers well into the twentieth century, combined traditional religious beliefs with an awakening ecological sensibility. White depicted nature as a well-ordered and harmonious system in which every species occupies a specific niche. The challenge for humans was to find their place within the God-given natural order and restore the organic unity of all life forms (Worster, 1994: 4–9). White's and the Romantics' vision of how humanity should relate to its natural environment was one of care and restraint. By contrast, the imperial version of ecology promoted a better scientific understanding of nature in order to enable humankind to control and dominate it. Already visible in the taxonomic work of Carl von Linné and the Linnaean tradition in biology, imperial ecology advanced an anthropocentric and utilitarian world view that advocates a more efficient use of nature through the application of scientific methods. It conceived of environmentalism as a rationalist project that provides tools for human advancement and economic progress, not restraint (Worster, 1994: 50–5).

The divide between arcadian and imperial versions of ecology had a profound impact on the development of the early environmental movement, particularly in North America where preservationism and conservationism came to form the two main strands of environmentalism (Meyer, 2017: 34–35; McCormick, 1989, 12–17). Romantic ideals and arcadian ecological thinking played a key role in preservationists' desire to protect the remaining landscapes that had not yet been transformed by the rapidly expanding US economy. John Muir, the founder of the Sierra Club and 'Father of the National Parks' in the United States, argued for the protection of areas of wilderness as a sacred refuge that provided humans with an antidote to modern industrial society. In his influential writings, most notably *My First Summer in the Sierra* (1911), Muir echoed Henry David Thoreau's famous dictum that '[w]e need the tonic of wildness' (*Walden*, 1968 [1854]: 280). For Muir, as much as for the Transcendentalists Ralph Waldo Emerson and Thoreau, environmentalism grew out of a desire to seek accommodation within, not dominance over, the natural environment. Muir's campaign for national parks sought to remove areas of outstanding natural beauty from commercial use, not manage them more rationally, in order to preserve the aesthetic value of an unspoilt nature (Steinberg, 2002: 139).

If Thoreau is the spiritual father of the preservationist tradition in American environmentalism, then George Perkins Marsh can be said to have laid the foundations for its conservationist alternative. Marsh's book *Man and Nature* (1865) provides one of the first systematic accounts of the ecological and economic damage caused by deforestation, which leads to soil erosion and reduces soil productivity. Marsh warned that unless the principles of nature conservation were respected and acted upon, the United States was set to repeat the devastating consequences of deforestation and desertification that had afflicted the ancient civilisations of the Mediterranean. By establishing the intellectual foundations for the efficient and sustainable use of forests and other natural resources, *Man and Nature* was to have a lasting influence on forestry policy in the United States. When President Theodore Roosevelt appointed Gifford Pinchot to head the newly created US Forest Service in 1905, Pinchot and his colleagues, who were trained in the latest developments in forestry, geology and hydrology, set out to put Marsh's ideas into practice and developed a more rational approach to the use of the nation's natural wealth. As Pinchot argued in his *The Fight for Conservation* (1910), with a clear nod to Jeremy Bentham's utilitarian philosophy: 'Conservation means the greatest good to the greatest number for the longest time' (quoted in Meyer, 1997: 270). For conservationists, the aim behind national parks and sustainable forestry was not the preservation of nature for its own sake, but the protection of nature's continued use for economic gain. Theirs was a 'gospel of efficiency' (Hays, 1959), based on sustainable use principles, that put anthropocentric values at the heart of environmentalism.

Preservationists and conservationists came to clash repeatedly over both the ends and means of environmental policy in the United States. The conflict over the so-called Hetch Hetchy project, which involved the building of the O'Shaughnessy Dam in a scenic valley in California (first proposed in 1890, completed in 1923), provided one of the most famous flashpoints in this struggle between rival environmentalist traditions. Muir and other preservationists opposed the building of the dam in a valley that was part of Yosemite National Park. Rejecting utilitarian arguments for the benefits of sustainable electricity production, they pointed out that the dam would destroy idyllic river valleys and breeding grounds of wildlife (Meyer, 1997: 272–3, 276–7). But despite such high-profile conflicts, the differences between the two main strands of environmentalism should not be exaggerated. Preservationist and conservationist arguments often overlapped, and in many cases both sides pursued the same agenda. Where they presented a united front, they could shape national and international policy in a lasting manner. Theodore Roosevelt, the first US president to establish nature conservation as a national policy objective, took inspiration from preservationists such as John Muir but relied on utilitarian conservationists such as Pinchot to implement his policy goals (Dorsey, 1998, 11). Domestic support for the first international conservation treaties that the United States signed with Canada came from across the environmental spectrum, and the success of those treaties depended on 'the ability of conservationists to justify [them] on both sentimental and economic grounds' (Dorsey, 1998: 16).

The debate between arcadian and imperial versions of ecology, and between preservationist and conservationist approach to environmental action, has continued to resurface in ever new disguises throughout the twentieth century. It can be found in philosophical debates between 'deep ecology', a perspective that is rooted in biological egalitarianism and places equal emphasis on all interrelationships among nature's organisms, and what their proponents describe as 'shallow ecology', the conventional focus on fighting industrial pollution and resource depletion in the interest of human well-being and prosperity (Naess, 1973; De Steiguer, 2006: chapter 16). It comes to the fore in the contrast between ecocentric and anthropocentric philosophical justifications for environmentalism, with the former emphasising the need to protect nature's intrinsic value and the latter viewing environmental protection through the lens of human utility (Eckersley, 1992). And it also informs the differences in political strategy between more radical environmentalists that seek to protect nature at all cost and more reformist environmentalists that promote environmental measures in balance with other societal objectives, such as economic growth.

Towards a Global Ecological Consciousness

With the exception of colonial scientists and administrators, environmentalism in the nineteenth century was predominantly focused on local ecological problems. The twin forces of industrialisation and colonialism certainly made sure that environmental degradation took on ever more transboundary forms, but it was not until the twentieth century that humanity was beginning to exert a dominant influence on the earth's geology and ecosystems. Awareness of the growing global environmental interdependencies also took its time to develop. The first step on this journey from local to global ecological consciousness was the emergence of ecology as a scientific paradigm. Building on the insights of global explorers, such as Darwin and Humboldt, ecology established itself as a scientific discipline by the early twentieth century. Over time, the study of ecology began to amass more and more evidence of humanity's growing impact on the natural environment and at different scales. Technological innovation helped improve scientists' ability to detect pollution, track its global spread and assess its impact on human health and ecological systems. These advances in ecological science were of critical importance to the emergence of a global ecological consciousness, and they also helped to transform environmentalism into a global political force with profound consequences for international society's normative structure.

The science of ecology grew out of efforts by eighteenth- and nineteenth-century naturalists and explorers to discover and record the planet's flora and fauna and to study geological and atmospheric conditions around the world. In his multi-volume masterwork *Kosmos,* published between 1845 and 1862, Alexander von Humboldt had already established the idea of the planet as a single, living entity (Wulf, 2015). Other researchers added further evidence of global ecological links that slowly gave rise to the new scientific world view of ecology. The term 'oecologie' was first coined in 1866 though it was Eugenius Warming's seminal book *Plantesamfund* (1895, published in English as *Oecology of Plants* in 1909) that produced the first key intellectual synthesis and allowed other scientists to develop ecology's conceptual framework, with American universities taking the lead in creating the new field's scientific foundations in the early twentieth century (Worster, 1994: 192; Nash, 1989: 55; Golley, 1993: 2–3). After the Second World War, ecological ideas slowly began to seep into public discourse and inform environmental politics. Barry Commoner's publications did much to popularise ecological thinking in the English-speaking world. His first law of ecology – 'everything is connected to everything else' (Commoner, 1971: 29) – still offers the most succinct summary of the new sense of global interconnectedness that was behind environmentalism's global turn from the 1960s onwards.

Governments in the leading powers after the Second World War played a key role in the growth of global ecological knowledge. Many of the technological advances that enabled scientists to develop a more sophisticated understanding of the planet's global ecology were funded by the two superpowers during the Cold War. Driven by the desire to develop new military capabilities, the US government in particular poured vast sums of public money into the environmental sciences, from oceanography to atmospheric science, physical geoscience and meteorology (Doel, 2003; Goossen, 2020). The space race between the superpowers led to new capacities for geo-monitoring, with the first satellites used for monitoring global weather patterns launched by the US Department of Defense. As part of its wartime experiments on nuclear weapons and post-war efforts to track Soviet nuclear tests, the US government promoted new radiocarbon dating techniques, which later became a key tool in tracking the movement of carbon from fossil fuels into the atmosphere. During the 1950s and 1960s, the US Navy also sponsored research on whale intelligence and acoustics, which helped substantiate environmentalists' claims that whales deserved special protection as uniquely intelligent mammals. (Weart, 2003, 21–35 and 110; Garfield, 2013: 240; Burnett, 2012, chapter 6).

Environmental Thinking Today: Between Unity and Diversity

By the second half of the twentieth century, environmentalism had evolved from a few disparate normative initiatives in the nineteenth century into a coherent set of ideas about nature protection and a political programme for political change. Built around a normative core (environmental stewardship), informed by a scientific programme (ecology) and driven by a social and political movement (environmental grassroots organisations and NGOs), environmentalism had morphed into a distinctive political ideology and movement with the desire to bring about lasting change in domestic and international politics.

At the same time, the environmental tradition has produced many different strands of thinking over the last two centuries. From an early stage, environmentalists were motivated by different, if not mutually exclusive, ethical positions: In the nineteenth century, preservationists argued for protecting nature's intrinsic value while conservationists justified environmental protection on the basis of its benefits for society and economy. Similar debates resurfaced during the twentieth century, with the more radical wing of the environmental movement demanding an ecocentric, or deep ecology, perspective and the more conservative wing arguing for an anthropocentric or reformist notion of environmental politics.

Environmental thinking also continues to show distinctive national roots and cultural biases. Conceptions of 'nature' and 'environmental protection' vary across different societies, and distinctive regional differences exist in terms of how the environmental agenda is defined. Environmentalists in industrialised countries have long focused on issues relating to the preservation of wilderness or pollution from industrialism, while environmentalists from the developing world have tended to emphasise the links between environmental protection, sustainable livelihoods and social justice. When analysing environmentalism's influence on the normative structure of international society, we therefore need to pay close attention to the ongoing tensions and debates within the environmental tradition and the ways in which different strands of environmentalism have come to dominate global debates at different times.

Environmental Visions for Global Order

In order to successfully introduce a new norm into international society, norm entrepreneurs need to meet two critical conditions: They need to identify a global problem that requires not just an ad hoc solution but a profound change in the international normative order; and they need to propose a new norm that addresses this problem and develops 'resonance' with relevant audiences, because of some affinity with the existing normative structure (Keck and Sikkink, 1998; Payne, 2001). As discussed earlier, environmentalism emerged in the nineteenth century in response to rapidly rising levels of environmental degradation. As industrialisation spread worldwide throughout the nineteenth and twentieth centuries, the ecological pressures created by the global industrial system became ever more apparent. Environmentalists may have put forward different conceptions of nature protection and justified environmental action with divergent ethical frames, but they successfully identified an overwhelming need to protect the environment. Norm entrepreneurs also need to propose a new norm that resonates with its target audience and is capable of winning enough support for it to be established. New norms usually come up against contestation and resistance, and out of the many normative proposals that are routinely made in political debates, only some manage to be adopted. The successful introduction of the environmental stewardship norm will be examined in later chapters, but the question for the remainder of this chapter is how the new norm of environmental stewardship could be made to fit into the international order.

Seeking to green international society raises several difficult questions for environmental norm entrepreneurs: how would the new environmental norm relate to the state-centric foundations of an international society that is based on sovereignty, territoriality, international law, diplomacy, war, the balance of power and great power management? Would the new environmental norm be consistent with these primary institutions, or would it challenge some or all of them? If some form of normative accommodation is needed to establish environmental stewardship as an international norm, how would the environmental norm – or other existing norms – need to adjust?

Furthermore, if the division of the world into territorially defined sovereign states is part of the problem of the global ecological crisis, as some environmentalists claim, would a greener international society need to move beyond its Westphalian roots? Would states have to pool sovereignty and establish powerful international authorities to keep the world within its ecological limits? Is indeed a green world government inevitable if humanity is to survive?

Finally, would the society of states have to engage more with world society, the realm of social and political interaction among non-state actors, in order to find viable responses to environmental problems? Would it eventually have to merge with world society in some larger and more complex system of global governance? Or is world society the primary site of effective environmental action? Does an effective global ecological rescue depend on circumventing state-centric international society altogether and building an ecological world society?

As we have seen, environmentalism is held together by a common concern for the natural environment but is made up of different ethical traditions and has endorsed different political strategies for resolving global environmental problems. Unsurprisingly, therefore, different proposals have emerged within the environmental tradition for establishing a green global order. Some of these proposals are directed at making international society more environmentally responsive. Others seek to establish a green global order out of the interactions between individuals and societal groups in world society. Some assume a high degree of decentralisation and value diversity within the new green order, others aim for the creation of a universal set of values that would govern a more

Table 3.1. *Models of green world order*

	Pluralism	Solidarism
International Society	*(A) Green Westphalia*	*(B) Global environmental governance*
World Society	*(C) Eco-localism*	*(D) Eco-globalism*

sustainable relationship between humans and their natural environment. The ES framework developed in Chapter 2 helps us bring some order to the often-bewildering array of proposals for a greener world system. To be sure, this is not the only way to structure the environmental debate, but in the context of the central question in this book – whether global environmentalism has become part of the normative structure of international society, and how it has influenced normative developments in international relations – the ES approach provides a useful logic for thinking about green global order models. Based on the core conceptual distinction between international and world society, and between pluralist and solidarist normative orders, we can distinguish four principal models, as depicted in Table 3.1:

- *Green Westphalia* (model A), which builds on the state-centric international order in a pluralist context, with a high degree of value diversity, decentralised decision-making and only thin levels of international cooperation.
- *Global environmental governance* (model B), which envisages a transition from a pluralist to a solidarist society of states, with greater normative convergence and deeper cooperation among otherwise sovereign states.
- *Eco-localism* (model C), which operates in a world society characterised by a high degree of value pluralism and decentralisation, with local communities seeking to realise their own versions of environmentally sustainable livelihoods largely on their own.
- *Eco-globalism* (model D), which sees individuals and societal groups form a world community that builds on a common set of values and pursues a shared vision of global environmental sustainability.

To be sure, these four models represent ideal types of a green world order. They abstract from the complex reality of relating norms to politics, and they ignore the fact that they often overlap in global environmental discourses and politics. Thus, we find that most of the international environmental regimes created over the last 50 years are rooted in the Westphalian principles of international society but some are motivated by a solidarist ambition to create global environmental governance. Likewise, state-centric and world society-oriented models of environmental politics have gradually come to overlap. International environmental regimes may be created and controlled by states, but they increasingly engage a wide range of non-state actors in the process of implementing international rules. And the myriad of environmental groups that operate in world society are creating new structure of environmental governance outside state-centric realms, but also engage with states in building hybrid structures of global environmental authority. These four models thus represent four different logics of developing a global green order, though in reality we would expect these logics to interact and overlap to a considerable extent. I will now discuss each of the models in more detail.

Green Westphalia

The first model of a green global order operates within the constraints of pluralist international society. It accepts and works with the core pluralist norms (e.g. sovereignty, territoriality, diplomacy, balance of power, war, great power management) that originate in the classic Westphalian order, hence the label *Green Westphalia*. It assumes that states pursue environmental objectives only where these correspond to their narrowly defined national interest. Maximising national power and wealth will usually override concerns for the natural environment, unless environmental protection can be achieved cheaply or environmental degradation impinges directly on those core state interests, for example where an ecological catastrophe reduces a nation's power and wealth or threatens its survival. In the Green Westphalian scenario, states follow an ecological *raison d'état* that prioritises domestic environmental objectives over global ones while defending the state against environmental harm emanating from outside its borders. Following the pluralist logic of coexistence, states will pursue global environmental objectives mainly insofar as they serve domestic objectives. If such objectives require working with other states, then a Green Westphalian order can include some degree of international cooperation, though this is bound to be weak, mostly short-lived and based on a narrow range of shared norms. Shared fate issues, for example where major ecological crises pose an existential threat to international society as a whole, may lead to more extensive international cooperation consistent with the pluralist logic of coexistence (Buzan, 2004: 145; Attfield, 2014: 183). In most other cases, however, states will interpret an international norm of environmental stewardship through the lens of environmental nationalism

(De-Shalit, 2006) while remaining largely indifferent about global ecological concerns.

In the Green Westphalian model, global environmental goals are to be achieved in line with core primary institutions. It involves a gradual expansion of diplomatic activities to achieve common environmental objectives, including the negotiation to a narrow range of secondary institutions. It engages existing norms of non-interference and dispute settlement in international law, and it leaves open the possibility of some form of great power responsibility for the pursuit of environmental objectives that are central to the stability and survival of international society. One area where a pluralist international society could give rise to more forceful international environmental action is the use of hegemonic power in the creation of a greener international order in line with Westphalian principles. If committed to environmental stewardship on a global level, a hegemon – be it a single power or a group of great powers providing collective hegemony (Clark, 2011) – could take it upon itself to create the conditions for more durable and meaningful international environmental cooperation. This form of environmental hegemony would instil a sense of ecological *raison de système* into an otherwise decentralised pluralist international order. As suggested by hegemonic stability theory, the hegemon could provide leadership in the creation of international environmental regimes, induce other states into global cooperation and enforce common international rules through sanctions or side payments or generally support global environmental action by providing the necessary economic or technological resources (Falkner, 2005: 589).

The use of power asymmetry in support of global environmental protection is not without historical precedent, though it is hampered by considerable legitimacy problems. A first manifestation of environmental hegemony can be said to have emerged during the nineteenth century, when colonial powers increasingly saw it as part of their 'civilising mission' to improve nature conservation in their African and Asian colonies (see chapter 4). By the early twentieth century, European environmentalists had launched the first campaigns for the protection of large mammals and the creation of natural parks in colonial territories, arguing that the wild fauna and flora in Africa and Asia belonged to humanity and needed to be protected for the benefit of future generations. Similarly paternalistic arguments were made by nature conservation organisations during and after the decolonisation process in the second half of the twentieth century. Unsurprisingly, the colonial legacy of nature conservation severely undercut later efforts to create international environmental trusteeships during and after decolonisation. Weak legitimacy is the

Achilles heel of any hegemonic project to create and impose a global green order.

Green Westphalia also works within the parameters of classical international law, which is state-centric in nature and focused on guaranteeing the coexistence of sovereign states. Developed in the nineteenth century, classical international law establishes only a minimal set of state responsibilities and is largely limited to dispute resolution. Applied to the environmental field, states' main legal duty is to prevent transboundary environmental harm being inflicted on their neighbours, leaving them free to act in their domestic realm as they see fit. Environmental resources that exist outside any state's national boundaries (e.g. high seas) have traditionally been assumed to be inexhaustible and politically unproblematic. In legal terms, they are referred to as *res nullius* – 'belonging to no one and therefore open to all' (Bodansky, 2010: 21). The first international environmental treaties that were negotiated before the 1972 Stockholm conference followed these legal principles. They respected national sovereignty but sought to manage the transboundary dimensions of environmental problems, such as the protection of migratory birds or whales or the prevention of oil pollution from international shipping (see Chapter 4). States created international environmental law via intergovernmental treaties, though this need not involve a more extensive development of legal norms to advance global environmental sustainability.

If applied consistently and with the full force of law, the *Green Westphalian* model has the potential to play an important role in regulating states' external environmental behaviour. In theory, the principle of state responsibility in customary international law could give rise to state liability for transboundary environmental damage to other states, which could act as a strong deterrent against transnational forms of pollution (Kettlewell, 1992: 438–9). Although the no harm principle was enshrined in the 1972 Stockholm Declaration (Principle 21), the foundational text of the environmental stewardship norm (see Chapter 5), in practice it has remained an ambition rather than a regulatory norm. This is partly because the no harm principle is restricted to cases where transboundary harm is not only substantial and serious but also attributable to a particular state (Tinker, 1995: 786). More importantly, its effective enforcement would require the kind of powerful international authority enforcing international law that is at odds with the pluralist nature of international society. For this reason, the regulative power of classical international law has remained extremely limited.

More generally, Green Westphalia falls short of environmentalists' demands for globally coordinated action because of the narrow conception of ecological *raison d'état* within a pluralist context. Many global environmental problems (e.g. species extinction, biodiversity loss, protection of marine ecosystems) do not represent an immediate and direct threat to core state interests but require sustained international cooperation based on shared values. Because ecological vulnerability and sensitivity are not equally distributed among the world's nations, a pluralist international society is unlikely to generate the kind of commitment by states that is needed to create and sustain deeper forms of international environmental cooperation.

Global Environmental Governance

The solidarist alternative to Green Westphalia is *global environmental governance*, the second pathway towards a state-centric green global order. Global environmental governance represents a more ambitious model as it includes the creation of strong international institutions based on a global consensus around humanity's universally shared environmental values and beliefs. Global environmental governance puts forward a cosmopolitan vision of global environmentalism, which is about protecting humanity's common environmental heritage, for both current and future generations. States assume a responsibility not only for their own natural environment but also for the global commons, working together to protect global ecosystems. While still operating within a state-centric international system, this solidarist model envisages the creation of an ever denser web of international rules and regulations (regimes, or secondary institutions) that can 'guide international behavior along a path of sustainable development' (Keohane, Haas and Levy, 1993: 4). Global environmental governance does not negate the existing primary institutions of sovereignty, territoriality, diplomacy and international law. Instead, it seeks to bring about a greening of these fundamental norms so that they support the objectives of global environmental sustainability. Creating global environmental governance involves a reform project that moves international society away from its pluralist origins towards recognition of humanity's shared interest in a globally managed environment.

Solidarist ideas and ambition have a long history in the global environmental movement. After the Second World War, when global ecological interdependencies came to the fore in ecological debates, environmentalists urged states to establish an international institutional framework, based on international law, for steering societies and economies towards a greener future. References to the planet as a fragile ecosystem that is shared by humanity increasingly came to frame debates about the environmental responsibilities of states and the need for a more expansive network of international rules. The notion of the planet as 'Spaceship Earth', which rose to prominence in the 1960s (Ward, 1966; Boulding, 1966), came to epitomise humanity's unity in the face of a worsening global environmental crisis. Adlai Stevenson, the US Ambassador to the UN, famously referred to humans as travelling like 'passengers on a little space ship, dependent on its vulnerable reserves of air and soil; all committed for our safety to its security and peace; preserved from annihilation only by the care, the work, and, I will say, the love we give our fragile craft' (quoted in Tal, 2006: 168). NASA's famous images of planet Earth taken from space, especially 'Earthrise' in 1968 and 'The Whole Earth' in 1972, reinforced this sense of global unity amidst a planetary ecological crisis (Cosgrove, 1994). For environmentalists, the solidarist implications were clear: Faced with the threat of collective destruction, humanity needed to overcome its division into nation-states and reimagine the world as 'a community sharing a common destiny' (Höhler, 2015: 5).

Global environmental governance fits well with the primary institution of diplomacy, indeed it creates the expectation of a significant increase in environmental diplomatic activity. It opens up a new agenda of creating a large number of secondary institutions and of equipping states with the foreign policy tools they need to engage with this new layer of intergovernmental governance. Great power management, although more commonly seen as a pluralist primary institution, could support a solidarist vision for green global order where great power responsibilities take on a more expansive interpretation, referring not only to minimal system maintenance in international relations but also to guardianship for the global environment.

Global environmental governance also works with and strengthens international law. The creation of more ambitious secondary institutions leads to an expansion of international environmental law and also pushes it into a more solidarist direction. One such area of legal development is the concept of 'common heritage of humankind'. First developed in the late nineteenth century (Milicay, 2015: 273–4) and formally introduced in the context of the Law of the Sea negotiations in 1967, the common heritage idea focuses on ecosystems that exist outside national jurisdiction and therefore require collective management by international society on behalf of humanity, including in the interest of future generations (Attfield, 2014: 189). It provides a blueprint for how international society could strengthen shared responsibilities and promote the collective management of environmental problems that cannot be solved by individual states. The common heritage idea has since become an established legal concept in the Law of the Sea and in the regimes to protect outer space and Antarctica (Wolfrum, 1983), though its extension to other areas of global environmental protection remains heavily contested. Solidarist treaty-making has also led to demands for strengthened compliance and dispute settlement mechanisms, which would further enhance the authority of international law.

The solidarist vision for global green order has a potentially more disruptive impact on other primary institutions, such as the market, nationalism and sovereignty. Together with the market norm, solidarist global environmentalism favours global over national rules. However, the need to bring about a radical restructuring of the global economy to tame its ecologically destructive tendencies stands in contrast to the decentralised nature of the market principle that puts severe limits on politically motivated interventions by the state or international authorities. Against nationalism, the global environmental governance model assumes universal environmental values and interests that can only be realised through the creation of strong international rules and governance mechanisms. It downplays the local and national character of environmental sentiment in favour of a form of green one-worldism, which is further reinforced by perceptions of a looming ecological crisis with existential consequences for all of humanity. Solidarist environmentalism similarly militates against a strict application of sovereignty and territoriality, both cornerstones of the Westphalian order. A system of global environmental governance can be created within a sovereignty-based international society but requires a reinterpretation of what it means to be a territorially based, sovereign nation-state. A move towards more effective global governance makes it difficult to maintain a position of strict non-intervention in cases where states commit gross acts of ecological harm.

The next logical step in this direction would be the creation of an ever-stronger international authority that can enforce global rules in the planetary interest: a green world government. For some environmentalists, the urgency of the global ecological crisis makes it necessary to overcome humanity's division into nation-states in order to organise an internationally coordinated ecological rescue. Neo-Malthusians, for example, have long argued that the planet is close to running out of sufficient food, energy and mineral resources to sustain a growing population, making it impossible for politics to carry on in a business-as-usual manner, based on small-scale and incremental reforms (Meadows et al., 1972). Only a world government would be equipped 'with authority and coercive power over sovereign states sufficient to oblige them to keep within the bounds of the ecological common interest of all on the planet' (Ophuls, 1977: 210). Should a worsening ecological crisis deepen the zero-sum-logic of the Westphalian order, setting off a competitive race for global resources, then only a world 'government of rescue' (Bahro, 1989) could ensure an orderly and fair allocation of scarce resources and force societies to live within ecological boundaries. To be sure, the idea of a world government has never developed much traction in international relations – 'world government is not around the corner' (Keohane, Haas and Levy, 1993: 4) – though it lives on in IR debates (Wendt, 2003; Craig, 2011). It has remained marginal even within environmental circles, raising fears of an overbearing form of environmental politics, or even eco-fascism. It is also coming up against the distinctly anti-statist instincts of large parts of the environmental movement. Given its utopian status, world government proposals have remained at the fringes of the debate on creating a green global order. I shall therefore restrict the discussion of solidarist order proposals to those that focus on creating global environmental governance.

Eco-localism

The two models of a green global order discussed so far are all focused on the society of states. They vary according to the degree to which pluralism versus solidarism characterises international society, and the extent to which it is possible to create effective and strong international institutions to govern the behaviour of states. But state-centric politics is only one route towards a greener global order. For many environmentalists, the main hope lies not in governmental action but the capacity of societal actors to chart a path towards greater sustainability. Global environmentalism is as much about creating a green world society as it is about greening interstate relations. As with international society, we can distinguish two principal types of world societal order: a pluralist version, hitherto referred to as *eco-localism*, which is characterised by a high degree of value diversity, with individuals belonging to local communities and local forms of environmental governance coexisting with only weak transnational links; and a solidarist version, or *eco-globalism*, which is built around a universal set of values, a high degree of cooperation and integration across societies and transnationally integrated forms of environmental governance.

To be sure, just as with pluralist and solidarist international society, pluralist and solidarist world society are analytical ideal types, they coexist and interact in reality. After all, 'think globally, act locally', the famous slogan of 1970s environmentalism, captures the seemingly paradoxical nature of a modern environmental movement that is rooted in local contexts but conscious of global ecological connections.

Environmentalists usually recognise the need to balance local and global dimensions, though they do not generally agree on how this balance can be achieved. For eco-localists, the dominant political focus is on the local level. Their politics is rooted in local grassroots campaigns and the solutions they seek to implement are found in local communities. They reject economic and political globalisation as one of the main sources of the global ecological crisis. Reflecting a profound belief in social and cultural heterogeneity, they resist the homogenising tendency of global capitalism and international governance. Their environmentalist vision is one of self-reliant communities that manage to live within the ecological constraints imposed by nature. These communities may all subscribe to a globally shared commitment to the principles of sustainability, but eco-localists expect each community to find its own path towards an environmentally sustainable future. Viewed through an ES lens, eco-localists thus represent the pluralist end of green world society.

Eco-localism has many different philosophical roots. It can be found in the communitarian belief in human identity as bounded in space and time, which views individuals as belonging primarily to local or familial communities. Communitarian environmentalism, which is a major source of inspiration for conservative environmentalism (Scruton, 2012), views local emotional ties as the main motivating factors for environmental action. It emphasises the fact that many of the environmental battles that are being fought against polluting factories, toxic waste dumps or deforestation are fought at the local level and depend on the ability to mobilise 'place-based activism' (Eckersley, 2006: 97). Communitarians belief that the most viable solutions for environmental problems exist not in international regulatory schemes but in locally organised responses. In similar fashion, deep ecology values decentralisation of power and local political autonomy as the basis for effective environmental action (De Steiguer, 2006: 187). By redirecting economic activities to the community level and developing new forms of shared ownership – from local currency schemes to agricultural cooperatives and car sharing clubs – environmentalists hope to inject a sense of both place-based belonging and ecological responsibility into economic exchanges.

Eco-localists are anti-statist and anti-international in the sense that they seek to return the locus of political action from the nation-state to those levels of societal organisation that are most in tune with ecological systems. Some focus on local politics as this is the level at which direct democracy and accountability are easiest to establish, while others advocate bioregionalism as the organising principle for political systems, with societies coming to respect the ecological relationships that exist in their relevant geographical space (Eckersley, 2006: 97–8). By living in bioregions, humans become more aware of their natural surroundings, understand the ecological limitations of their consumption patterns and are able to adjust their lifestyles to suit their ecological surroundings (Sale, 1985). Ernest Callenbach captures the essence of such an eco-local vision in his futuristic novel *Ecotopia* (1975), which sold over a million copies and became a 'cult classic' in the genre of environmental utopian thinking (Allitt, 2014: 83). Ecotopia is the name of a fictional new state, formed out of the US states North California, Washington and Oregon, which has turned its back on a materialist and growth-obsessed American society. It consists of self-contained local communities that use simple technologies and consume locally produced food as part of a no-growth economy. Like Ernst Schumacher's *Small Is Beautiful* (1973), another classic of 1970s environmentalism, Callenbach advocates devolved decision-making for politics and the economy, allowing small-scale communities to work out their own path towards environmental sustainability.

Eco-localism is widely debated and practiced in environmentalist circles around the world, wherever local communities resist the globalising logic of global capitalism, but it has proved difficult to carve out the social and political spaces to develop it as a more viable political alternative. Eco-localists' belief in pluralism also comes up against the arguments of those that only a globally connected environmental movement is able to challenge the power structures that keep the global political economy on its ecologically destructive path. Moreover, while it may be possible to create a sustainable economic model in specific local contexts that carefully manages local environmental resources, it might still fail to address the global challenge that transboundary ecological problems, such as climate change and marine protection, pose. A myriad of local environmental solutions may not add up to an adequate solution for global environmental problems. As eco-globalists argue, it is not enough to 'think globally', the environmental movement also has to organise and 'act globally'.

Eco-globalism

Eco-globalists view the ecological crisis as a global threat that has the potential to unite humanity behind a common cause. Their search for a political solution revolves around the creation of a global community of individuals and groups that act on the basis of a shared understanding of the ecological challenge and a common vision for a global green order. As Newell (2020: 2) notes: 'Green articulations of political community are intrinsically global'. The environmental movement may comprise many strands of environmental thinking and practice, but it is its underlying unity of purpose and normative outlook that creates the conditions for a successful transition to green global order. Eco-globalism thus represents the solidarist, cosmopolitan, ideal within world society.

Eco-globalists view world society as the appropriate political space for finding a globally coordinated response to the ecological crisis. Operating outside the state-centric society of states, they engage other world society actors in the creation of transnational networks that seek to advance the cause of environmental sustainability. These non-state actors comprise environmental campaigners but also business actors and other societal groups that have the capacity for environmental problem-solving. Environmental action in a world society context comprises a wide range of political action at different scales of political organisation, from the local to the subnational, transnational and global. As such, eco-globalism is a multifaceted phenomenon that is difficult to pin down. It can range from ecological knowledge creation and sharing of scientific knowledge to environmental awareness raising and the creation of environmental norms and rules: private environmental governance, in other words.

International environmental NGOs are at the forefront of creating the global ecological consciousness and concern that defines the eco-globalist model. Whereas the older conservationist or preservationist organisations were predominantly rooted in local contexts and only slowly formed transnational linkages, many modern environmental organisations founded from the 1960s onwards have adopted an explicitly global outlook. Modern environmental NGOs usually espouse an ethic of global citizenship. When appealing to citizens for support in their global campaigns, they do this not with reference to local identities but on the basis of a cosmopolitan belief in the value of protecting the global environment, be it endangered species or ecosystems (Attfield, 2014: 184). Greenpeace is the leading example of an environmental NGO that puts global environmental threats (e.g. nuclear weapons testing, whaling, global warming) at the centre of its transnational campaigns (Wapner, 1996: 48). More than many other NGOs, Greenpeace has adopted a centrally organised approach to transnational campaigns, operating 'as a hierarchy with power centralized in the international office in Amsterdam' (Doherty and Doyle, 2013: 9). Its campaigns, which generally speak to a worldwide audience, reflect the globalist tradition of environmentalism that emphasises human solidarity in a world of ecological interdependencies.

Environmental NGOs are the leading agents of eco-globalism, but the solidarist vision of a green global order also encompasses the activities of other world society actors. Scientists have played an important role in globalising environmental awareness. By highlighting the planetary scale of ecological interdependencies, environmental sciences have prompted humans to rethink their place in the world and to see themselves as part of larger ecological systems that transcend national boundaries (Warde, Robin and Sörlin, 2018: chapter 4). Subnational actors, such as cities and municipalities, are also connecting with each other across boundaries in global action networks, thereby circumventing national governments as the sole representatives of their societies in the international context (Betsill and Bulkeley, 2006; Bansard, Pattberg and Widerberg, 2017; Van der Ven, Bernstein and Hoffmann, 2017). Furthermore, companies have emerged as important transnational actors in the search for environmental solutions outside the state-centric international system. Often reacting to civil society and consumer pressure, a growing number of businesses have engaged in private governance initiatives that seek to limit or reduce the environmental burden of economic activities (Falkner, 2003; Green, 2013; Neuner, 2020).

Eco-globalism is thus about organising a global response to the ecological crisis within a world society that is characterised by common values and transnational governance networks. Understood as the 'great society of humankind' or *normative* world society (Buzan, 2018: 127–9), eco-globalism expresses a globally shared identity that is cosmopolitan in nature, with humankind as its referent. It exists and operates apart from international society but can often be found to be engaged in close interaction with the state-centric world. Understood as a *political* world society (Buzan, 2018: 129–30), eco-globalism injects its ideological resources into the operations of the society of states, lobbying states to build global environmental governance. Where it intersects with international society, advocacy (Buzan, 2018: 135) serves as the main primary institution of world society. Depending on the intensity of interplay between ecological world society and international society, we can also imagine a situation in which the two enter into a state of ever closer interaction, creating integrated global governance mechanisms involving both state and non-state actors. This would amount to what Buzan (2018: 136) refers to as the eventual emergence of an 'integrated world society based on acknowledged and legitimate functional differentiation amongst different types of units'. To date, we can find evidence of autonomous world society-based environment action and some degree of interplay between world and international society. Full integration of international and world society based on functional differentiation remains as yet only a theoretical possibility (see Chapter 11).

Environmental Stewardship as an International Norm

In this chapter, I have traced the origins and evolution of environmental thinking and identified the different ways in which environmental ideas map onto the international/global realm. The next two chapters (Chapters 4 and 5) examine the origins, emergence and social consolidation of environmentalism as a primary institution of global international society, from the nineteenth century up to the 1970s. The focus in these chapters will gradually narrow from the broad tradition of environmentalism as it grew into a transnational movement to the specific international environmental norm (which I label 'environmental stewardship') that was imported into international society's normative structure. Before I can take up the historical narrative, a brief discussion of the nature of this international environmental norm – its content and boundaries – is warranted.

As noted in chapter 2, to qualify as a primary institution a new norm needs to be based on a single value or principle that applies across regional or global international society (Falkner and Buzan, 2019: 136). This value or principle is not fixed in eternity. Just like other fundamental norms, it is malleable in response to the changing collective purposes that it represents. However, malleability and change in the long run as such do not negate the existence of a core value that is distinctive and recognisable over time. Much like political ideologies (Freeden, 1996: 77), international norms combine core and peripheral concepts. The normative core gives a norm its stable identity – what remains constant throughout the maelstrom of historical change. Peripheral concepts, in contrast, give norms a degree of flexibility. They provide a bridging function between the normative core and empirical reality, they allow political principles to be applied to specific historical contexts and they relate the normative core to other fundamental values and principles.

As discussed earlier, environmentalism needs to be understood as a broad church that combines a range of different philosophical and political positions. At its heart is, put simply, an environmental awareness and concern. Environmentalists are united by an ecological consciousness, an awareness of the threat that humanity poses to the health and diversity of its natural environment. They also share an ecological concern, that is an ethic of care for the natural environment, which gives rise to notions of environmental stewardship or trusteeship, whether they are rooted in religious of secular beliefs (Attfield, 2014: 21–3). Beyond this shared inner normative core, environmentalists hold diverging views on key aspects of their environmentalist belief systems. They differ over whether environmental concern is motivated by human-centred values (anthropocentrism) or should be based on an ecocentric perspective that attributes intrinsic value to all living beings and ecosystems. They disagree over questions of political strategy, whether effective environmental protection can be achieved with gradual adjustments to existing political–economic systems (reformists) or whether it requires a fundamental transformation of these very systems (radicals). And they hold sharply diverging views on the role of the nation-state, and by implication international society, whether it is an important agent in the greening of society and economy or is itself deeply implicated in the global ecological crisis.

The environmental norm that emerged as a primary institution in international society also has an identifiable core and more variable – and contested – peripheral elements. Its emergence and evolution in international relations, which I trace in subsequent chapters, is thus a story about how environmentalism inserted itself into the international normative order and how interaction with international society in turn shaped the environmental norm and gave it specific shape and form.

At its core, environmental stewardship posits a fundamental responsibility of the state, and of international society, to protect the natural environment. States are expected to act as guardians of the environment not just within their own territory but also in a regional and global context. Guardianship in this context expresses an ethic of care for the environment, not in the sense of a religious belief system but a secular recognition of the interdependence between humanity and ecology. It is a form of enlightened self-interest that seeks to transcend short-termism in favour of a more long-term sustainability perspective – 'living today as if tomorrow will happen' (Berry, 2006: 2). It is rooted in a profound expansion of the underlying purpose of the state during the twentieth century, from a more narrowly defined national security and industrial growth imperative to a larger responsibility for social welfare and environmental protection (Reus-Smit, 1996; Eckersley, 2004a; Meadowcroft, 2005). Environmental stewardship reflects the expansion of the social purpose of the state in modern society, and in this sense, it is an extension of states' domestic environmental responsibility to the international realm. It is also a predominantly anthropocentric norm. International society has taken on an environmental mandate in response to societal demand, reflecting humans' interest in (and potentially a human right to) a healthy environment.

Environmental stewardship can thus be thought of as a single norm, and I shall subsequently refer to the rise of 'the environmental norm' or just 'environmentalism' in international relations. However, much like other fundamental international norms (e.g. nationalism, human rights) it is important to remember that it represents a complex mix of core and peripheral norms, and its emergence and evolution has given rise to distinct 'norm complexes' as the basis for global governance arrangements (Bernstein, 2001: 6). Elements of the norm, usually its peripheral concepts, have remained contested even as its core value has strengthened in international relations. As I shall discuss in Chapter 7, social consolidation and norm contestation have gone hand in hand in global environmental politics. Such contestation has focused on the question of how environmental stewardship relates to other primary institutions of international society (e.g. sovereignty, market, development). It concerns specific environmental principles and norms that may, or may not, be part of the international environmental norm, from the precautionary principle in risk regulation to the concept of common heritage of humankind. Contestation has also occurred over the role that equity concerns play in environmental stewardship, and specifically with regard to how the burden of global environmental protection should be distributed between rich and poor nations.

The specific meaning of the international environmental norm has also been challenged by radical ecologists. In their view, it fails to address the deep-rooted causes of the global ecological crisis and can therefore only lead to dysfunctional policy responses. Environmental ideas may have been successfully exported from world society to international society, but radical ecologists would argue that in the process of norm transfer the original environmentalist message has been tamed and converted into 'global managerialism' (Newell, 2020: 142). Undoubtedly, the successful rise of new international norms involves a certain degree of normative adaptation and accommodation. It is one of the central premises of the norms literature in IR that in order to be successfully adopted, new norms need to offer a certain fit with existing cultural preferences and normative structures at international and/or domestic levels (Finnemore and Sikkink, 1998; Acharya, 2004: 243). As we will see subsequently, the environmental norm, although in many ways challenging established principles in international order, has not been exempt from this pressure to 'fit in'. Only a narrow selection of the wide range of ethical and political positions that make up the environmental tradition were able to develop the kind of resonance in international society that allowed environmental stewardship to grow into a primary institution. This selection and filtering process will be subject to closer scrutiny in the subsequent empirical chapters.

Conclusions

As I have shown in this chapter, environmentalism has developed from a disparate set of ideas about the relationship between humans and nature into a political ideology in its own right. All political ideologies are complex assemblages of values and beliefs that show some variation across time and space, and environmentalism is no exception in this regard. With diverse roots in religious beliefs, organicist thinking, aesthetic values and scientific insights, environmentalism emerged in the nineteenth century as a multifaceted movement that argued for the need to protect humanity's natural environment, be it landscapes, natural monuments, forests or individual species. Environmentalists have held different views about how humans should go about protecting nature, and how to balance nature protection with other societal goals, most notably economic development. While some have argued for preventing any human use of at least some parts of nature, others advocate the sustainable use of nature for humanity's benefit. Divergent notions of what needs protecting, and how, have thus given rise to a range of different views about environmental activism and reform. What unites these different strands, however, is a common notion of care for nature. Environmentalists share a fundamental ecological sensibility, a concern for nature's integrity and survival, and a desire to protect the natural world and its diversity for future generations: environmental stewardship, in short.

When it comes to applying environmentalism to the international realm, different models exist for creating a greener global order. Using the ES's conceptual dyads of pluralism/solidarism and international/world society, I identified four principal models of achieving a greener future worldwide. Within the state-centric context of international society, environmentalists have sought to establish environmental stewardship as a fundamental international norm. Many environmentalists have pushed for a solidarist vision of greening international society, which is based on universalist ecological sentiment and values and aims at creating strong international institutions and law (*global environmental governance*). Alternatively, global environmental goals can also be pursued in a more minimalist fashion, within the constraints of pluralist international society, which accepts and works with the diversity of values to be found in different societies and pursues national environmental policies with only limited forms of international cooperation (*Green Westphalia*).

Within the context of a world society that exists and operates independently from state-centric international society, two further models for green global order exist. At the solidarist end of the spectrum, environmental action can be organised by societal actors that work through transnational networks and establish global norms and rules based on cosmopolitan environmental values (eco-globalism). At the pluralist end, environmental action is rooted more in local and national contexts, carried out with regard to local and national conceptions of environmentalism, thus reflecting a fundamental value diversity in world society (eco-localism).

As the discussion in this chapter has shown, environmentalism understood as a fundamental norm can be mapped onto the international/global realm in different ways. To be sure, the four models identified above are ideal types; in reality they often overlap and interact. As will be discussed in subsequent chapters, environmentalists operate in both international and world society, they seek to create both state-centric and non-state-centric rules and regulations. Global environmental politics is characterised by an unusually high degree of social interaction and overlap, if not merging, between international and world society. But it is important to establish these four ideal-type models not least for analytical purposes, in order to be able to trace the different ways in which environmentalism has impacted on societal and normative development in international relations. The subsequent Chapters 4 and 5 trace the origins of environmentalism as a global movement and the emergence of environmental stewardship as a primary institution. Chapters 6 and 7 examine the social consolidation of the new environmental norm in international relations and the normative conflicts and processes of contestation that have accompanied its growth into a fundamental norm for *global* international society.

Part II

History

4 The Origins of Global Environmentalism

The origins of organised environmentalism can be found in nineteenth-century Europe and North America. The first environmental campaigners responded to mainly local concerns about the prevention of cruelty against animals, preservation of areas of natural beauty and protection against urban pollution. With its roots in local experiences of place and community, and of loss and degradation, the early environmental movement paid little attention to the transboundary dimensions of environmental problems. However, from its early days, environmentalism was a global phenomenon. Environmental organisations and campaigns sprang up seemingly spontaneously in different parts of the industrialised world, but they all reflected an emerging environmental awareness that had grown out of a transnational exchange of ideas between naturalists, philosophers and scientists. By the end of the century, environmental campaigns also started to address the transboundary dimensions of environmental problems. The first international NGOs were formed in the early twentieth century to campaign on environmental issues that required international cooperation, such as the protection of migratory birds in Europe or iconic mammals in Africa. Slowly but steadily, the contours of a green world society were beginning to emerge, though it took until the second half of the twentieth century for a nascent global environmental movement to leave its mark on international relations.

International society took much longer to respond to the emergence of a global environmental agenda. During the nineteenth century, some states began to work together to manage shared resources (e.g. rivers, lakes) and European powers initiated the first international conservation policies in their African and Asian colonies, but these forays into international environmental management did not translate into a more general environmental role for international society. The growth of an international environmental agenda was held back not least by the fact that well into the twentieth century states themselves did not accept a domestic responsibility for environmental protection. The greening of the nation-state was a necessary, if not sufficient, condition for the greening of international society.

Towards the end of the nineteenth century, persistent lobbying by environmental campaigners started to have an effect, and leading industrialised countries began to introduce the first environmental policies. By the early twentieth century, the first international environmental treaties were being negotiated, though repeated efforts, before and after the First World War, to convene a global conservation conference and create an international environmental body came to nothing. The 1919 Paris Peace Conference provided campaigners with a unique opportunity to embed environmental objectives in the mandate for the newly created League of Nations, but yet again leading powers rejected any international responsibility in the environmental field. Efforts to green international society proved to be a frustrating experience for an environmental movement that had gradually expanded its campaigning horizon to the international level. The outbreak of the Second World War in 1939 not only resulted in unprecedented levels of war-induced environmental destruction, but also crushed pre-war hopes for the creation of a new international environmental order. This chapter traces the roots of organised environmentalism from the origins of environmental campaigning in the nineteenth century to the largely failed efforts to establish an international environmental agenda in the first half of the twentieth century.

The Emergence and Transnationalisation of Organised Environmentalism in the Nineteenth Century

The sprouting of local environmental initiatives in nineteenth-century Europe and North America reflected so many different moral, political and aesthetic concerns that it would have been difficult for contemporaries to interpret them as being part of one coherent social movement. In countries that experienced rapid industrialisation and urbanisation, local communities organised protests against air and water pollution as part of a wider effort to improve the often appalling living conditions in densely populated cities. Campaigns for stricter rules against cruelty to animals drew on support from the urban bourgeoisie, which saw animal welfare issues as part of a wider effort to improve the 'moral temper' of society (Kean, 1998: 36). Organisations that sought to preserve national monuments and areas of outstanding natural beauty were driven by a mix of romantic and patriotic sentiment. And the rise of sustainable forestry in the nineteenth century owed a great deal to the confluence of economic motives and the nation-state's interest in preserving strategic natural resources. With hindsight, however, it is possible to identify these different strands of early environmental thinking and practice as part of a general shift in societal attitudes towards nature. In short, they represent a collective social response to the nineteenth century's 'global transformation' (Buzan and Lawson, 2015), the disruptive changes brought about by the twin forces of industrialisation and urbanisation.

Diverse Origins of Organised Environmentalism

Animal protection became one of the earliest rallying cries for environmental campaigners in the nineteenth century. In Britain, the Society for the Prevention of Cruelty to Animals, founded in 1824 and renamed the Royal Society for the Prevention of Cruelty to Animals (RSPCA) in 1840, launched moral campaigns against the poor treatment of urban workhorses and stray dogs (Li, 2000). The French Société Protectrice des Animaux (Society for Protection of Animals) was established in 1845, with further animal welfare groups springing up in different regions and cities in subsequent decades (Matagne, 1998: 362). The beginnings of organised animal advocacy in the United States can be traced to the American Society for the Prevention of Cruelty to Animals (ASPCA), founded in 1866, which spawned the creation of nearly 700 smaller organisations across the country (Beers, 2006: 3). In Germany, some of the earliest forms of organised environmentalism were also focused on animal rights. Counting over 150 local societies as members, the Verband der Tierschutzvereine des Deutschen Reiches (Association of Animal Protection Societies of the German Empire) of 1881 became one of the largest environmental umbrella organisations in Wilhelmine Germany. It later grew into a major vehicle for engaging citizens in nature protection throughout the twentieth century (Radkau, 2011: 70). A particularly prominent concern was bird protection, which allowed environmentalists to build alliances with agricultural interests. Founded in 1899, Germany's Bund für Vogelschutz (League for Bird Protection) acquired a particularly large following that provided the basis for a sustained lobbying campaign for national and international bird protection (Markham, 2008: 62–3). In Britain, campaigners against the trade in decorative bird feathers formed the Society for the Protection of Birds in 1889 (renamed in 1904 as Royal Society for the Protection of Birds, RSPB), which was to grow into Europe's largest wildlife conservation organisation in the twentieth century (Rootes, 2007: 35).

The 'urban environmental crisis' (Radkau, 2011: 63–4) of the late nineteenth century saw a myriad of local campaign groups appear in the industrial heartlands of Europe and North America. Due to the widespread siting of factories close to residential areas and use of coal-fired power generation, urban conglomerations were severely affected by air and water pollution (Uekötter, 2004: 118–19). In the United States, where the federal government had no authority to introduce air quality standards for cities until well into the twentieth century, anti-pollution organisations sprang up in several cities, from the Anti-Smoke League of Baltimore and the Society for the Prevention of Smoke of Chicago to the Smoke Abatement League of Cincinnati and the Citizens' Smoke Abatement Association of St Louis (Uekötter, 2004: 118–19; Shabecoff, 1993: 63). In Britain, efforts to reduce pollution and improve hygiene in urban areas often took on an explicitly moral tone, becoming part of a Victorian campaign for public health and moral rectitude that viewed the fight against pollution as a condition for achieving other moral and social reform objectives (Wohl, 1983).

Protecting natural habitats and areas of cultural significance became a further focal point for environmental campaigning in Europe and North America, though the meaning of what deserved protection varied from country to country. In the United States, areas of wilderness took on a special significance for the preservationist movement, which sought to set aside unspoiled nature and exclude it from human use. American wilderness was celebrated in nineteenth-century art and literature and became a core part of American identity (Nash, 2001). A rise in nature tourism starting in the 1820s and 1830s helped build public support for nature protection measures, and by the late nineteenth century environmental groups started to lobby for the creation of national parks. The Sierra Club, created in 1892 by John Muir and a group of Californian nature lovers and mountaineers, originally set out to campaign for the preservation of 'forests and other natural features of the Sierra Nevada Mountains' (Sierra Club, 1892) but soon developed into a nation-wide movement. The cause of wilderness protection was helped by growing concerns over the loss of natural habitats for wildlife, and especially birds. After the first Audubon Society was established in Massachusetts in 1896, a further thirty-five state-level groups of the bird protection society were formed by 1901, followed by the National Audubon Society in 1905 (Merchant, 2010: 17–18), which promoted bird watching but also campaigned for the creation of bird reserves.

The protection of wilderness played only a marginal role in European environmental campaigns, not least because few wild ecosystems were left by the time industrialisation took hold across Europe.

European environmentalists were more interested in the preservation of culturally significant aspects of nature, such as natural monuments and landmarks as well as cultivated landscapes. Informed by Romantic ideals and nationalist sentiment, European conservation thinking linked nature protection with wider concerns about the need to strengthen national identity.

Several nation-wide organisations emerged in Europe around the turn of the century that presented nature preservation as part of a wider homeland protection effort. In Britain, the National Trust for Places of Historic Interest or Natural Beauty (1895) grew into the country's pre-eminent organisation for the preservation of historic buildings as much as for the conservation of landscapes and natural habitats. In France, the Société pour la Protection des Paysages et de l'Esthétique de la France (Society for the Protection of the Landscape and Aesthetics of France, 1901) was established as an umbrella body for nature protection groups, with the stated mission 'to conserve sites and natural objects in their primitive beauty, to defend them against useless degradation by industry and bill-posters, and publicly denounce any act of vandalism' (quoted in Matagne, 1998: 360). In Germany, the Bund Heimatschutz (League for Homeland Protection, 1904) became a major organisation campaigning for the protection of culturally significant natural monuments and landscapes as well as native species (Markham, 2008: 54–5). The cultural significance of nature may have differed from country to country, but the link between environmental protection and cultural identity was equally strong across the continent.

One of the most important outcomes of the nature preservation movement was the creation of national parks and other protected areas. The United States had led the way with the creation of the Yellowstone National Park in 1872 and the Yosemite National Park in 1890 (Shabecoff, 1993: 60). Environmentalists around the world were eager to learn from this 'American invention' (Nash, 1970) and soon embarked on their own campaigns to designate natural habitats as protected areas. Britain's Society for the Promotion of Nature Reserves (SPNR) was formed in 1912 and soon emerged as the country's fourth major conservation organisation after the RSPCA, RSPB and the National Trust (Evans, 1997: 41–5). Australia's first national park, the Royal National Park of Sydney (1879), and the first Swiss national park in the Engadin Alps (1909) both drew on the experience with America's Yellowstone Park. Other countries (e.g. Japan, Sweden, Spain, Italy, Iceland, Romania and Greece) were inspired by US national parks even if their nature protection schemes were 'not a direct copy of the American model' (Frost and Hall, 2012: 44). The Association des Parcs Nationaux de

France et des Colonies (Association of National Parks in France and Its Colonies, 1913) similarly sought to copy the American example and successfully lobbied for the creation of France's first national park in the Alpine region of L'Oisans (Pincetl, 1993: 85). Europe's colonial powers also exported this policy to some of their colonies, with the Dutch creating the Ujung Kulon reserve in Java in 1915, and Belgium creating the Albert National Park in Congo in 1925.

A Green World Society in the Making

As the diffusion of the national parks idea shows, environmental organisations from different countries increasingly learned from each other's campaigns and lobbying successes. Throughout the nineteenth century, scientists and naturalists had already been exchanging ecological observations and ideas for some time, and by the second half of the nineteenth century more durable networks of environmental expertise began to emerge. As the natural sciences underwent a process of professionalisation and globalisation, environmentalists and conservation experts from around the world began to meet more regularly at international scientific congresses. The first international botanical congress was held in 1864, the first ornithological congress in 1884 and the first zoological congress followed in 1889. In an age where regular international travel and communication was still limited to a wealthy elite, international scientific meetings served as a leading platform for organising transnational environmental campaigns. In a sense, they became the essential infrastructure of a nascent green world society.

Colonial empires provided another important context within which ideas about ecology and environmental policy circulated internationally (Anker, 2002; Tilley, 2011; Ross, 2017). Faced with unfamiliar and often precarious ecological conditions in tropical regions, colonial administrations themselves had long promoted the exchange of knowledge about ecological science, sustainable forestry practices and soil preservation. In this way, colonial policies had become an incubator for new ecological thinking and conservation practices that were later applied in Europe and North America. By the end of the nineteenth century, environmental campaigners also started to turn their attention to environmental problems in colonial territories. In Britain, France and Germany, the protection of wildlife in colonial territories became the rallying cry for some of the first explicitly international environmental campaigns.

The cause of wildlife protection brought together an unusual mix of conservationists, hunters and colonists who were united by the desire to protect the iconic mammals that could mainly be found in African and

Asian colonies. Large mammals, such as elephants, lions and tigers, had been hunted by indigenous communities for a long time, either for food or to protect farmlands, but when the growing popularity of big-game hunting among colonial elites led to a dramatic decline in their numbers, hunting groups began to lobby colonial powers to impose tighter hunting restrictions and establish game reserves (MacKenzie, 1988: 232). The Society for the Preservation of the Wild Fauna of the Empire (SPWFE), established in 1903, became the driving force behind a transnational campaign to prevent the extinction of wild animals within the British Empire (Prendergast and Adams, 2003; Neumann, 1996). By the turn of the century, concern about the extinction of large mammals in Africa was already well established, and British and German colonial authorities had been creating game reserves since the 1890s (Prendergast and Adams, 2003: 252). At the London Conference of African colonial powers in 1900, Britain, France, Germany, Portugal, Spain, Italy and Belgian Congo agreed a Convention for the Preservation of Wild Animals, Birds and Fish in Africa. The convention, one of the first international conservation treaties ever to be signed, was innovative in that it introduced a system of schedules, which placed animals in different categories of protection. It never entered into force, however, and it was not until 1933 that the European powers replaced it with the Convention Relative to the Preservation of Fauna and Flora in the Natural State (Boardman, 1981: 34).

Early State Responses

As we have seen, by the turn of the century European and North American governments had been confronted with at least some societal demand for environmental protection. Their responses to this awakening environmental sentiment were, on the whole, slow and ineffective. Well into the twentieth century, national governments tended to treat citizens' complaints about air and water pollution, lack of adequate waste treatment and the destruction of natural habitats as local concerns. For this reason, municipal authorities in Europe and America were initially at the forefront of developing the first policies on pollution prevention and the sustainable use of nature, though they usually lacked the resources and enforcement powers to put in place effective measures (Markham, 2008: 46). The situation began to change somewhat in the early twentieth century, when several countries began to introduce the first legislative acts to promote nature protection at the national level. Such initiatives were still ad hoc measures, however, not yet part of a systematic approach to environmental policy. In any case, most states remained committed to a programme of national economic development that prioritised industrialisation and urbanisation over environmental sustainability.

In some areas of conservation practice, the United States was at the forefront of developing more forceful national responses. The main breakthrough came in 1901, when Theodore Roosevelt succeeded William McKinley as the twenty-sixth President of the United States. Unusually for heads of state at the time, Roosevelt made nature conservation a national priority during his eight years in the White House. Although Roosevelt professed to have been influenced by John Muir's preservationist ideas, his administration followed a decidedly utilitarian approach. The appointment of Gifford Pinchot to head the newly created Forest Service in 1905 signalled a focus on principles of 'wise use' in natural resource management. Roosevelt also gave conservation a distinctively progressive political agenda. The country's vast natural resources were to be used for the public benefit, rather than private gain, and nature conservation was framed as a question of democracy and popular sovereignty, not simply the result of either romantic sentimentality or commercial interest (Hays, 1959; Tyrrell, 2015). The first blossoming of environmentalism in American politics was short-lived, however. Conservation ideals had reached their nadir in US politics with the end of Roosevelt's presidency. William Howard Taft, who succeeded Roosevelt in the White House in 1909, reverted to business as usual as he did not share his predecessor's enthusiasm for nature conservation (Tyrrell, 2015: 236–7).

Although European governments did not make conservation a national priority, they nevertheless began to introduce various laws on nature protection, mostly in response to specific causes that environmental organisations had campaigned on. The British government responded to widespread public concern over animal welfare by creating the Wild Birds Protection Act (1904), the Grey Seals Act (1914) and the Importation of Plumage Act (1921) (Evans, 1997: 47–8). Unlike in the United States, however, conservation never played a prominent role in British electoral politics. In Germany, several states had created agencies to identify and preserve monuments and buildings of historical significance during the nineteenth century. By the early twentieth century, these state offices gradually absorbed nature protection as a further objective. Prussia's Staatliche Stelle für Naturdenkmalpflege (State Office for Natural Monument Preservation), a body that was established in 1904 by the influential botanist Hugo Conwentz (Lekan, 2004: 21), became a key driver behind Germany's emerging approach to nature protection at the time. As in other European countries, the French state also stepped up its response to rising environmental concern, but most measures were half-hearted and poorly resourced. The one area where French conservation policy benefited from the long-standing tradition of state intervention was forestry. As part of its mission to safeguard essential resources, the French state pursued a comprehensive policy on forest conservation throughout the nineteenth and early twentieth centuries (Ford, 2004: 176–80).

A small number of states also began to deal with transboundary environmental problems, although such international initiatives remained the exception to the rule. In a few isolated cases, civil society pressure succeeded in pushing governments to negotiate the first international environmental treaties. Years of campaigning by bird protection groups, which were among the then most internationally connected NGOs, led to the adoption of the Convention for the Protection of Birds Useful to Agriculture, agreed in 1902 by twelve European countries, and the Migratory Birds Treaty, a bilateral treaty between the United States and Britain (on behalf of Canada) from 1916. Bird protection campaigners subsequently strengthened their international links even further and in 1922 created one of the first international environmental NGOs, the International Council for Bird Protection (ICBP) (Meyer, 2017: 41). The United States, Britain (on behalf of Canada), Russia and Japan also signed the North Pacific Fur Seal Convention in 1911, which was designed to regulate the commercial hunting of fur-bearing mammals in and around the Bering Sea (Dorsey, 1998).

As we have seen, by the turn of the century a number of states had begun to introduce the first environmental policies at local, national and even international levels. However, most states still lacked a national environmental strategy and treated environmental issues mostly as resource management or local administrative issues. The global dimensions of the various environmental problems were still poorly understood, and the few international environmental treaties that had been negotiated before the First World War amounted to little more than isolated cases of transboundary environmental coordination. The path to internationalising environmental protection went through the nation-state, and for as long as states refused to recognise environmental protection as a core state responsibility, it would prove difficult for environmental campaigners to establish environmentalism as a responsibility for international society.

Towards a Global Conservation Agenda before the First World War

By the early twentieth century, environmentalism had made the first inroads into national policy-making in key industrialised countries, and several great powers had started to address transboundary issues through international treaties. This was still a far cry from the kind of international environmental diplomacy that was to emerge during the 1970s, but the moment seemed to have come for a major push to take conservation to the international level. Two such efforts to create a global conservation agenda were launched before the First World War, one in the United States and one in Europe. In the end, both ended in failure. Environmental campaigners renewed their efforts after the war, at the Paris Peace Conference in 1919 and later at the League of Nations. Despite growing awareness of transboundary environmental problems and a gradual expansion of international environmental negotiations during the interwar years, the leading powers rejected demands for the League to take on an environmental mandate. International society was unwilling to accept a general duty for global environmental protection.

Roosevelt's International Conservation Agenda

Theodore Roosevelt was the first US president, and indeed the world's first head of state, to make nature conservation an integral part of his domestic political programme. Forest protection and the creation of wildlife preserves featured prominently in his first State of the Union Address to Congress in 1901. Soon after, the Roosevelt Administration began turning this conservationist vision into practice, and with considerable success: by the end of his presidency, Roosevelt had signed into existence 5 national parks, 18 national monuments, 55 national bird sanctuaries and wildlife refuges, and 150 national forests (Dant, 2016: 102). Throughout his presidency, however, Roosevelt had to overcome considerable resistance in Congress. In an effort to institutionalise his conservation policy, Roosevelt called the governors of all US states to a White House Conference on Conservation in 1908, a key event in the emergence of America's national conservation movement (Shabecoff, 1993: 68). The conference led to the adoption of a landmark declaration on conservation and the creation of the National Conservation Commission, which conducted an interdisciplinary stocktaking exercise and produced a three-volume report on all aspects of conservation policy (Tyrrell, 2015: 5).

Roosevelt was also the first head of state to attempt to internationalise conservation policy. Soon after hosting the US conservation conference, Roosevelt went one step further and convened a North American Conservation Conference in order to promote continental cooperation on nature protection (Tyrrell, 2015: 5). Delegates from Canada, Mexico and Newfoundland met in the White House in February 1909 and debated the threat from the rapid depletion of forests, coal and water resources in North America. In his invitation to the Canadian government, Roosevelt had highlighted the need to think of nature conservation as a transboundary task: 'It is evident that natural resources are not limited by the boundary lines which separate nations, and that the need for conserving them upon this continent is as wide as the area upon which they exist' (quoted in Dorsey, 1995: 407). Both the Canadian and Mexican delegates expressed support for the American proposal and supported Roosevelt's suggestion to include other states in the initiative. The conference passed a concluding Declaration of Principles, which suggested that 'all nations should be invited to join together in conference on the subject of world resources and their inventory, conservation and wise utilization' (North American Conservation Conference, 1909). The US Secretary of State submitted the proposal for such a world conservation conference to be held as soon as possible, stating that it was the duty of all states to protect the environment in the interest of future generations: 'As to all the great natural sources of national welfare the peoples of to-day hold the earth in trust for the peoples to come after them' (Scott, 1923: 129).

Roosevelt's proposal for the world conservation conference signalled a shift in his foreign policy away from a muscular imperialism to a new form of conservation internationalism. It included the idea of an international conservation body, a council that would promote research, establish a global inventory and advise governments on 'conservation, development and replenishment' (Tyrrell, 2015: 213). Although a modest proposal by contemporary standards of global governance, the idea was a departure from the then strictly Westphalian approach to conservation matters. The American initiative was initially met with support by several countries in Europe and Latin America. By the end of 1909, twenty-three countries had declared their interest in attending the conference, and the Netherlands agreed to host the world gathering in The Hague (Tyrrell, 2015: 225). The positive response from around the world signalled the growing attention that was being paid to conservation matters in both industrialised and developing countries. As a 1908 editorial of the journal *Conservation* noted, somewhat optimistically, the 'cult of conservation' was 'by no means wholly American', it was in fact 'altogether cosmopolitan' (quoted in Tyrrell, 2015: 225).

In the end, however, Roosevelt's proposal came to nothing. As soon as Roosevelt's second term in office had ended, his successor in the White House, William Howard Taft, abandoned Roosevelt's National Conservation Commission and lost interest in his predecessor's international initiative. With the US government's support formally withdrawn in 1911, the Netherlands decided not to go ahead with plans to host a world conservation conference (Tyrrell, 2015: 227).

The 1913 Berne Conference and the Consultative Commission

At around the same time that Roosevelt pushed for a global conservation conference, European environmentalists were lobbying their own governments to promote greater international coordination. At the International Congress for the Protection of Nature, held in Paris in 1909, a wide range of conservation experts, scientists and campaigners from across Europe met to discuss the need for greater cooperation across boundaries to make conservation a global priority. By this time, conservationists had established several transboundary threats that could not be addressed within individual states or even within colonial empires. States were still considered the primary actor to deliver international conservation strategies, in line with green Westphalian thinking, but conservationists were now beginning to think more creatively about the conditions for effective and lasting international action. If the common biological inheritance of humankind was to be preserved for future generations, then a new international institution was needed to establish an international knowledge base and monitor the safekeeping of wild creatures and their habitats (Ross, 2015: 225). The 1909 Paris Congress thus became the first ever international conference to issue a call for the creation of an international environmental body (Boardman, 1981: 29).

A year later, environmentalists met again at the Eighth International Congress of Zoology in the Austrian town Graz. Again, delegates debated how to establish an international conservation agenda and formed a committee to prepare the ground for an intergovernmental agreement on nature protection. Paul Sarasin, one of Europe's leading environmentalists and co-founder of the first national parks in Switzerland, began to lobby Swiss government officials about the need to host such an international gathering. The initiative fell on fertile political ground. Much like its neighbours, Switzerland had started to adopt conservation policies to protect natural landmarks and resources in the late nineteenth century. After suffering an accelerated phase of deforestation in the early nineteenth century, which had contributed to several severe floods in the 1830s and 1850s, Switzerland first experimented with local forestry measures before introducing a national forest protection law in 1876 (Mather and Fairbairn, 2000: 408). By the early twentieth century, Swiss authorities were beginning to consider the introduction of national parks. The Swiss government endorsed Sarasin's proposal for a world conservation conference not least as it saw an opportunity to boost its international diplomatic standing by championing the cause of international conservation. With nature protection increasingly seen as a common interest of all 'civilised' nations, hosting the conference would underline the country's capacity to convene international conferences and reaffirm its leading role in what Swiss politicians perceived to be a global civilisational project (Wöbse, 2008: 522).

A total of seventeen countries attended the first ever international conservation conference, held in Berne in 1913, including the then leading powers (the United States, Britain, France, Germany, Austria, Russia). The Berne conference had a broad remit and focused on some of the main conservation issues of the time. The British delegate, for example, urged governments to establish more nature reserves (Rothschild, 1913), while others sought international measures to protect migratory birds and whales, or simply to promote information exchange about conservation practices. The most significant outcome of the conference was the creation of an international environmental organisation, the Consultative Commission for the International Protection of Nature. The agreement was signed by delegates from seventeen countries (fifteen European countries, the United States and Argentina) (Acte De Fondation d'une Commission Consultative Pour La Protection Internationale De La Nature, 1913), and Article VI stated as the new international body's main task the creation of a knowledge base about the current state of international nature protection and dissemination of 'propaganda for the international protection of nature' (Boardman, 1981: 29).

The discussions about the purpose of the new Commission witnessed the first clash between solidarist and pluralist conceptions of international environmental politics. Environmental campaigners, such as Switzerland's Paul Sarasin, argued for the creation of an international institution with authority to promote nature conservation across the world. In the ambition that he outlined for the new international body, Sarasin espoused an early form of environmental internationalism, though he left the institution's powers and mandate ill-defined (Wöbse, 2011: 334–5). In contrast, other delegates continued to view nature protection as a national, not international, duty. Hugo Conwentz, the Prussian Commissioner for Natural Monuments and an influential figure in international conservation circles, spoke on behalf of those who rejected the idea of states taking on a formal responsibility for the global environment. Conwentz's reasoning was a classic case of Green Westphalian logic: He identified the diverse nature of environmental problems around the world and the national sovereignty principle as the main obstacles to creating an international body that would develop and apply uniform rules across the world. The only scope for international environmental action was in the area of 'informal exchange of information among experts developing nonbinding recommendations' (Meyer, 2017: 38).

To bridge the gap between these two positions, the Swiss hosts of the conference proposed a compromise solution, which supported the creation of a new international body but limited the scope for international action to the most urgent issues that were beyond national regulation, the 'high seas, the deserts, the steppe' (quoted in Meyer, 2017: 38). International cooperation was also presented as a measure of last resort, after voluntary and private efforts by NGOs and national state action had been tried. Thus, it was Conwentz' pluralist minimalism, rather than Sarasin's solidarist ambition, that came to define the Consultative Commission's mission. Established with a limited mandate to collect, review and disseminate all available information on international nature protection, but without requiring member states to provide any funding for its activities (Meyer, 2017: 38–9), the Commission had more in common with the international technical unions of the nineteenth century (Boardman, 1981: 30) than the global environmental governance institutions of the late twentieth century.

The success of this initiative was short-lived. Due to the outbreak of the First World War, a further conservation conference planned for August 1914 had to be cancelled, and despite fourteen countries nominating their delegates to the new body, the Commission never took up its work. After the end of the war, some delegates tried to resuscitate the Berne agreement, but the deep divisions that the war had caused in Europe prevented the new Commission from taking up its role (Meyer, 2017: 39; McCormick, 1989: 22–3).

Had it developed into an international environmental body with significant support from the great powers, the Consultative Commission could have signalled at least a first move away from the de-centralised approach of the past. It certainly was the first notable attempt to create a secondary institution in the field of environmental protection, which could have promoted the emergence of environmentalism as a primary institution of international society. Lobbying by world society actors had provided the initial impetus for this institutional innovation, a pattern that was to repeat itself in international environmental politics throughout the twentieth century. But conservationists were unable to overcome the deep fissures in the international order and persuade major powers to accept a general responsibility for the global environment. Their campaigning networks, although increasingly operating in a transnational context, were still thin by contemporary standards. Conservation groups were based on a small social and political elite, drawn mostly from scientific and bureaucratic circles, which relied on connections in high places of government and diplomacy. Without a broader social movement underpinning their demands for a green international order, they relied on the goodwill of state leaders to implement their demands. Unsurprisingly, therefore, a green Westphalian logic focused on states' sovereign rights and limited international information exchange prevailed. The time was not yet ripe for international society to embark on elevating environmental stewardship to the status of a fundamental international norm.

An Environmental Mandate for the League of Nations?

International peace conferences after major wars can be seen as 'important conjunctures for the practice of legitimacy' (Clark, 2005: 8) in international society. They provide an opportunity for states to re-think the fundamental rules that define international order and lay the foundations for a new order. World society actors have repeatedly used such historical episodes to influence the normative development of international society. Clark (2007) identifies several critical junctures at which new international norms were introduced, such as the peace conferences after the Napoleonic wars in Europe and the abolition of the slave trade, and the post–Second World War settlement at the San Francisco conference and the creation of the human rights agenda. The post–First World War negotiations on a new international order were no different in this regard. Environmental campaigners saw the Paris Peace Conference of 1919 as a new opportunity to lobby state leaders with proposals to embed environmental objectives in the peace settlement. To be sure, environmentalists were not the only ones seeking to bring about international normative change. A variety of civil society organisations lobbied their national representatives at the Paris Conference to promote global values, from racial equality to social justice, drug trafficking and the protection of ethnic minorities (Clark, 2007: chapters 4 and 5; Davies, 2013: 81–2; Wöbse, 2008: 521). The prospect of a new international organisation, the League of Nations, had fuelled activists' hopes that the new body would become a champion for deeper international cooperation, that it would assert humanity's universal values over narrow national interests. Such hopes were short-lived, however, as environmental campaigners were soon to find out for themselves.

Environmental lobbying at the Paris Conference and later at the League of Nations was dominated by a small number of actors mostly from European and North American organisations. From its origins in the nineteenth century, the conservation movement was predominantly made up of members of a small social and political elite, usually drawn from the urban middle and upper classes. What they lacked in popular support they made up for in access to high-ranking diplomats and government officials. In making their case for a green mandate for the League of Nations, many conservationists were able directly to lobby the key architects of the new post-war order. In Britain, for example, the RSPCA sought to convince Britain's foreign secretary Arthur Balfour that animal rights ought to be included as a mandatory objective in the peace treaties. A Scandinavian coalition of animal rights campaigners similarly appealed to the US President Woodrow Wilson to propose an international treaty on animal protection as part of the post-war peace settlement (Wöbse, 2012: 136–7).

The most comprehensive proposal for an international conservation agenda came in the form of a submission to the international peace conference by the Austrian zoologist Eduard Paul Tratz. In his 1919 pamphlet *Entwurf für ein internationales Naturschutzgesetz* (draft proposal for an international nature protection law), Tratz argued for the creation of a new body of international law to protect migrating birds. The proposal, which was focused on the creation of nature reserves as a form of environmental trusteeship, combined solidarist one-worldism with a heavy dose of environmental paternalism, an attitude that was widespread in international conservationist circles at the time. Backed up by international law and policed by a new international organisation, such nature reserves would remove areas of environmental importance from the exclusive control of nation-states and protect them as part of humanity's natural heritage (Wöbse, 2008: 521). If implemented, Tratz' proposal for a comprehensive system of international environmental law would have pushed international society well beyond the strictures of Green Westphalia. In the end, none of these proposals were taken up at the Paris Conference. Despite having given the League of Nations a set of new international responsibilities – from the protection of minority rights to economic development – the victorious powers did not include environmental protection in the League's mandate.

Undeterred by this setback, environmentalists resumed their internationalist campaign as soon as the League of Nations had taken up its work. For the first time, environmentalists had an international organisation to focus their lobbying efforts on. They hoped that despite the lack of a formal mandate, the League might be persuaded to become a champion for international nature protection efforts, from preventing oil pollution from shipping to protecting large mammals in Africa and regulating ocean fisheries and whaling (Meyer, 2017: 40–2). Initially at least, their arguments fell on fertile ground within the League of Nations Secretariat. Paul Sarasin, the driving force behind the 1913 Berne Conference, approached the League Secretariat with a proposal for the dormant Consultative Commission for Nature Protection to come under the aegis of the new international organisation. Article 24 of the Covenant of the League of Nations provided for a mechanism whereby any international organisation established by an existing treaty could be placed under the direction of the League, provided it had the consent of the parties. The Secretariat's first response was encouraging. Eric Drummond, the League's first Secretary-General, expressed his belief that the League was on a civilising mission and that environmental protection fitted well into the rapidly expanding list of global governance challenges that it sought to address (Wöbse, 2008: 523). Drummond also signalled his approval of a proposal by Britain's RSPCA for an 'International Charter for the Prevention of Cruelty to Animals', though he urged the NGO that it needed to get its own government to make a formal submission to the League if the Secretariat was to act on the proposal (Wöbse, 2012: 138–9).

While environmentalists could draw comfort from the League Secretariat's solidarist ambition of creating a system of global governance based on international law, their hopes for a new green international order were short-lived. The deterioration of great power relations during the interwar years put an end to the League Secretariat's own ambition for a global environmental role. Member states were reluctant to give the League a mandate, let alone formal legal authority, for initiating environmental protection measures. And as Sarasin discovered for himself, the original coalition of scientists and government officials behind the creation of the Commission for Nature Protection could not be restored after the war. When asked about their views on resuscitating the Commission, German delegates rejected the proposed association of the Commission with the League, while Belgian and British delegates refused to be part of a body that included the erstwhile enemies of Germany and Austria (Wöbse, 2012: 56–7). On the advice of Nitobe Inazō, Under-Secretary General of the League, Sarasin then lobbied the Swiss government to formally propose the establishment of nature protection as one of the functions of the League. The Swiss parliament debated the matter in 1922 but reached the conclusion that the time was not ripe for a revival of the pre-war Commission. Other urgent political matters took precedence over nature conservation (Wöbse, 2012: 59).

Only a few years after the end of the war, it had thus become clear that international society was still unwilling to accept global environmental protection as part of its international responsibilities. The League Secretariat continued to encourage environmentalists to expand their transnational networks and keep up their lobbying effort but could not itself take the initiative to introduce global environmental measures. Lacking a functional mandate for environmental protection, the League passed environmental issues to other technical sections of the international organisation, such as transport and economic affairs (Wöbse, 2008: 525).

Environmental campaign groups continued to direct some of their lobbying campaigns at the League of Nations, but their focus increasingly shifted to other international initiatives with a more limited ambition. The hopes for creating an international environmental organisation never went away, but with international society refusing to endorse this proposal, environmentalists tried to establish it outside the intergovernmental framework of the League. By the late 1920s, Dutch, Belgian and French conservationists sought to breathe new life into the Consultative Commission that had been agreed at the 1913 Berne conference. In 1928, they re-founded the Commission, first under the name International Office of Documentation and Correlation for the Protection of Nature, and again in 1934 under the new name International Office for the Protection of Nature (IOPN) (Anon., 1935). Based in Brussels, IOPN was tasked with mainly scientific service functions: to collect scientific data and studies, to facilitate international cooperation between institutions and individuals in the environmental field, to undertake technical studies on nature protection and to organise propaganda activities for global nature protection (Bergandi and Blandin, 2012: 17–18). The Second International Congress for the Protection of Nature of 1931 expressed its hope that governments would support the new body and presented it as an interim step towards an official international organisation. Even the modest hope of securing state support for an independent IOPN was misplaced. Against a worsening international situation in the 1930s, IOPN received only limited financial support from a small number of countries (France, Poland, the Netherlands) and was unable to develop any significant international role. It was a case of 'too little, too late'.

Frustrated by their lack of success in engaging with international diplomacy, conservationists redirected their efforts to national campaigns and pursued international objectives through other forums, including minilateral ones. Whaling was one such issue where transnational conservation campaigns were slowly beginning to develop traction in international diplomacy. During the 1920s, various League of Nations committees had already been considering proposals for a comprehensive marine protection regime, which would have based conservation efforts on the progressive notion of shared international environmental responsibility. The negotiations failed to reach a consensus, however, on what would have been a significant step towards a more solidarist form of global environmental governance. Instead, a more limited convention to regulate whaling was passed under the auspices of the League in 1931 (Convention on the Regulation of Whaling). The convention – a landmark development in international law, in that it was based on the principle of common use of global resources – failed to have the desired effect of reducing the killing of whales. Frustrated with its inadequacies, a smaller group of countries came together in London in 1937 to develop a separate whaling treaty outside the League system, though again without managing to establish a working regulatory system (Wöbse, 2012: 237–9; Leonard, 1941: 100–1).

Preventing marine pollution from shipping became another focus of transnational environmental campaigning during the 1920s, which led the great powers to be confronted once more with demands for global environmental responsibility. After the First World War, commercial shipping companies increasingly switched from coal to oil to power their ships. At that time, it was common for ships to fill their empty oil tanks with water to stabilise the vessel. Before refuelling, they would simply deposit the oil–water mixture back into the sea, causing large amounts of oil to float on the surface and damage not only coastal wildlife but also harbour installations and tourism (Wöbse, 2012: 70–4). Under pressure from coastal communities and environmentalists, the US and British governments soon moved to press for an international solution, an early example of the now familiar pattern of environmental frontrunners seeking to internationalise domestic environmental protection (DeSombre, 2000). However, with other states not experiencing similar levels of coastal pollution or domestic pressure, it proved difficult to reach an international agreement on how to regulate the shipping industry.

The United States was the first country to seek to create an international regulatory framework for shipping. At the insistence of Congress, which had passed a domestic Oil Pollution Act in 1924, the US government convened an international conference in 1926, with delegates from thirteen nations in attendance at the meeting in Washington, DC. After running into strong resistance by Sweden, Germany and the Netherlands, as well as domestic lobbying by US shipping companies, the US government watered down its original regulatory proposal. Instead of calling for the mandatory installation of separators that would prevent the release of oil into oceans, the United States agreed on the more limited solution of banning the disposal of oil within a 50-mile zone from the coast and 150 miles in exceptional cases. Despite this concession, the weakened draft treaty failed to gain sufficient support to enter into force, and a voluntary agreement to accept the 50-mile discharge prohibition zone by the International Shipping Conference (representing national shipowner associations) became the default situation instead (M'Gonigle and Zacher, 1981: 81–2; Wöbse, 2012: 93–5).

Dissatisfied with their own government's response, British bird protection campaigners kept up the pressure and rallied support among the public for a renewed international regulatory effort. They reached out to Dutch and American conservation groups to join forces in an international campaign. Their efforts were helped not least by the ICBP, founded in London in 1922 and one of the first international environmental NGOs, which served as a coordination mechanism for the international campaign. In response to strong civil society pressure, the British government renewed its support for an international solution. With the United States having lost interest in the matter, London in 1934 urged the League of Nations to produce a new draft convention for adoption at a future international conference. As in the 1920s, the League was still willing to take up environmental matters and prepared a draft convention based on the 1926 Washington treaty. Yet again, determined lobbying by shipping interests held up progress, and the rapid deterioration in great power relations made sure that the conference, originally planned in 1937, never took place. It was not until the 1950s that renewed international efforts were made to regulate oil pollution from international shipping (M'Gonigle and Zacher, 1981: 82–3; Wöbse, 2012: 101–6, 121–5).

Thus, despite the emergence of more sustained transnational lobbying campaigns by environmentalists and repeated efforts to establish international environmental treaties, little progress was made during the interwar years towards establishing states' general environmental responsibility. With the League of Nations losing legitimacy as a multilateral forum for regulating interstate affairs, progressive states and environmentalists had no other international institutional framework to fall back on. Despite the creation of the first international environmental NGOs, the nascent green world society was too weak to offer an alternative to state-centric efforts. As before the First World War, any hopes for an imminent breakthrough in establishing environmental protection on the international agenda were short-lived. With limited domestic mobilisation behind environmental action and deteriorating relations between the great powers, it was unlikely that international society would come to accept environmentalism as an emerging fundamental norm.

Conclusions

The nineteenth and early twentieth centuries witnessed the transformation of the environmental movement from a few scattered initiatives at the local level to an increasingly transnational movement that pressured states into environmental action. As such, environmentalism was a classic case of a new norm that emanated from world society and awaited adoption by the society of states. During this time, a pattern had begun to emerge that saw NGOs lobbying leading states to establish international cooperation on conservation matters. Although still limited in its scope and success, the early forms of engagement between the nascent environmental movement and states foreshadowed the high degree of interaction between non-state and state actors in global environmental politics towards the late twentieth century.

Before the Second World War, environmentalism was an elite concern that lacked wider mass appeal, which severely limited its impact on national and international politics. The membership of environmental organisations was drawn from a small social and political elite, usually wealthy individuals with connections in scientific and governmental circles. Their wealth and status gave them privileged access to state representatives and diplomats, allowing them to lobby for conservation objectives at the highest level. But without wider societal support, environmental organisations were unable to exert the kind of political pressure on states that could make environmental protection a priority in international politics.

By the early twentieth century, a small group of industrialised countries had already adopted domestic environmental protection measures and was now beginning to make first forays into international environmental diplomacy. The first international environmental treaties signalled some willingness to address transboundary issues, though still within the constraints of a Green Westphalian approach. Proposals for a stronger international authority to guide and monitor national action were roundly dismissed, in line with international society's reluctance to accept a fundamental responsibility for the global environment. Despite a significant uptick in environmental conferences and initiatives during the interwar years, the greening of international society remained a distant vision. This was to change only after the Second World War.

5 The Emergence of Environmental Stewardship as a Primary Institution

At the end of the Second World War, a fresh opportunity arose for the environmental movement to establish environmental stewardship at the heart of the new international order. The war had heightened fears about resource scarcity and technology's destructive potential, leading to renewed efforts to address conservation concerns at the international level. Yet again, though, international society rejected a general responsibility for the global environment. Environmental protection did not become one of the core objectives of the newly created United Nations. It was not until the 'environmental revolution' of the 1960s, which transformed environmentalism from an elite concern into a mass movement with wider electoral consequences for governments, that international society began to accept environmental stewardship as a new primary institution. Within a short space of time, from the mid-1960s until the early 1970s, leading industrialised economies established environmental protection first as a comprehensive domestic duty of the state and then as a general responsibility for international society. The 1972 Stockholm conference, the first UN conference on the environment, became the equivalent of a 'constitutional moment' in the greening of the international normative order. This chapter traces the process through which world society actors successfully transmitted environmentalism into international society, with leading powers such as the United States providing critical leadership along the way.

Rebuilding Global Environmentalism after the Second World War

The Second World War and the need to rebuild international order after the defeat of Germany and its allies provided a second major opportunity in the twentieth century to redefine the key principles of international society (Clark, 2005). As with the Paris Peace Conference of 1919, the San Francisco conference in 1945 might have given world society a chance to feed environmental values into a reconfigured concept of international legitimacy. After all, the conference did become the key event in the creation of a global human rights regime, thanks not least to energetic interventions by NGOs (Vincent, 1986: 98; Sikkink, 2014). By contrast, lobbying by conservationists was much more muted during the San Francisco conference, and environmental objectives ended up being relegated to the margins of the newly created United Nations. Yet again, the leading powers rejected the idea that environmental protection ought to be elevated to a fundamental norm of international relations, and individual states continued to deal with domestic and transnational environmental issues in a mostly ad hoc fashion.

Six years of global warfare had severely weakened the nascent global environmental movement. Having lost the pre-war momentum in building more lasting transnational networks and institutions, environmentalists' main concern in 1945 was to rebuild the remnants of an embryonic green world society. Unlike after the First World War, however, the post-1945 conditions were more auspicious for a profound and lasting expansion of global environmental ambition. Even before the war had ended, the Allied powers embarked on a major planning exercise to redesign the international order. This included the creation of a raft of new international organisations that would replace the defunct League of Nations and strengthen global governance for the post-war global economy (Mazower, 2012: chapter 7). Once domestic reconstruction got underway, leading industrialised countries also faced an expanded reform agenda at home. As will be discussed subsequently, the societal and political shifts that occurred in the post-1945 era were to lay the foundations for the successful greening of international society from the 1970s onwards.

The devastation wreaked by the Second World War itself played a major role in raising global environmental awareness. Advances in military technology had made this the most devastating global war in human history, leading to over sixty million casualties and creating deep scars in the topography of urban and rural landscapes. Both sides in the global conflict had mobilised natural resources and caused pollution on an unprecedented scale, leading to the wholesale militarisation of the environment (Pearson, 2015). Above all, it was the invention of the atom bomb that came to epitomise humanity's new power over nature and the existential threat that continuous technological progress posed to humanity's survival. The images of huge mushroom clouds rising above the Japanese cities Hiroshima and Nagasaki in August 1945 were to provoke a profoundly emotional response in the post-war years, with nuclear fear emerging as one of the main forces behind the resurgence of the environmental movement in the 1960s (Weart, 2012: 194).

Initially at least, the post-war focus on environmental issues revolved around technical questions of natural resource management and conservation. Amidst international efforts to rebuild war-torn territories, the UN, responding to a request from the United States, convened the United Nations Scientific Conference on the Conservation and Utilization of Resources (UNSCCUR) in 1948. The link between resource shortages, economic hardship and international conflict was well-established in post-war American thinking, and US President Harry S. Truman had urged the scientific community to consider how 'Conservation can become a major basis of peace' (quoted in Warde, Robin and Sörlin, 2018: 36). UNSCCUR took place in parallel to a second UN-sponsored meeting of experts, the International Technical Conference on the Protection of Nature (ITCPN), which had been initiated by the UN Educational, Scientific and Cultural Organization (UNESCO). The two conferences, which were held at Lake Success in New York, the temporary headquarters of the UN, represented the main strands in environmentalist thinking: while UNSCCUR emphasised the need for better and more efficient exploitation of natural resources, ITCPN promoted discussions about how to strengthen the preservation of natural habitats. The long-standing divide between conservationist and preservationist thinking had thus resurfaced in international environmental debates (Warde, Robin and Sörlin, 2018: 39–41; on conservationism and preservationism, see chapter 3), though neither of the two conferences had a lasting effect on post-war international environmental policy.

UNESCO, the sponsoring organisation behind ITCPN, was the only UN body to have gained at least a partial mandate for environmental matters, as part of its mission to promote scientific exchange and cultural heritage (the Food and Agriculture Organization (FAO) focused only on sustainability in forestry). That UNESCO should play this environmental role was largely a reflection of the influence that key individuals had over the direction of the new international organisation, most notably Julian Huxley, the organization's first Secretary-General (Meyer, 2017: 46–7; Wöbse, 2011: 338). Huxley, a biologist by training, saw nature conservation as part of a bigger project of promoting humanity's progress and unity, on the basis of scientific rationality and cosmopolitan values (Sluga, 2010; Wöbse, 2011). His vision for UNESCO was that of a 'scientific world humanism, global in extent and evolutionary in background' (Huxley, 1946: 8). Building on the legacy of the League of Nations' International Committee on Intellectual Co-operation, which had considered coordinating national policies to establish national parks, UNESCO sought to promote knowledge exchange about conservation policies alongside its wider educational, cultural and scientific objectives.

UNESCO's major contribution to the revival of international environmental politics after 1945 was as an orchestrator. It sponsored the first major international conferences on nature protection in 1948, with ITCPN alone attracting an estimated 150 experts and activists (Wöbse, 2011: 331). UNESCO also became a key vehicle for engaging world society actors in the workings of the UN, going as far as to facilitate the creation of new NGOs at the international level (Martens, 2001: 397–8; Radkau, 2011: 104–7). Under Huxley's leadership, UNESCO thus played a significant, if limited, role in re-establishing an international environmental agenda after 1945. Given its limited mandate and lack of policy-making authority, Huxley was keen to mobilise as many non-state actors as possible to create a global web of environmental action. His most notable success came in 1948 with the creation of the International Union for the Protection of Nature (IUPN, changed in 1956 to International Union for Conservation of Nature and Natural Resources, IUCN) at an UNESCO-sponsored international environmental conference in Fontainebleau, France (Wöbse, 2011: 340). But with Huxley's term in office already coming to an end that year, the international organisation lost its main champion for a strong environmental role. In the end, UNESCO was unable to evolve into a champion for environmental issues within the UN. Without a more fundamental commitment by international society to global environmental protection, charismatic leadership by individuals, even when operating from within the UN, could only go so far in pushing environmental matters onto the international agenda.

Much like during the interwar years, international society continued after 1945 to deal with global environmental issues in an ad hoc manner, focusing on the management of transboundary pollution or resource use issues. One of the first notable outcomes of post-war diplomacy was the 1946 International Convention for the Regulation of Whaling (ICRW), which led to the creation of the International Whaling Commission (IWC) as its governing body. The new whaling regime had grown out of earlier treaties and protocols agreed under the auspices of the League of Nations, starting with the 1931 Convention on the Regulation of Whaling, a 1937 agreement and further protocols in 1938 and 1945 (Gillespie, 2005: 4), all of which shared the characteristic of being based on a cartel of leading whaling nations. Despite its explicit reliance on conservation science, the ICRW's primary purpose was the protection of the whaling industry, not whales (Meyer, 2017: 49). The new Commission failed to get a grip on the problem of declining whale populations, largely due to various opt-out clauses in the treaty and lack of implementation controls. By the time of the UN environment conference in Stockholm in 1972, global whale stocks had become so thoroughly depleted that most countries – with the exception of Japan, Norway and the Soviet Union – had by then given up commercial whaling (Dorsey, 2013: 44).

Other international efforts to manage shared global commons were similarly piecemeal and ineffective. Much like the whaling treaty, the Convention for the High Seas Fisheries of the North Pacific Ocean, which was signed in 1952 by the United States, Japan and Canada, ended up protecting national fishing industries rather than fish stocks (Flippen, 2008: 618). Only two years later, the Convention for the Prevention of Pollution of the Sea by Oil was adopted, which built on similar efforts during the interwar years to limit the rights of ships to discharge oil into oceans. The convention prohibited the discharge of oil into the sea within 50-mile coastal exclusion zones and enabled coastal states to take enforcement action against all ships within their territorial seas. Critically, however, the convention left the punishment of offences on the high seas under the law of the flag state where the offending ship was registered (Legault, 1971: 212). The treaty underwent several amendments in 1962, 1969 and 1971 before it was incorporated into the International Convention for the Prevention of Pollution from Ships (MARPOL) in 1973 (Mitchell, 1994).

None of these post-1945 treaties signalled any desire among the most powerful states to move beyond the limitations of the Green Westphalian approach. The only conceptual innovation that pointed towards a more solidarist understanding of global environmental responsibilities was the growing recognition of the 'common interest' or 'common concern of humanity' in international environmental management (Shelton, 2009). The Whaling Convention of 1946, for example, recognises in its preamble the 'interest of the world in safeguarding for future generations the great natural resources represented by the whale stocks' and declares that it is in the 'common interest' to achieve the optimum level of whale stocks as rapidly as possible. The High Seas Fisheries Convention of 1952 similarly expresses 'the common interest of mankind, as well as the interests of the Contracting Parties, to ensure the maximum sustained productivity of the fishery resources of the North Pacific Ocean' (International Convention for the High Seas Fisheries of the North Pacific Ocean, 1952). An even stronger nod in the solidarist direction came with the 1959 Antarctic Treaty (Stec, 2010: 365). Although the first explorations of Antarctica led to various territorial claims for the exploitation of mineral and other resources, and potentially the creation of military bases, the great powers were able to pull back from geopolitical competition on the frozen continent and declared it a global commons. In a unique show of international unity, the 1959 treaty sought to remove the Antarctic from international economic and military competition and to preserve the territory for scientific cooperation, though it was not until 1991 that the Madrid Protocol on Environmental Protection added an explicit environmental mandate to the treaty (Blay, 1992).

As far as the global environmental movement was concerned, it struggled to be heard internationally as the post-war reconstruction phase gave way to the Cold War. While most conservation organisations in North America and Europe were still overwhelmingly focused on national issues, a small group of environmental campaigners tried to rebuild the international networks that they had begun to create during the interwar years. In 1946, the Swiss League for the Protection of Nature hosted a meeting for a small group of experts from Britain, France, Belgium, the Netherlands, Czechoslovakia, Norway and Switzerland to discuss the revival of the Berne Conference's Consultative Commission of 1913. Swiss campaigners hoped to combine this initiative with efforts by Dutch conservationists to resuscitate the International Office for the Protection of Nature (IOPN) of 1934, which according to Dutch conservationist Pieter Van Tienhoven 'still exists, although in a dormitory state' (quoted in Holdgate, 1999: 19). The first breakthrough came at a follow-up conference in Brunnen, Switzerland, in the summer of 1947, which drew a larger crowd of about eighty conservationists from twenty-three countries, this time also with delegates from the United States, India and Guatemala, as well as representatives of UNESCO, FAO and the UN Trusteeship Council (Holdgate, 1999: 24).

The Brunnen conference considered various models for a new international environmental body and debated the extent to which conservationists should seek UN and governmental involvement in the new institution. While some environmentalists sought to establish a new global environmental body outside the states system, as part of an emerging green world society, others felt that in order to be effective the new body needed the backing of international society. Unable to resolve this chasm in environmental thinking, the conference concluded with the adoption of a compromise solution, based on a hybrid form of state and non-state representation. A year later, the Brunnen compromise led to the creation of the IUPN (IUCN from 1956).

IUPN's early years, which were marked by severe financial difficulties, were inauspicious (Holdgate, 1999: 53–6). Nevertheless, over the years the new international organisation, constructed as a Governmental and Non-Governmental Organisation (GONGO), was to become one of the early success stories of the post-war international conservation movement. IUPN's purpose, as defined in Articles 1 and 2 of its constitution, was the facilitation of international cooperation between governments and national and international organisations on nature protection, the promotion of national and international action, the collection, analysis and dissemination of information about nature protection and the international exchange of policy-relevant ideas and information. In this, IUPN came remarkably close to the remit that UNEP was given in 1973, though IUPN was deliberately designed to operate at arm's length from the UN system. In the late 1940s, governments had willingly accepted the IUPN's new model of semi-autonomous existence, not least as this reduced their financial liabilities and avoided a difficult re-negotiation of existing environmental mandates that UNESCO and FAO had taken on.

IUPN's main function was to provide an international network that facilitated the gathering and exchange of scientific information relevant to nature protection. Much like pre-war conservationist networks, it relied heavily on existing personal links between scientists, government officials and politicians. Although acquiring consultative states with the UN's Economic and Social Affairs Council (ECOSOC) in 1951, it failed in its ambition to achieve a more co-equal position alongside UNESCO and could only pursue a limited agenda of non-political advice and coordination behind the scenes. Given the profound weakness of global environmental norms in the post-war international order, it is hardly surprising that IUPN's early lobbying efforts had little impact on governmental policies around the world. As Jean-Paul Harroy, Secretary-General of IUPN, noted in 1954: 'Most of the time, the letters written by the Union are officially acknowledged with a polite note to the effect that the letter would be "forwarded to the competent departments"' (International Union for the Protection of Nature, 1955: 27).

IUPN's weakness was mirrored in the parochial nature of most environmental groups after 1945. Since the second half of the nineteenth century, a plethora of conservation organisations had come into existence across the industrialised world, but most were still preoccupied with local issues and failed to make emerging global environmental threats the subject of their campaigns (Tucker, 2013: 567). Despite some success in establishing more durable transnational links between campaigners, a green world society was still a distant vision. As can be seen from the distribution of topics that international NGOs (INGOs) campaigned on after the war, environmental concerns were still underrepresented in post-war global civil society. While thirty-three INGOs focused on promoting human rights in 1953, fourteen on international law and eleven on peace, only two were concerned with environmental issues. Even Esperanto fared better than global ecology, with eleven INGOs championing the international language (Keck and Sikkink, 1998: 11). The founding of the IUPN may have laid the foundation for a future growth in global environmental networking, but its mode of operation – working as a small club of well-connected conservationists, scientists and bureaucrats – meant that it had more in common with the nineteenth century-style elite networks that sustained the early conservation movement than the grassroots campaigns and large membership-based organisations that were to propel modern environmentalism to global prominence in the 1970s.

The Environmental Revolution of the 1960s/1970s

The social and political revolution that was to transform an elite concern for nature conservation into a mass movement with far-reaching consequences for international society can be dated to the late 1960s and early 1970s. In his history of global biodiversity protection, Holdgate (1999: 101) speaks of an 'environmental explosion' between 1966 and 1975. Radkau (2011: 135–6) suggests 1970 as the key turning point (see also Kupper, 2003), though he recognises that the societal developments in the 1960s prepared the ground for the eruption of environmental activism in the 1970s. Even observers writing at the time sensed that the new awakening of environmental awareness signalled a profound shift in public attitudes, at least in the industrialised countries. Max Nicholson's influential book *The Environmental Revolution*, first published in 1970, points to the broadening appeal of the conservation agenda and a transformation of environmentalism from a minority concern into a mass phenomenon. Opinion polls in the United States and other industrialised countries support this notion of a step change in public attitudes. Membership figures in environmental organisations shot up dramatically and new organisations emerged with a more explicitly political orientation. When an estimated twenty million US citizens attended the events marking the first Earth Day on 22 April 1970, politicians across America's political spectrum saw with their own eyes that environmentalism had become an electorally significant force (Shabecoff, 1993: 111–28). US President Richard Nixon captured this sense of a broad environmental awakening in his State of the Union address in January 1970:

> The great question of the seventies is, shall we surrender to our surroundings, or shall we make our peace with nature and begin to make reparations for the damage we have done to our air, to our land, and to our water? Restoring

nature to its natural state is a cause beyond party and beyond factions. It has become a common cause of all the people of this country. (quoted in Kalb, Peters and Woolley, 2007: 667)

The nature of the socio-economic changes that made the environmental revolution possible has been at the centre of a long debate in the social sciences. Ronald Inglehart (1977) famously identified the experience of two decades of uninterrupted economic growth after the Second World War as the trigger for the rise of post-material values in Western societies. Once people's basic economic needs had been met, they began to prioritise non-material values, such as beautiful landscapes and clean air. With the state still focused on promoting industrial growth and material welfare, citizens began to organise in grassroots campaigns to question the dominance of the modern industrial system. Inglehart's theory has been challenged at both empirical and conceptual levels: not all environmentalists are economically well off; other factors, such as education and political orientation, may have played a more important role than income levels; and subsequent fluctuations in support for environmentalism don't match shifting economic fortunes (Duch and Taylor, 1993; Graaf and Evans, 1996). Other critics, such as Ramachandra Guha and Juan Martinez-Alier (1997), point out that environmentalism in the developing world has different roots, stemming from struggles of poor communities against 'environmental destruction which directly affects their way of life and prospects for survival' (Guha and Martinez-Alier, 1997: xx) rather than a post-material value shift. Undoubtedly, the environmental revolution in the last third of the twentieth century is a multifaceted phenomenon that is difficult to capture in mono-causal theories. Still, as far as the industrialised world is concerned, the post-war economic boom clearly created the material conditions for a wider societal re-evaluation of the relationship between humanity and nature. In this sense, the link between the dramatic expansion of post-war economic prosperity and societal change remains an important starting point in accounting for the emergence of modern environmentalism (McNeill, 2000: 336–7).

The remarkable rise of environmentalism from the late 1960s may have reflected societal changes, but it also came with a shift in environmental thinking itself. Whereas early twentieth-century environmentalists were strongly influenced by preservationist thinking that sought to protect nature from human usage, the modern environmental movement was more unequivocally driven by a desire to improve humanity's condition within a healthy environment (Radkau, 2011: 147). 'You're next, people!' was the key message of environmentalists demonstrating against the pollution of New York City's Hudson River on Earth Day

(Shabecoff, 1993: 113). Romantic notions of the loss of wilderness and species extinction continued to resonate in environmental circles, but it was the new anthropocentric focus on the dangers of overpopulation, resource constraints and industrial pollution that gave modern environmentalism its mass appeal and enabled it to shape modern politics in a way that traditional conservationists had failed to do. Nuclear weapons testing during the early Cold War and the growing number of environmental disasters caused by industrial accidents – from the release of mercury into Japan's Minamata Bay (1960s) to the Torrey Canyon oil spill off the coast of England (1967) and dioxin spills in the United States and Italy (1970s) – reinforced ecologists' warnings of the threat that industrial growth posed to human well-being. Rachel Carson's book *Silent Spring* (1962), widely regarded as the foundational text of modern environmentalism, was particularly successful in directing public attention to the dark side of industrial progress and turning environmentalism into a mainstream concern (Kinkela, 2013: 118–19). By highlighting the unintended consequences of pesticide use in agriculture, which eventually found their way back into the animal and human food chain, Carson helped popularise ecology's core tenet that 'everything is connected to everything else' (Commoner, 1971: 16).

The rise of modern environmentalism also transformed the nature of environmental activism. Established conservation groups, such as the Sierra Club, Audubon Society and Wilderness Society, had been reluctant to directly challenge political and economic elites over politically charged issues of environmental pollution. By contrast, the new breed of environmental NGOs, such as Friends of the Earth (1969) and Greenpeace (1971), adopted a more overtly political campaigning style and targeted big business and the military as sources of global environmental degradation. David Brower, for example, who had led the Sierra Club for 16 years as its executive director, tried to shift the organisation's focus from non-political mountaineering to environmental action. His tireless efforts turned what had been a sleepy 'gentlemen's club' into a lobbying and campaigning organisation, though he grew increasingly frustrated by the organisation's refusal to campaign against nuclear power. After being forced out of the Sierra Club in 1969 over claims of financial mismanagement, Brower went on to create Friends of the Earth, which developed a more confrontational campaigning style with an explicitly global focus (Shabecoff, 1993: 100–1). Whereas the traditional conservationist movement emphasised the need to nurture close links with bureaucratic and political elites in order to effect political change from within the state, the new environmental groups took environmental issues to the street and sought to bring about political change from the bottom up. By turning environmental protection into a matter of open political contestation over the future of industrial society, they forced a profound re-thinking of the core purpose of the state (Reus-Smit, 1996) in an era when untrammelled economic growth could no longer guarantee improvements in human well-being.

The modern environmental movement also succeeded in reframing environmental degradation as an essentially *global* problem, thereby inevitably turning the spotlight on states' international responsibilities. This is in stark contrast to much of the environmental debate in the nineteenth and early twentieth centuries, when the transnational dimensions of nature protection played only a marginal role. At a time when the prevalent focus was on protecting scenic landscapes and landmarks – be it North America's unspoiled wilderness or Germany's mythical forests – environmentalist discourses were often couched in nationalist rhetoric. By contrast, the environmental NGOs that grew out of the new social movements of the 1960s embraced globalism and benefitted from the new opportunities for transnational networking and campaigning. Thanks to the attention that television and print media paid to their publicity-friendly stunts, NGOs such as Greenpeace operated in a political space that increasingly transcended national boundaries (Wapner, 1996: chapter 3). By the 1970s, it was becoming a matter of routine for environmental organisations to develop linkages across national boundaries, to campaign on global issues and to mobilise supporters worldwide (McCloskey, 2005: 157–69). Even traditional conservation groups, such as the National Audubon Society and Humane Society, were starting to team up internationally to demand international action, as can be seen in the internationally coordinated 'Save the Whales' campaign to end commercial whaling (Dorsey, 2013: 54).

The emerging global focus of the environmental movement was informed not least by a similar shift in ecological science after the Second World War. Ecology had always promoted a more holistic understanding of nature that emphasised interdependencies between living organisms and their environment. By the mid-twentieth century, thanks to technological advances in environmental monitoring, global interdependencies had occupied a more prominent place in ecological research, and it had become increasingly common for scientists and environmentalists to describe the earth as an integrated ecological system. The beginnings of planetary surveillance can be traced back to what Davis (2002: 217) calls the British Empire's emerging 'world climate observation system', which relied on the growing network of colonial observation stations connected by telegraph and undersea cables in the second half of the nineteenth century. However, it was the advances in aviation and satellite technology after 1945 that finally gave scientists the means with which to collect more reliable data on transnational environmental flows and global ecosystems. The launch of the International Geophysical Year in 1957, originally driven by US military needs and involving an unprecedented level of cooperation by scientific organisations from across the East–West divide (Goossen, 2020), led to the creation of new 'World Data Centers' freely available to all scientists. A year later, the Mauna Loa Observatory in Hawaii began to monitor atmospheric carbon dioxide. Over time, scientific research into planetary ecosystems provided growing empirical evidence of a planet imperilled by modern technology (Aronova, Baker and Oreskes, 2010). When the Apollo 17 crew beamed the famous 'blue marble' image of the planet back to mission control in 1972, the spectacular picture merely reinforced the then emerging view of the planet as a fragile ecosystem under threat.

These societal and ideational shifts produced a conducive context within which a truly global environmental movement emerged. However, to have any lasting impact on international relations environmentalists needed to engage with state-centric politics. Most importantly, the modern nation-state had to be greened first before international society itself could undergo a similar normative shift. In this sense, shifts in the legitimacy of the nation-state were the critical link between the rise of a green world society and the emergence of environmental stewardship as a primary institution in international society. Leading states first had to accept a core environmental responsibility at the national level before they could collectively accept a similar responsibility at the international level. Fortunately, the timing was right for such a normative reorientation. The modern environmental movement arose at a time when governments in advanced industrialised countries were developing an ever more comprehensive role in steering the economy and society, to achieve economic growth and full employment but also to address wider social problems associated with rapid economic expansion, technological change and mass consumption. State activism and interventionism were at an all-time high just as environmentalists embraced the state and sought to expand the modern state's core purpose to include environmental sustainability (Hunold and Dryzek, 2002: 22). The visible hand of the state could also be felt at the international level. The UN and the Bretton Woods institutions were created to establish a multilateral framework for regulating the expansion of the post-war global economy while ensuring UN members with sufficient domestic autonomy to pursue both economic growth and social stability. John Gerard Ruggie's notion of 'embedded liberalism' (1982) captures the essence of the compromise that Western powers struck between the creation of a liberal rules-based international order and the need to accommodate an expanded role for the state in guiding domestic economic and social development.

The gradual greening of the state took on many forms and proceeded at different speeds around the world. Early successes in establishing environmental policies and institutions in the United States were to prove particularly important, both because they became a source of international policy diffusion and because the United States provided critical leadership in the international context. To be sure, the United States was not alone in promoting the internationalisation of the environmental agenda in the 1970s. After all, it was Sweden that came to propose the first UN environment conference in 1968. But it was America's pioneering role in establishing comprehensive environmental legislation and institutions that provided other industrialised countries with a blueprint for integrating environmental objectives into national policy. Already in the 1960s, the Lyndon B. Johnson administration had introduced a wide range of environmental laws, from the Wilderness Act of 1964 to the Land and Water Conservation Act of 1965, Solid Waste Disposal Act of 1965, Water Quality Act of 1965, Endangered Species Preservation Act of 1966 and Air Quality Act of 1967. Many of these laws proved inadequate and had to be amended or replaced in later years, though bipartisan support for environmental policy ensured that even after Johnson's departure the United States continued on the path to strengthening federal authorities in environmental matters. Indeed, the passing of the National Environmental Policy Act (NEPA) and the creation of the US Environmental Protection Agency (EPA) in 1970 proved to be two of the most significant environmental policy legacies of the Nixon administration (Flippen, 2008: 616).

Nixon's environmental policy may have been swayed more by electoral calculation than a deep-seated commitment to nature protection (Flippen, 2008: 614; Hopgood, 1998: 71), but his institutional and legislative initiatives locked in bureaucratic support for environmental stewardship as a national policy priority in the United States. They also set in motion a process of horizontal policy diffusion to other countries and vertical policy export to the international level. From the late 1960s onwards, other industrialised countries started to develop their own legal and institutional infrastructure for national environmental protection, and many looked to the United States as a role model. In 1969 Germany established a federal environmental policy competence based on the US model, while Britain and France created their first environmental ministries in 1970 and 1971, respectively

(Hünemörder, 2004: 154–5; Szarka, 2002: 71; Weale, 1992: 15). Even socialist countries began to experiment with some environmental policies behind the Iron Curtain, despite officially claiming that industrial pollution only afflicted capitalist systems. Starting in 1968 and continuing through the 1970s, the Soviet Union created laws on specific environmental aspects, on land, water, minerals, forestry, air quality and wildlife, though it only moved to prepare a comprehensive environmental protection law in the late 1980s (Pryde, 1991: 7). It even entered into bilateral agreements with the United States on scientific and environmental cooperation as part of détente policy (Josephson et al., 2013: 195). By the early 1970s, the domestic conditions had thus been created for leading countries in the West, where domestic environmental pressure had been strongest, to move towards a global definition of environmental responsibility.

The 1972 Stockholm Conference and Global Environmental Stewardship

The early 1970s were the critical juncture that allowed environmental values to enter the normative structure of international society. To be sure, the momentum for this profound change had been building up slowly since the end of the Second World War, but with the emergence of the modern environmental movement in the late 1960s it now reached a decisive tipping point. A combination of ideational shifts, political mobilisation in world society and political leadership within international society made it possible to establish environmentalism as an international norm. International scientific organisations had amassed growing evidence of a looming environmental crisis, with rising populations, resource extraction and economic growth putting ever more pressure on vulnerable ecosystems. Ecological notions of a globally interconnected biosphere with limited carrying capacity were being disseminated in environmental discourses and campaigns and began to seep into public discourse in major industrialised countries. And as more and more politicians on all sides of the political spectrum adopted the new language of the planet as an endangered 'spaceship', calls for a new cooperative spirit across national boundaries became commonplace. The time was right, at least in the most advanced industrialised economies, to recognise environmental stewardship as a duty for states and international society.

One measure of the growing momentum behind the greening of international society was the integration of environmental concerns into the mandates of ever more international organisations (Borowy, 2019).

UNESCO and the FAO were the original UN bodies that developed work programmes on environmental matters. While the FAO focused mainly on sustainable forestry issues and never aspired to play a more comprehensive environmental role, UNESCO gradually expanded its conservation agenda, which had revolved around science promotion, heritage protection and information exchange. In 1961 UNESCO established a new Division for Natural Resources and in 1965 proposed a world conference to promote scientific cooperation on environmental matters (so-called Biosphere Conference of 1968). By the 1960s, the Organisation for Economic Co-operation and Development (OECD, known until 1961 as Organisation for European Economic Co-operation, OEEC) also started to discuss pollution problems as part of its emerging work on the qualitative aspects of growth. The OECD set up two working groups within its Committee for Research Cooperation, on water pollution (1967) and air pollution (1968), and in 1970 became the first international organisation to establish its own environmental committee (Schmelzer, 2012: 1008). Similarly, the Council of Europe initiated discussions on cross-border air pollution in the early 1960s and set up a Committee of Experts on Air Pollution in 1966, which helped launch the Declaration of Principles on Air Pollution Control in March 1968 (Meyer, 2017: 51). The 'rising tide of active environmentalism' (Holdgate, 1999: 111) had thus begun to engulf various international organisations within and outside the UN.

Despite the uptick in environmental debates at the international level, the society of states had yet to make an explicit and firm commitment to pursuing global environmental stewardship. As the final report to UNESCO's Biosphere Conference rightly noted, the world still 'lacked considered, comprehensive policies for managing the environment' (quoted in Caldwell, 1996: 54). To address this gap, Sweden submitted a proposal to the UN's ECOSOC in July 1968 to hold an international conference on the environment. This was the first time that the UN would formally serve as a high-level diplomatic forum for addressing global environmental challenges. Until then, various UN bodies had merely facilitated expert-level meetings that operated at the margins of international diplomacy. By contrast, the proposed UN conference was intended to establish environmental protection at the heart of the international agenda, with explicit buy-in from state leaders. It was to become the constitutional moment in the creation of environmental stewardship as a primary institution of international society.

The Swedish proposal quickly garnered the support of a number of influential states, most notably the United States. After the UN General Assembly had endorsed the proposal, Maurice Strong, a businessman and the first head of Canada's International Development Agency, was tasked with leading the preparatory process. Strong established a consultative process and commissioned a report, the *World Report on the Human Environment*, which was to provide the main intellectual blueprint for the UN's new environmental agenda. Produced by a Committee of Corresponding Consultants, consisting of 152 scientists from over sixty countries (Javaudin, 2017: 83–4), the report became a global bestseller after it was published in book form under the title *Only One Earth* (Ward and Dubos, 1972). Right from the outset, the authors of the report made it clear that the UN conference would have to pursue a decidedly anthropocentric form of global environmentalism: Stockholm 'is not focused on abstract problems of theoretical ecology' but 'the characteristics of the environment which affect the quality of human life' (Ward and Dubos, 1972: 24). The report reiterated the by then commonplace observation that 'environmental problems are becoming increasingly world-wide' and that they therefore required a 'global approach' (Ward and Dubos, 1972: 28). With humanity having to 'accept responsibility for the stewardship of the planet' (ibid., 25) but individual governments being unable to deal with many environmental problems on their own, a new capacity for 'global decision-making and global care' (ibid., 270) was required. In short, international society needed to make 'a new commitment to global responsibilities' and put this commitment into action (ibid., 270).

Only One Earth included an ambitious call for a new 'planetary order' (ibid., 263) that had strong solidarist undertones. The authors noted that recognition of global environmental interdependencies could foster 'a sense of global community, of belonging and living together, without which no human society can be built up, survive and prosper' (ibid., 297), and that the required expansion of international cooperation would inevitably weaken strong notions of national sovereignty. The authors went as far as expressing the hope that 'we may find that, beyond all our inevitable pluralisms, we can achieve just enough unity of purpose to build a human world' (ibid., 297). But their vision for a greener international society was based on an extension of existing forms of international cooperation, not a radical reform of the international system or a move towards a green world government. Environmental one-worldism was tempered by the pluralist reality of late twentieth-century international relations: 'world action for pollution control and an enhanced environment would simply be seen as logical extensions of the practice of limited intergovernmental cooperation, already imposed by mutual functional needs and interests' (ibid., 298). Thus, by the end of their report Barbara Ward and René Dubos returned to the more pragmatic theme of working with, not against, the existing Green Westphalian logic of international society.

The global environmental movement, although broadly supportive of the UN process, was divided in its assessment of what to expect from the Stockholm conference. On the one hand, environmentalists had long pushed for states to cooperate more fully on global environmental issues and now saw the first tangible opportunity to create the institutional infrastructure for a global environmental rescue. From its early days, IUCN (IUPN until 1956) was focused on shaping state practices at the international level, and in 1970 its General Assembly restated as one of its central purposes the need 'to encourage and assist in the making of coordinating legislation and international conventions' on environmental issues (quoted in Macekura, 2015: 103). Many leading conservationists had developed close links with state bureaucrats and political leaders and now saw an opening at the highest level to insert environmental values into the normative fabric of interstate relations. On the other hand, the environmental movement had experienced far too many setbacks in the past when dealing with governments. Some campaigners feared that state leaders' newly found enthusiasm for the environment would be short-lived. US President Nixon was not the only head of state whose advocacy of global environmental action was influenced more by electoral calculation than a deep desire for nature conservation. Even IUCN, which had worked more closely with the political establishment than many other environmental organisations, held little hope that the Stockholm conference 'would produce any basically new approach to the problems of the environment', as its 1971 annual report stated (Holdgate, 1999: 112). Some environmental purists operating within the preservationist tradition were also concerned that Stockholm would focus more on the benefit that humans could derive from environmental politics than on nature protection for its own sake. NGO suspicions about states' intentions were further deepened when many non-state actors found themselves excluded from the main proceedings of the Stockholm conference and were only offered participation in a parallel event at some distance from the main diplomatic gathering (Holdgate, 1999: 113).

Despite the constraints that the conventions of diplomacy imposed on environmental campaigners, Stockholm turned into a unique opportunity for world society actors to shape the emerging international environmental agenda. Partly as a result of the novelty of environmental issues as a high-level diplomatic concern, environmentalists were able to exert unprecedented influence over the conference proceedings. Many scientists and conservation experts were intimately involved in advising diplomatic missions at UNCHE. In some cases, leading environmentalists were appointed as members and even heads of country delegations. The US delegation was led by Russell E. Train, a prominent conversation campaigner. As Vice-President of the World Wildlife Fund (WWF, renamed in 1986 as World Wide Fund for Nature) and President of The Conservation Foundation in the second half of the 1960s, Train had occupied influential positions in the American environmental movement. Having been made Chairman of the Task Force on the Environment for President-elect Richard Nixon in 1968, he was in a uniquely influential position to shape the American approach to the Stockholm conference. Train benefitted from the fact that President Nixon cared little about environmental issues beyond their electoral value and 'assigned a low priority to the actual content of environmental policies' (Macekura, 2015: 109). America's new international environmental policy attracted considerable bureaucratic infighting in Washington (Hopgood, 1998: 66–75), but as delegation leader at the Stockholm conference, Train was able to provide IUCN and WWF with unprecedented influence over the US delegation's approach (Macekura, 2015: 109).

It was clear from the beginning of the preparatory process for UNCHE that deep divisions existed in international society over how strongly environmental protection should be pursued internationally. For one, not all industrialised countries shared a common vision of the global environmental agenda. Sweden had proposed the UN conference not least to initiate international action against acid rain caused by sulphur dioxide (SO_2) and nitrogen oxide (NO_x) emissions from the burning of fossil fuels, a problem that largely originated abroad and therefore required an international solution. Many of the main emitters in Europe, however, opposed intrusive regulatory measures, citing a lack of scientific evidence. France and Britain, in particular, were reluctant to agree to ambitious new rules that would harm their industrial interests, be it energy companies or supersonic air transport (Brenton, 1994: 41). Many countries were also wary of creating new international institutions. The US position paper for Stockholm included opposition to a new environmental body as a critical US interest (Hopgood, 1998: 95). British diplomats similarly expressed concern about the UN Secretary-General's proposal for the creation of a specialised environmental agency, advising their delegation to resist the growing pressure for a 'new machinery' at the international level (Arculus, 1970; Sterling, 1970).

An even deeper rift existed between the West and the group of communist countries led by the Soviet Union. According to official communist doctrine, industrial pollution was strictly a capitalist problem (Cole, 1993: 35), and the Stockholm conference should thus have been of limited importance to the Soviet Union and its allies. Soviet representatives nevertheless participated in the preparatory meetings only to stage a boycott of UNCHE when the German Democratic Republic (GDR), at the time not yet a member of the UN, was refused a seat at the table in Stockholm. American diplomats suspected that the Soviet leadership was 'using the environment issue as a political football in trying to achieve international recognition for the GDR' (United States Department of State, 1971). Of all Eastern bloc countries, only Romania sent an official delegation to Stockholm, while the Soviet Union and GDR delegations participated as observers (Macekura, 2011: 503–4).

The most troubling divide in international society, however, which threatened to derail the Stockholm conference preparations, was between countries of the Global North and Global South. Developed economies had called for the UN conference to be convened and sought to focus its agenda on typically Northern environmental problems, such as protection of natural habitats, marine pollution and air pollution. Alarmed by Northern environmentalists' neo-Malthusian concerns over population growth (e.g. Ehrlich, 1968) and calls for a limit on global economic growth (Meadows et al., 1972), many developing countries challenged Northern countries' agenda for the Stockholm conference. Brazil took the lead in formulating an alternative Southern perspective that sought to protect national sovereignty and the development imperative. At various G77 meetings in the run-up to Stockholm, Brazilian delegates took a tough stance that held the North responsible for environmental problems and demanded financial compensation from the North for past pollution and future development needs (Macekura, 2015: 113–16). In the Declaration and Principles of the Action Programme adopted at its 1971 summit in Lima, the G77 declared that 'no environmental policy should adversely affect the possibilities for development, either present or future, of developing countries', and that environmental measures adopted in the North should not 'create additional obstacles such as new non-tariff measures' (quoted in Nicholls, 1973: 67). These points were to be reiterated throughout the preparatory process for Stockholm, and some developing countries even talked of boycotting the UN conference altogether (Manulak, 2017: 106). It was thus clear from well before the Stockholm conference had opened that any attempt to create a global consensus behind the new norm of environmental stewardship would fail if it did not take on board developing countries' concerns around development and global justice.

Much of the UNCHE preparatory process was focused on bridging this fundamental divide between the global North and South. At the root of the conflict were contrasting views not just on the urgency of establishing international society's global environmental responsibility but also on what this responsibility entailed. Northern countries tended to define environmental stewardship as a general, collective, responsibility of all states, focused on managing the effects of transboundary industrial pollution, marine pollution, depletion of global resources such as fish stocks and whales, and overconsumption of resources. In contrast, developing countries stressed the special responsibilities of Northern countries, both for causing resource depletion and industrial pollution and for helping Southern countries meet their own environmental objectives, mainly through international aid and technology transfers (Brenton, 1994: 39–40; Manulak, 2017: 105–7).

Maurice Strong, who had identified this conceptual divergence early on in the process, convened an international expert group to debate the relationship between environment and development. When the committee of twenty-seven development economists from developed and developing countries met in Founex, outside Geneva, in June 1971, two principal conceptions of this relationship came to the fore. On the one side were liberal economists who viewed environmental and development policy as being locked into a zero-sum game, with poor societies needing to prioritise their economic development and taking up environmental concerns only later, after they had grown wealthier. On the other side, economists of a more interventionist persuasion countered that environmental problems in the South were of a different kind, about how human well-being was affected by poor sanitation, inadequate housing, soil erosion and vulnerability to natural disasters. The latter pushed for a broader concept of economic growth, beyond gross domestic product (GDP), and for government intervention in the development process to reduce environmental disruption and increase human well-being (Manulak, 2017: 110–12).

In the end, the intellectual gap between liberals and interventionists was never fully bridged. The committee only achieved a 'manufactured consensus' (Manulak, 2017: 115). It was based on a compromise position that emphasised the intricate links between environment and development, a theme that was to resurface in new form in the late 1980s as part of the sustainable development discourse. Significantly, environmentalists' warnings of an impending ecological crisis and the need to rein in economic growth played only a marginal role at the Founex meeting. If environmental stewardship was to become a fundamental norm in *global* international society, it had to be constructed so as to fit in with developmentalism and national sovereignty, the two key primary institutions that developing countries prioritised in their approach to global environmental issues.

In a significant nod to developing country concerns, the Founex Report expressed the expectation that 'the global concern for the environment may reawaken the concern for elimination of poverty all over the globe'. Rich nations' emerging understanding of 'the indivisibility of the earth's natural systems' would help 'strengthen the vision of a human family', even to the extent that it would 'encourage an increase in aid to poor nations' efforts to improve and protect their part of the global household' (Founex Report, 1971: 4.12). Going into the Stockholm conference, industrialised countries thus had a clear understanding of the expectation that increased international aid would be the price to pay for developing countries to accept environmental stewardship. Already in 1970, British diplomats had noted 'strong indications that the UN Conference [UNCHE] will give rise to increased pressure for aid to be spent for environmental purposes; and that the LDCs will not give priority to environmental factors unless assisted financially' (Wheeler, 1970). The G77 had been demanding a redistribution of global resources for some time, as part of its campaign for a 'New International Economic Order' that was meant to restructure global economic relations (Mayall, 1990: 132–9). Much to the frustration of donor countries, environmental aid was soon added to the list of global redistributive demands. While Sweden's social–democratic government was prepared to put financial transfers on the new environmental policy agenda, other industrialised countries, most notably the United States and Britain, initially tried to fend off demands for large additional financial commitments (Macekura, 2015: 110–11). They were unlikely to get their way, however, as a British diplomat noted sarcastically in the run-up to Stockholm: 'Whether or not the UK contribution to such a fund should be made within existing ceilings, we shall inevitably have to pay financial homage to the new environmental god' (Wheeler, 1970).

The outcomes of the Stockholm conference confirmed that global environmentalism had firmly emerged as a new component of international society's normative order. Agreed by 113 on 16 June 1972, the Stockholm Declaration (United Nations, 1973) became the first international agreement to declare unambiguously that '[t]he protection and improvement of the human environment is [...] the duty of all Governments' (Preamble). In a nod to the preservationist objectives of the early environmental movement, the Declaration identifies 'a special responsibility to safeguard and wisely manage the heritage of wildlife and its habitat' (Principle 4). This may appear to make the non-human environment the key referent object of environmental stewardship, but such an ecocentric conception of environmental protection was far from state leaders' minds. The remainder of the Stockholm Declaration is on firm anthropocentric ground: humans have a right 'to freedom, equality and adequate conditions of life, in an environment of a quality that permits a life of dignity and well-being' (Principle 1; see also Principles 2 and 3). Both traditional conservationists and modern environmentalists could thus find elements in the Declaration that reflected their philosophical beliefs, though the anthropocentric focus on humans' right to live in a clean environment was to prove the dominant motive behind establishing environmental stewardship as a primary institution in international society.

The Stockholm Declaration also combines pluralist and solidarist understandings of environmental stewardship. Many principles reflect the constraints of an international normative order built on sovereignty and non-intervention, such as the commitment to prevent transnational environmental harm (Principle 21). Others allude to a more ambitious agenda of protecting common areas (e.g. preventing pollution of the seas, Principle 7) and integrating environmental considerations into development planning (Principle 13). Yet others point strongly to a solidarist future of global environmental cooperation, as in the commitment to develop international law on environmental liability and compensation (Principle 22).

As with many other international declarations, the Stockholm Declaration represents a careful balancing act that seeks to satisfy the demands of different constituencies. Clearly, to be acceptable to most countries, and particularly the large number of developing countries, environmental stewardship had to be framed as fitting in with, rather than posing a fundamental challenge to, the existing international normative order. International society did not endorse the kind of radical break with past state practice that some environmentalists had called for. Recognition of global ecological interdependencies and solidarist notions of 'Spaceship Earth' and 'common heritage of humankind' did enter the deliberations during the preparatory meetings for the Stockholm conference (Ward and Dubos, 1972), but few states were willing to consider the possibility of ceding regulatory authority to a new international body representing the planetary interest. Far from endorsing the more radical calls for a profound reorganisation of the international order (e.g. Falk, 1971), the Stockholm Declaration embeds new environmental duties within an unambiguous reassertion of the principles of national sovereignty and development (Principles 21 and 24). Simply put, environmental stewardship needed to go with the normative grain of international relations if it was to become a viable, and universally accepted, primary institution in international society.

Conclusions

Despite rejecting a general duty to protect the global environment as part of the post–Second World War international order, from the late 1940s onwards international society gradually expanded its engagement with environmental problems through a series of international conferences and treaty negotiations. While this largely mirrored the limited scope and pace of environmental cooperation during the interwar years, it nevertheless laid the foundation for a dramatic escalation of international society's involvement in environmental policy-making from the 1970s onwards. Spurred by the growing realisation of the *global* dimensions of many environmental problems and a groundswell of domestic demand for greater environmental protection, governments in North America and Europe began to put in place the building blocks for a more active state role in the environmental field – both domestically and internationally.

The key breakthrough in the greening of international society came in the early 1970s, as a consequence of decisive leadership by a small group of countries that put environmental stewardship on the international agenda. The 1972 Stockholm conference, proposed by Sweden and actively supported by the United States, provided the critical forum for taking stock of the global environmental crisis and determining what role international society should play in solving it. The conference took place within the state-centric strictures of international diplomacy, but it was the emergence of the modern environmental movement in the preceding years that had successfully challenged the existing understanding of states' domestic and international legitimacy. Acting as part of political world society, the environmental movement successfully established the idea that collective action by states was needed, to prevent catastrophic environmental degradation and to secure humanity's collective interest in self-preservation.

Although driven by a solidarist logic of realising humanity's shared interest in a clean environment, the Stockholm conference did not frame environmental stewardship in isolation from the pluralist context of international relations. It certainly did not endorse the kind of radical break with the Westphalian tradition that some environmentalists had called for. The cosmopolitan rhetoric of 'Spaceship Earth' and 'common heritage of humankind' may have motivated the push towards recognition of international society's environmental responsibility, but to become successfully established as a new norm, environmentalism could not directly challenge the existing international order. In order to be acceptable to most leading powers as well as to the developing world, environmental stewardship was framed to be consistent with the principles of national sovereignty and developmentalism. The Stockholm Declaration reflects solidarist ambition but is built on pluralist foundations. The key question now was how these two competing logics were to be reconciled as international society began to furnish its new normative commitment to environmental stewardship with secondary institutions and governance mechanisms.

6 The Globalisation of Environmental Stewardship

On 3 December 1968, the United Nations General Assembly agreed to convene the UN Conference on the Human Environment, the first ever UN environmental summit. Few diplomats taking part in the UN decision could have imagined that they had set in motion the process that would establish environmental stewardship as a fundamental norm in global international society. In the late 1960s, environmental protection was still regarded as a marginal topic on the international agenda. By the time UNCHE was held in June 1972 in Stockholm, societal pressure had grown on governments in leading industrialised countries to act against the 'continuing and accelerating impairment of the quality of the human environment' (United Nations General Assembly, 1968). The Stockholm conference turned out to be the key turning point in the gradual greening of GIS. After five decades of failed attempts to establish states' collective responsibility for the global environment, the Stockholm Declaration declared unambiguously in its preamble that '[t]he protection and improvement of the human environment is [...] the duty of all Governments' (United Nations, 1973: 3). The majority of states had finally accepted what environmental campaigners had been arguing for since the late nineteenth century, namely that individuals and societies have a fundamental duty of care towards their natural environment. By defining this environmental responsibility in terms of a human right to environmental protection, state leaders may not have followed ecologists down an ecocentric path, which accords nature and humans a co-equal ethical status. Yet they opened the door to a still ongoing debate about how states and international society are to discharge their duty to protect the global environment.

Not all newly proposed norms in international relations become a broadly accepted standard of international behaviour. Norm entrepreneurs have sought to change many aspects of the international normative order, but only a few manage to bring about lasting ethical transformations in international relations. Some norms emerge and disappear again after some time, others remain contested to such an extent that they never reach universal levels of recognition. The ES's solidarist wing has long agonised over the fate of human rights as a progressive international norm, hoping that it would curtail the violence and injustice perpetrated by sovereign states and create a justification for humanitarian intervention in international society (Vincent, 1986; Wheeler, 2003). The end of the Cold War increased expectations that a new era of multilateral cooperation could strengthen the universal applicability of liberal values, but such hopes have been confounded more recently, leading some to conclude that human rights never acquired the status of a primary institution of *global* international society (Buzan and Schouenborg, 2018: 231).

How do we know when normative innovations have become fully established as part of GIS's normative structure? As discussed in Chapter 2, we need to look for evidence of significant social consolidation, across time and space, if we are to make the empirical claim that a new norm qualifies as an (emergent) primary institution. Two principal mechanisms are involved in this process of social consolidation: the creation of secondary institutions, which build on and reproduce a new primary institution, thereby developing, strengthening and entrenching it; and observable patterns of change in states' behaviour and identity, which are in conformity with the new norm. In this chapter, I explore these two mechanisms of social consolidation, tracing the evolution of international environmental politics from the Stockholm conference in 1972 to the Rio Earth Summit in 1992 and beyond.

The first part of this chapter reviews the creation of secondary institutions, or environmental regimes, and how dynamics of international environmental rule-setting reinforced but also reinterpreted the underlying environmental norm. The second part focuses on indicators of changing state behaviour and identity, particularly with regard to environmentalism's impact on diplomatic practices and multilateralism as a mode of international cooperation. The third part completes the story by examining the spatial dimension of environmental stewardship's social consolidation, with a focus on how environmental ideas and practices spread worldwide and how the Global South came to adopt but also redefine global environmental responsibilities.

The Creation of Environmental Secondary Institutions

Secondary institutions are important indicators of the social consolidation of primary institutions: they provide a measure of how quickly new norms emerge, how they strengthen and evolve over time, but also how they weaken and eventually decline. Deliberately created to serve specific functions of international cooperation, secondary institutions turn primary institutions into issue-specific regulative rules and principles. They give a more concrete and observable expression to fundamental norms. They help to reproduce and also change them. By providing an institutional context in which states are socialised into international society's core norms and practices, they also create a link between fundamental norms and state behaviour. At the same time, secondary institutions are sites of normative contestation and resistance. Because secondary institutions are continuously created and re-negotiated by states, they offer opportunities for political resistance to those that oppose new norms. As sites of normative innovation, reproduction, contestation and decay, secondary institutions thus play an important role in the life cycle of primary institutions.

The Post-Stockholm Boom in Multilateral Environmental Agreements

Global environmental politics has proved to be a particularly fertile area for the creation of secondary institutions. Estimates of how many multilateral environmental agreements (MEAs) have been signed vary, but the widely used *International Environmental Agreements (IEA) Database Project* at the University of Oregon offers the most reliable and comprehensive record, listing a total of just under 2,300 bilateral, over 1,300 multilateral and 250 other environmental agreements as of 2020 (Mitchell, 2002–2020). The first such agreements date back to the mid-nineteenth century, and the number of newly adopted treaties, protocols and amendments started to rise steadily (though at a modest rate) during the first half of the twentieth century. After the 1972 Stockholm conference, two significant upward shifts occurred in the rate of adoption of new environmental treaties. Both during the 1970s and 1990s the number of newly created bilateral environmental agreements shot up dramatically, while the signing of multilateral treaties peaked in the first half of the 1990s (Mitchell et al., 2020: 106).

At first sight, the Stockholm conference seems to have had only a modest impact on the frequency with which states created new multilateral accords, though there has been a notable increase in protocols and amendments to existing MEAs from the early 1970s onwards. In any case, a purely numerical interpretation of these trends fails to account for the shift in emphasis that UNCHE produced, in terms of the focus, ambition and scale of international environmental policy-making. Before Stockholm, only a relatively small number of states had been involved in negotiating bilateral or plurilateral environmental treaties, and these were largely ad hoc efforts to deal with relatively isolated environmental problems, mostly in relation to animal protection and transboundary resource management. As noted in Chapter 4, several European countries agreed to the 1902 Convention for the Protection of Birds Useful to Agriculture. In 1909, the United States and Canada signed the USA–Canada Boundary Waters Agreement, which included a provision that prevented the pollution of shared lakes and rivers. The United States, Britain, Russia and Japan agreed to the 1911 North Pacific Fur Seal Convention to regulate the commercial hunting of fur-bearing mammals, and European colonial powers adopted the 1933 Convention Relative to the Preservation of Fauna and Flora in the Natural State in order to strengthen nature conservation in African colonies. None of these treaties can be said to have been part of a more systematic effort to expand global environmental policy-making, and no international organisation was created to oversee the implementation of internationally agreed environmental objectives. Moreover, many of the early environmental treaties either did not enter into force or lacked the institutional infrastructure to turn them into effective regulatory instruments. They were often symbolic efforts at best, and they certainly did not reflect a commitment to global environmental responsibility among the parties.

By contrast, the post-Stockholm conference era witnessed a sustained effort to develop a more holistic approach to global environmental governance and to provide it with a more appropriate institutional support base. The agreements reached at UNCHE included an Action Plan for the Human Environment, which put forward 109 recommendations to promote global scientific cooperation and environmental impact assessment. Most significantly, Recommendation No. 4 included a commitment to establish within the United Nations a single body to coordinate global environmental protection efforts. The UN General Assembly followed up this recommendation in 1973 with the creation of UNEP. After several failed efforts to establish such an international body in the interwar years (see Chapter 4), GIS had finally accepted the need for a dedicated UN body with an explicit environmental mandate, even though UNEP was given only limited powers and financial resources (Ivanova, 2010). Still, UNEP came to play an important role in stimulating scientific knowledge exchange, setting the international environmental agenda and facilitating the negotiation of new environmental treaties. As the international environmental agenda expanded in subsequent years, several other international environmental bodies were added to complement the work of UNEP (e.g. UN Commission on Sustainable Development) and to manage the increasing flows of international environmental aid (e.g. Global Environment Facility; Green Climate Fund).

The post-Stockholm years also saw an ambitious expansion of the focus of international treaty-making. Some of the first international treaties to be signed after UNCHE still reflected environmental concerns of the immediate post-war era. This was the case with the 1973 Convention on International Trade in Endangered Species (CITES), which followed in the tradition of early twentieth-century conservationism. The 1972 London Convention on Dumping at Sea and the 1978 MARPOL Convention on Pollution from Ships were also rooted in long-standing efforts to limit environmental harm to the marine environment from commercial shipping, going back to the 1920s and 1930s (see Chapter 4). However, the adoption of the Convention on Long-range Transboundary Air Pollution (CLRTAP) in 1979 signalled industrialised countries' willingness to tackle industrial pollution at the regional and even global level. By the early 1980s, the focus had shifted to global atmospheric pollution, with the 1985 Vienna Convention for the Protection of the Ozone Layer and its 1987 Montreal Protocol dramatically expanding the scope of global environmental governance (Benedick, 1991; Parson, 2003). Emboldened by the success of the ozone regime, international society then set about to create two further mega-regimes that took the ambition and scope of international environmental policy-making to a new level: the UN Framework Convention on Climate Change (UNFCCC) and the Convention on Biological Diversity (CBD). Agreed at the 1992 UN Conference on Environment and Development (UNCED) in Rio de Janeiro, both the UNFCCC and CBD represented GIS's new-found confidence that it was now ready to manage some of the most complex ecological threats on a planetary scale. By the end of the twentieth century, GIS had thus come to establish a comprehensive institutional framework for dealing with an ever-wider range of global issues, covering animal and plant species, genetic resources, natural habitats and ecosystems, pollution of air, water and oceans, as well as the Earth's atmosphere.

From International Coordination Towards Global Environmental Governance

Few areas of global policy-making have witnessed such a rapid and comprehensive creation of global governance mechanisms as happened in the environmental field. The sheer scale of institution-building underscores the strengthening commitment to environmental stewardship as a primary institution in the post-Stockholm era. To be sure, this process of institutionalisation played itself out over several decades and experienced many setbacks along the way. Questions also need to be asked about whether the environmental secondary institutions were equipped with the required authority and funding to fulfil their objectives (on the limitations of global environmental governance, see chapter 8). In fact, after decades of multilateral efforts to create MEAs, some observers have pointed to a widening gap between the proliferation of international legal instruments, on the one hand, and the lack of coordination and capacity to deliver on international environmental obligations, on the other. Complaints about treaty congestion, summit fatigue, implementation gaps and institutional fragmentation have become commonplace among diplomats and analysts (Anton, 2012; Chambers, 2008; Wapner, 2003), but few would question the degree to which international environmental policy-making has become a routine practice in GIS. States' repeated expressions of support for environmental stewardship are more than diplomatic rhetoric or cheap talk. The normative shift that first manifested itself at the 1972 Stockholm conference – and again at subsequent UN summits – was not simply an outgrowth of 1970s zeitgeist but a reflection of a normative commitment that has since deepened and widened.

It would be misleading to assume that the growth of international environmental governance is the result of functional necessity rather than political design and struggle. With hindsight, we can see that the rise of GEP coincided with what environmental scientists and historians refer to as the 'Great Acceleration' of humanity's impact on the environment since the mid-twentieth century. During this period, human populations expanded their energy consumption and resource throughput, and also increased pollution levels and rates of environmental degradation, to such an extent that humankind has now 'emerged as the most powerful influence on global ecology' (McNeill and Engelke, 2016: 1). Unsurprisingly, therefore, societies and their governments have reacted to worsening environmental conditions by increasing environmental protection efforts. Given global ecological interdependencies, it was inevitable that demand for international environmental cooperation would also increase in the late twentieth century. But such a functionalist reading of GEP ignores the many obstacles to creating global environmental governance that had to be overcome. Although closely reflecting scientific advances and environmental needs, the creation and expansion of the international environmental agenda followed a distinctive logic of power. The creation of major new international environmental treaties depended on a favourable combination of interests, norms and power in international relations. Environmental norm entrepreneurs and powerful states had to champion new causes of global environmental action, international coalitions had to be formed to support new regulatory proposals and political compromises had to be struck to overcome resistance and draw environmental laggards into multilateral agreements.

Standard IR explanations of international cooperation based on hegemonic power have proved to be a poor guide to explaining the creation of secondary institutions in the environmental field (Falkner, 2005). Yet, the important role of leadership by major powers, whether or not they can exercise hegemonic power, is widely acknowledged in the literature (Skodvin and Andresen, 2006). Important environmental regimes were often created at the instigation of leading and powerful states that sought to internationalise their own national environmental policies, not least to ensure a level playing field for their domestic manufacturers that were competing in global markets (DeSombre, 2000; Kelemen, 2010). In the immediate aftermath of the Stockholm conference, for example, the United States was behind much of the expansion of the international environmental agenda (Ivanova and Esty, 2008; Kelemen and Vogel, 2010). Well into the 1980s, the United States routinely initiated or backed new global environmental initiatives, especially when they reflected US legislative mandates to seek international support for domestic policy objectives. Having introduced some of the most advanced endangered species protection legislation (Endangered Species Act of 1966 and later revisions; Marine Mammals Protection Act of 1972), which directed the US government to negotiate an international treaty to protect endangered species worldwide, the United States played a major role in the CITES negotiations that concluded at a conference held in Washington, DC, in 1973 (DeSombre, 2000: 52–8). The United States also developed pioneering national laws on air pollution that served as the basis for combatting transboundary air pollution. The first federal-level Clean Air Act was adopted in 1963, and the 1970s Clean Air Act Amendments became the basis for subsequent efforts in North America and internationally to combat sulphur dioxide and nitrogen oxide emissions (DeSombre, 2000: 81–9). The United States was also one of the first countries to regulate chemicals that were suspected of destroying the stratospheric ozone layer, and it was US pressure on other industrialised countries that forced the pace of the negotiations on the Vienna Convention and Montreal Protocol in the 1980s (Benedick, 1991).

Growing domestic polarisation and the loss of bipartisan support for environmental policy pulled the rug from underneath US environmental leadership during the 1990s. By that time, European countries had empowered the European Union (EU, known as European Economic Community until 1993) with sufficient decision-making authority for it to play a more active role in international environmental politics. Against the background of a strong uptick in organised environmentalism, including the arrival of the first green parties in European parliaments, the EU eventually emerged as a new international leader to drive forward environmental regime-building (Kelemen and Vogel, 2010). Having pioneered a new generation of domestic environmental regulations, the EU began to export domestic norms and policies to the international level, most notably on issues relating to sustainable development (Lightfoot and Burchell, 2005), climate change (Vogler and Bretherton, 2006), biosafety (Falkner, 2007) and chemicals regulation (Selin and VanDeveer, 2006). Just as was the case with US environmental leadership, the EU sought to globalise domestic environmental rules in order to create a regulatory level playing field for its industries. It was able to use its economic might as the world's largest import market to impose some of its environmental standards on other countries (Damro, 2015; Bradford, 2020). In this sense, environmental leadership and the asymmetrical distribution of power have been closely linked in GEP.

As the ambition of the international environmental agenda expanded, a growing number of countries were drawn into the process of international institution-building. By the 1990s, near-universal participation in multilateral negotiations had in fact become the norm in GEP. This is in stark contrast to the early days of environmental diplomacy, before and even after the Stockholm conference. Well into the 1970s, it was common for environmental treaties to be negotiated by just a small number of countries, often on a plurilateral basis or in regional settings. And even during the 1980s, when developing countries were beginning to play a more active role, many international environmental processes were still dominated by industrialised countries, especially when it came to initiating and creating new international rules. The ozone regime, for example, started out as a small club and only gradually grew into a fully multilateral framework (see subsequent discussion). By contrast, when the negotiations on a climate regime started in the late 1980s, many more developing countries were involved in the international process from the outset. One hundred and fifty-four countries signed the UNFCCC in 1992 (Bodansky, 1993: 454), and the total number of ratifications stands at 197 countries as of 2020. Other treaties that were launched in the 1990s (e.g. CBD; UN Convention to Combat Desertification, UNCCD) also achieved near-universal levels of ratification.

The growing support for environmental secondary institutions in the Global South is one of the remarkable achievements of environmental multilateralism. It stands in contrast to the first two decades after Stockholm, when developing countries are best described as rule takers rather than rule makers in GEP. Initially sceptical about what they perceived to be a Northern bias in how the environmental agenda was framed, developing countries were slow to rally behind their own agenda for global environmental governance. As discussed subsequently, this only began to change in the run-up to UNCED in 1992, which provided developing countries with an opportunity to firmly embed Southern priorities in the UN environmental agenda (Najam, 2005). Since then, developing countries have not only engaged more fully in multilateral processes but also initiated new international regimes. In the case of toxic waste, it was political pressure from the developing world that prompted UNEP's Governing Council to approve the Cairo Guidelines in 1987, a set of voluntary standards for regulating waste trade. Developing countries then successfully pushed for the creation of the 1989 Basel Convention on the Control of Transboundary Movements of Hazardous Wastes and their Disposal, a legally binding regime to regulate the international trade in toxic waste (Selin, 2013: 111; Clapp, 2001). The demand for global rules on trade in genetically modified organisms (GMOs) similarly originated in the developing world. Concerned about the inherent risks of novel genetic engineering techniques, developing countries first proposed the inclusion of biosafety rules in the CBD before demanding a separate biosafety protocol to the biodiversity treaty (La Vina, 2002). Both proposals were initially resisted by industrialised countries, but developing countries kept up the pressure and eventually succeeded in launching a negotiation process that led to the adoption of the Cartagena Protocol on Biosafety in 2000 (Falkner, 2000).

State power and leadership are essential ingredients in the creation of environmental secondary institutions, but few areas of global governance are as open to non-state actors as environmental protection. Indeed, scientists, activists and companies have played an unusually influential role in shaping the outcomes of international environmental negotiations. Expert knowledge provided by transnationally connected scientific communities has been critical to both the identification and framing of the environmental issues that states have taken up at the international level (Haas, 1995; Kohler, 2019). Their role in international negotiations has increasingly been institutionalised through the creation of scientific advisory bodies that provide authoritative assessments of available scientific and technical knowledge, such as the Intergovernmental Panel on Climate Change (IPCC) and the Subsidiary Body on Scientific, Technical and Technological Advice (SBSTTA) of the CBD. Environmental activists and NGOs similarly identify and frame the issues that states address through international governance processes, but their influence reflects a more normative and political role in international politics, namely advocacy. They have sought to influence outcomes in interstate bargaining by mobilising domestic public opinion, exerting moral pressure on states and lobbying diplomats during international negotiations (Betsill and Correll, 2008; Park, 2013). Business actors have also engaged more directly with environmental regime-building processes as the scope of international environmental regulation has expanded. Their role has evolved from initial opposition to the rise of global environmentalism to more nuanced international engagement, either to ensure business-friendly outcomes overall or shape international regulation to suit particular company or sectoral interests (Falkner, 2008; Clapp and Meckling, 2013).

By the end of the twentieth century, non-state actors had not only managed to insert themselves into international environmental negotiations, they were also beginning to create new forms of global environmental governance that operate outside intergovernmental realms. The rise of non-state or transnational environmental governance since the 1990s has led to a dramatic expansion of global environmental rule-making, with a myriad of transnational actors and initiatives establishing diverse governance mechanisms, from voluntary codes of conducts and emissions trading schemes to third-party sustainability certification schemes (Bulkeley et al., 2014; Roger and Dauvergne, 2016; Hale, 2020). As will be discussed in chapter 10, these new forms of environmental governance may appear to operate outside international society but in fact are often closely interwoven with state-centric processes. State authority looms in the background of private governance, either because states encourage voluntary initiatives by companies and NGOs or reinforce them through legal incorporation (Falkner, 2003). Strong national environmental regulation is often associated with greater participation by non-state actors in transnational initiatives (Andonova, Hale and Roger, 2017), and states and international organisations also increasingly orchestrate transnational governance initiatives in an effort to expand global governance capacity (Abbott, 2012; Abbott et al., 2015a). Environmental stewardship has thus strengthened among both state and non-state actors. It also provides a normative driver for integrating international and world society (see discussion in Chapter 10).

Greening State Behaviour and Identity

The creation of a dense web of secondary institutions is only one indicator of how the new norm of environmental stewardship established itself in international relations. As international society takes on a new normative commitment, we should also expect to see a change in the behaviour, and ultimately identity, of states. As can be observed in other global policy fields, the social consolidation of international norms goes hand in hand with the emergence of regularised patterns of behaviour that respond to the altered normative context (Crossley, 2016). This change is unlikely to follow a steady and linear path; it is more likely to occur slowly and unevenly, and it may take some time before it applies to all members of international society. Indeed, the pace and geographical scale of this change provides insights into the strength of social consolidation that a new international norm has experienced.

There are two principal ways in which such a change at the level of the state manifests itself. The first concerns observable behavioural change that involves states putting in place the domestic conditions for discharging their international environmental duties. This can take the form of creating domestic environmental agencies, adopting domestic environmental laws and regulations and integrating environmental objectives into other domestic policy areas. In the field of foreign policy, we would expect states to integrate environmental concerns into their diplomatic machinery and practices. Such forms of domestic policy change have attracted a great deal of scholarly attention, which have produced a wealth of evidence that indicates the diffusion of environmental policy practices from state to state, widening participation in international treaty-making and the growing number of ratifications of MEAs (Bernauer et al. 2010; Busch, Jörgens and Tews, 2005; Holzinger, Knill and Sommerer, 2008; Roberts, Parks and Vásquez, 2004).

The second manifestation of change at the state level concerns the identity of the state and involves a sociological understanding of how a change in understandings of appropriateness and expectations about 'normal' behaviour has a constitutive effect on the members of international society. In this view, the rise and consolidation of environmental stewardship can be expected to affect states' shared understanding of international legitimacy – what it means to be a rightful member of international society and to behave in accordance with mutually accepted rules (Clark, 2005: 5). Empirically tracing such shifts in legitimacy has proved a more complex challenge and may therefore have received less attention in the GEP literature, but recent scholarship on the rise of the 'green state' has begun to provide some evidence to this effect (Dryzek et al., 2003; Eckersley, 2004a; Meadowcroft, 2005).

The Diffusion of National Environmental Policy

The most pervasive change in environmental state behaviour can be found at the level of national legislation and institution-building. Although several industrialised countries had begun to establish nature conservation functions as early as the late nineteenth century, usually with small offices for forestry management or the protection of natural monuments attached to existing ministries, it was only from the late 1960s onwards that they broadened these policies into what is now recognised worldwide as a separate environmental policy field. In the years leading up to the Stockholm conference, the United States, together with Sweden and Japan, were the first major industrialised economies to institutionalise a broad range of environmental policy functions at the national level (Jörgens, 1996: 74). Some of these first initiatives occurred independently and in response to domestic political imperatives, but America's pioneering role in creating modern environmental laws and institutions served as an important model for other industrialised countries that also sought to establish environmental protection as a comprehensive national policy objective.

The Federal Republic of Germany (known as West Germany until unification in 1990) was one of the first major economies to learn directly from the American experience in setting up its own environmental institutions. Despite a long history of protection of natural monuments and natural resources, often at the sub-national level, it was only in 1969 that the German government established a unit within the Interior Ministry that was tasked with developing a new programme for *Umweltschutz*, a neologism in the German language based on the American concept of environmental protection. The unit's aim was to integrate previously dispersed environmental responsibilities under one roof, and its first success came in 1971 with the introduction of the first federal environmental programme for Germany. A wave of new environmental laws and regulations followed in subsequent years, leading to the creation of a federal environment agency (*Umweltbundesamt*) in 1974. In establishing this new institution, the German government drew heavily on the model of the US EPA (created in December 1970), though without giving it comparable powers of enforcement (Engels, 2006: 275–6).

Other industrialised countries followed a similar path of creating new agencies and introducing legislation that established environmental protection as a separate national policy domain. In 1971 and 1972 alone, fourteen industrialised countries started to institutionalise environmental policy in this way. Britain was the first country to create a separate environmental ministry, a model that was soon repeated in France, Canada, Australia, Denmark, Norway and Austria as well as in Poland and the East German state (Jörgens, 1996: 80). The 1973 oil crisis and subsequent downturn in the global economy temporarily dampened the emerging enthusiasm for environmental policy innovation (Weale, 1992: 30), but as soon as the economic turmoil of the 1970s subsided the environmental institutionalisation trend picked up again.

By the 1980s, a series of major environmental accidents (e.g. Bhopal gas leak in 1984; Sandoz chemical spill in 1986; Chernobyl nuclear accident in 1986) and novel environmental threats (e.g. ozone layer depletion) renewed momentum behind the global environmental agenda (Jörgens, 1996: 81–3).

The environmental policy diffusion process took much longer to take root in the Global South. As discussed in Chapter 5, leading developing countries had expressed severe reservations at the 1972 Stockholm conference about what they saw as a Northern environmental agenda. With governments firmly set on prioritising economic development and only limited domestic mobilisation on environmental issues, most developing countries did not establish domestic environmental institutions until the 1980s, and in some cases it wasn't until after the 1992 Rio Earth Summit that environmental policy was formally recognised as an integral part of national policy-making.

Participation in the Stockholm conference had at least encouraged some developing countries to take the first tentative steps towards establishing a domestic environmental policy focus. Two months before UNCHE opened in Stockholm, India set up a National Committee on Environmental Planning and Coordination within the Department of Science and Technology. The committee had no regulatory authority, however, and served mainly as an advisory body on how to develop environmental policies and integrate them into the country's developmental strategy (Dwivedi, 1997: 55). A year after Stockholm, Brazil established a Special Secretariat of the Environment within the Ministry of Interior (Hochstetler and Keck, 2007: 27). And in China, the Stockholm conference is credited with initiating a government review of how to establish an environmental policy competence. A year after the UN gathering, the country convened its first national conference on environmental protection, which produced a set of policy recommendations that were later adopted by China's State Council and led to the creation of the first environmental protection institutions (Heggelund and Backer, 2007: 421; He et al., 2012: 29).

One reason why the environmental policy diffusion effect of UNCHE was more limited in the developing world was the relatively low profile of environmental issues in domestic politics at the time. In India, environmental issues did not feature prominently in national electoral politics until the late 1970s, and a separate Department of Environment was only established in 1980. The country did pass several environmental laws after Stockholm, but weak implementation prevented them from having any significant impact (Dwivedi, 1997: 56). Brazil's response to Stockholm was similarly half-hearted. As Hochstetler and Keck (2007: 27) note, the

Brazilian government failed to provide adequate resources for its national environmental programme. The newly created environmental secretariat within the Ministry of Interior began its work with just three members of staff and virtually no budget. And in China, the first domestic environmental policy initiatives of the 1970s struggled to get off the ground at a time of profound political turmoil and economic upheaval. While the Cultural Revolution of 1966–76 led to the repudiation of environmental regulations as bourgeois revisionism (Economy, 2004: 54), Deng Xiaoping's subsequent economic reforms promoted rampant materialism and prioritised economic growth at all cost (Shapiro, 2012: 90–1).

There is broad agreement in the GEP literature that in countries where governments faced little or no domestic environmental pressure, international factors such as participation in the Stockholm conference at least created an opening for the gradual adoption of national environmental policies (Jörgens, 1996; Frank, Longhofer and Schofer, 2007). However, the spread of domestic environmental institutions is best understood as a policy diffusion process based on the twin drivers of international harmonisation and transnational communication (Busch and Jörgens, 2005; Holzinger, Knill and Sommerer, 2008). Participation in the Stockholm conference and subsequent environmental negotiations created a functional need for states to establish domestic institutions and define environmental priorities. In this sense, growing international enmeshment in environmental regimes served as a powerful external stimulus for domestic policy developments (Haas, 2002). At the same time, domestic policy-makers and regulators also began to learn from each other, exchanging ideas and best practices among them, thereby establishing horizontal links that eventually grew into transgovernmental networks (Slaughter, 2004).

Regional institutions provided a particularly powerful boost to environmental policy diffusion, especially in Europe. During the 1970s and early 1980s, environmental policy authority in Europe was still concentrated in the hands of national governments, and policy diffusion during this time was based primarily on horizontal policy learning. Once the EEC (renamed European Community and integrated into the EU in 1993) had gained environmental policy competence in the Single European Act in 1986, the gradual Europeanisation of environmental policy-making became the key driver of policy diffusion in Europe and beyond (McCormick, 2001; Andonova, 2004). EU institutions facilitated the greening of state practices mainly through the 'uploading' and 'downloading' of policy models (Padgett, 2003: 227–8). In some cases, the EU sought to harmonise environmental policy standards upwards, as was the case with the Dutch National Environmental Policy Plan of

1989 that served as a model for both EU-level and member state policy developments. In other cases, the EU itself became the main instigator of environmental policy developments. Having adopted sustainable development as a constitutional principle in 1997, the EU sought to promote environmental policies both internally and externally. Its Fifth Environmental Action Plan of 1992, for example, became an important reference point in the diffusion of environmental policy practices to Central and Eastern European countries as part of the EU's enlargement process after the end of the Cold War (Busch and Jörgens, 2005: 870).

The Greening of Diplomatic Practice

Apart from creating domestic environmental policy capacity, states also had to reform their foreign policy institutions and instruments to be able to engage more comprehensively with international environmental negotiations. The rise of environmental diplomacy in the late twentieth century provides another important indicator of environmental stewardship's deepening impact on interstate relations and state behaviour. For many states, especially in the developing world, it was the Stockholm conference of 1972 that first created a need to build institutional capacity for conducting environmental diplomacy in a more systematic fashion. The bureaucratic revolution in environmental diplomacy did not happen overnight, of course. As discussed in chapter 4, a small number of European and North American states had been negotiating environmental treaties since the beginning of the twentieth century and had thus gained some experience in this area. However, such early forms of 'conservation diplomacy' (Dorsey, 1998) were mostly ad hoc efforts to manage transboundary environmental problems, and states were able to rely on outside experts to work out international regulatory solutions. When negotiating the inland fisheries treaty of 1908, for example, United States and Canadian diplomats merely created a framework agreement on fisheries conservation – the American draft text had only four articles – and left it to the scientists that made up the two-member International Fisheries Commission to develop the treaty's protective regulations (Dorsey, 1998: 55–73). This division of labour between diplomats creating general political framework agreements and scientific or regulatory experts developing the details of regulatory regimes carried on throughout the twentieth century. As the demand for international environmental regulation increased, however, foreign ministries and their diplomatic missions had to build the capacity to conduct multilateral environmental negotiations on an ever-larger scale, and as a matter of routine. The rise of environmental stewardship thus prompted a lasting adjustment in diplomatic practice worldwide.

The proliferation of international environmental conferences and negotiations required some changes to the organisation and staffing of foreign ministries. Most diplomatic services have traditionally been organised around geographical knowledge and responsibilities. With the rise of economic diplomacy (Bayne and Woolcock, 2017) as well as other cross-cutting foreign policy issues (global health, energy, science), new issue-specific expertise and units had to be created alongside or within existing bureaucratic units. To some extent, foreign ministries have been able to draw on relevant expertise in specialised domestic departments (e.g. secondments to foreign ministries) and include their representatives as part of international delegations. Still, with environmental issues gaining in political salience on the international stage, foreign ministries had to develop their own ability to interpret complex scientific and environmental issues in the context of bilateral and multilateral relations (Murphy, 2018: 36).

The US Department of State was one of the first foreign ministries to reform its institutional setup after Stockholm. In response to legislation passed by the US Congress in October 1973, the State Department created a separate Bureau of Oceans and International Environmental and Scientific Affairs, which has since taken a leading role on international environmental policy (United States Department of State, 1997: 32). Other countries, too, began to expand their foreign ministries' environmental expertise during the 1970s and 1980s but progress towards fully fledged environmental diplomacy was slow and uneven. Swedish diplomat Lars Björkbom (1988: 134) summed up the experience of many countries in the late 1980s when he observed that environmental diplomacy 'has yet to be established and accepted as a household word in the ministries for foreign affairs'. Despite the advances that had been made in reforming existing diplomatic bureaucracies, the '[o]ffices for international environmental affairs are not only small and understaffed but often organizationally hidden as subdivisions of departments, which have their main attention elsewhere' (ibid., 134). This was the case in not only developing countries but also leading powers such as Britain. In the 1980s, the Foreign and Commonwealth Office (FCO) still included environmental issues in its science and energy portfolio. Only in 1990, as preparations for the 1992 Rio Earth Summit got underway, did the FCO upgrade environmental policy as part of the newly created Environment, Science, and Energy Department, and it took another decade for this to be restructured into a separate Environment Policy Department (Brenton, 2010).

For many countries, especially in the developing world, the Rio Summit in 1992 provided an important stimulus for strengthening environmental policy expertise within their diplomatic service. As Bo Kjellén (2008: 132), Sweden's lead diplomat at UNCED, noted: 'The preparations for the Rio conference engaged a new generation of negotiators, who learned the necessary skills of the new diplomacy on the job. They were traditional diplomats and their knowledge of the many complex issues of sustainable development was limited; they had to learn fast, and they did.' Rio helped create environmental diplomacy expertise, but the subsequent increase in the number of high-level UN sustainability summits and environmental negotiations also put an increasing burden on smaller countries, stretching their already thin diplomatic capacity. It is no exaggeration to suggest that today 'barely a fortnight goes by without an environmental meeting somewhere on the planet' (Depledge and Chasek, 2012: 19). The UN climate regime alone has created an endless pattern of negotiation rounds and additional meetings in between formal sessions. In 2009, for example, the UNFCCC Secretariat scheduled meetings for nearly 140 days (Depledge and Chasek, 2012: 21). Small countries and developing countries, in particular, have faced severe resource constraints when dealing with the competing demands of participation in MEA negotiations, conferences of the parties, UNEP meetings and UN summits. Most smaller nations would normally send not more than five officials, and often only one or two, to international environmental meetings (Chasek, 2001: 169). With MEA negotiations requiring ever-more scientific and technical expertise and often being conducted in several parallel working groups or contact groups, small delegations are at a severe disadvantage when trying to represent their country and influence outcomes. Coalition-building with other countries, reliance on advice from non-state observers and capacity-building for diplomats are thus critically important for overcoming such resource constraints (Chasek, 2001: 170–1).

While the 'environmental mega-conferences' of the 1990s and 2000s provided a major impetus for states to upgrade their environmental diplomacy capacity, it was the growing political attention to the threat from climate change that arguably had the most lasting impact on diplomatic practices. Between the 1992 Rio conference and the 2009 Copenhagen climate conference, climate change moved from the margins to the centre of the international diplomatic agenda. Encompassing an ever-wider array of related policy issues, from energy security to public health and biodiversity, climate change has come to be defined as one of the top global human security challenges of our time. This in turn has helped to transform it into a more routine element of international diplomacy. By the early 2000s, climate change had moved up the list of foreign policy priorities among environmental leaders, such as the United Kingdom (Lane, 2007). Both the UK and German governments commissioned major scientific reviews of climate change science as a security and economic threat (Stern, 2007; German Advisory Council on Global Change, 2008), and the UN Security Council debated climate change for the first time in 2007. Despite the setback of the 2009 Copenhagen conference to agree a post-Kyoto climate treaty, climate change had become internationally recognised as a high priority concern in foreign policy. By the early 2010s, climate change was established on the agenda of major international forums (e.g. G7/8, G20), and at least 70 per cent of states have identified climate change as a national security threat (Scott, 2015: 1330).

The Normative Power of Environmental Multilateralism

The ever-expanding agenda of international environmental rule-making not only led to a significant uptick in environmental diplomacy but also established a regular pattern of multinational negotiation on environmental issues. As Falkner and Buzan (2019: 146; see also Dimitrov, 2005: 3) argue, this has resulted in the emergence of a norm of good citizenship in GIS, whereby all states are expected to participate in multilateral environmental policy-making. At least by the 1990s, participation in international environmental negotiations was no longer considered a choice, it had become standard behaviour in international relations. To be considered a legitimate member of GIS today, all states, including some of the most recalcitrant powers in GEP, must participate in multilateral environmental processes.

To be sure, it took some time for this norm to establish itself. In the immediate aftermath of the Stockholm conference and well into the 1980s, participation in international environmental negotiations was still far from universal levels. At that time, it was not uncommon for a small set of countries to initiate a new environmental treaty and then invite others to join the agreement. For example, the Ramsar Convention (Convention on Wetlands of International Importance especially as Waterfowl Habitat) was agreed in 1971 in the Iranian city of Ramsar by just eighteen countries, with a further five countries in attendance as observers (International Conference on the Conservation of Wetlands and Waterfowl, 1972: 963–4). Over time, the treaty managed to acquire the ratifications of the vast majority of UN member states (171 as of 2019). Some global pollution issues were initially addressed by a select group of industrialised countries. Only twenty-four countries attended the first UNEP workshop on ozone layer depletion in 1982, for example. The conference to agree the 1985 Vienna Convention on ozone layer depletion was attended by forty-three countries, and the number eventually rose to 60 (by this time half from the developing world) during the 1987 Montreal Protocol negotiations (Benedick, 1991: 42, 44, 74). In the end, the ozone treaty reached a universal level of ratifications (198 as of 2019). Other issues, however, were of broader global concern and attracted a large number of negotiating countries from an early stage. CITES was agreed at a meeting of representatives of eighty countries in Washington, DC, in 1973 and reached 153 ratifications by 2020 (CITES, no year).

By the time of the 1992 UNCED, which played a key role in broadening the support for environmental stewardship across the developing world (discussed subsequently), the majority of developing countries had become fully engaged in the international environmental agenda. The high-profile climate and biodiversity negotiations were thus launched with the active engagement of developing countries right from the start. A total of 154 countries signed the UNFCCC (Bodansky, 1993: 454) and 157 signed the CBD at the Rio Earth Summit (McGraw, 2002: 7). Both conventions have since reached universal rates of ratification (the United States failed to accede to the CBD). Thereafter, international environmental negotiations have routinely achieved a balanced level of participation by industrialised and developing countries.

Undoubtedly, the strength of the environmental multilateralism norm varies from country to country, and not all countries that routinely negotiate MEAs necessarily support the objectives behind them. In any case, it is within a sovereign nation's right to participate in international negotiations and refuse to ratify the internationally agreed outcome. For this reason, environmental multilateralism is 'more a procedural than a substantive norm' (Falkner and Buzan, 2019: 146). The fact that all countries now routinely participate in international environmental negotiations tells us relatively little about their commitment to creating strong secondary institutions. It is rooted in the primary institution of environmental stewardship and reflects the widely shared expectation that states should collaborate to safeguard the planet's health. However, this leaves open the option for states to engage in international regime-building and to seek to slow down or weaken international environmental policy-making. The UNFCCC, for example, enjoys universal support, but some parties have shown little interest in legally binding emission reduction targets (e.g. China, Russia, the United States) while others have developed a pattern of openly obstructionist behaviour (e.g. Saudi Arabia).

The strength of the environmental multilateralism norm can be seen from the fact that on many occasions reluctant powers that resist strong environmental action nevertheless have felt compelled to participate in multilateral negotiations. The case of the United States is instructive in this regard. Having initially championed global environmental causes, the United States abandoned environmental leadership on many issues from the early 1990s onwards, from climate change to biodiversity and biosafety. Indeed, the United States has failed to ratify any significant new environmental treaty since the 1992 Rio Summit, owing largely to the US Senate's resistance to a strengthening of international legal obligations (Brunnée, 2004). Yet, even when it opposes proposals for international environmental regulation, the United States has continued to engage in multilateral environmental negotiations. Despite a sharp turn towards unilateralism under President George W. Bush (Eckersley, 2007), the United States never abandoned the UNFCCC process. Under the Obama administration, the United States returned to engage more fully with the multilateral climate negotiations and even became a key architect of the new logic of the Paris Agreement (Falkner, 2016b), only to declare its departure from the Paris Agreement (though not the UNFCCC) under President Donald J. Trump. To some extent, Trump's renewed unilateral shift represents the strongest test case yet for the survival of environmental multilateralism, not least as it threatens to encourage other countries also to walk away from international environmental efforts that they object to (Pickering et al., 2018).

International pressure to conform with the environmental multilateralism norm has also been directed at emerging powers. China, for example, has intensified its engagement with global environmental policy-making in the twenty-first century, reversing the country's longstanding resistance against international regulations that might impede its economic growth agenda. This shift, which has been notable in climate politics (Hilton and Kerr, 2017), reflects international expectations but is also the result of the growing salience of environmental issues in domestic politics. China's environmental record has sharply deteriorated as it experienced over three decades of record-breaking economic growth, and the country now counts as the world's largest polluter on many fronts. Ever since it overtook the United States as the world's biggest emitter of carbon dioxide emissions in 2006, world leaders have urged the country to make a stronger contribution to the global climate mitigation effort. President Barack Obama made an explicit link between China's great power status and climate responsibility when he asserted at the 2014 UN Climate Summit that the United States and China had a 'special responsibility to lead' the global fight against global warming (White House, 2014). For China as well as for other emerging powers, the environmental stewardship norm has thus created expectations of 'responsible behaviour' that they find increasingly hard to ignore. At the same time, this has also created opportunities for aligning domestic with global priorities that allow them to burnish their international image (Zhang, 2016: 812, 814–15).

In sum, the rise of environmental stewardship as a fundamental norm has changed behavioural patterns in international society. States created the required domestic institutional capacity to pursue environmental protection as an integral part of national policy-making. They also established environmental expertise within their foreign policy machinery and adapted diplomatic practices to support the rapidly expanding network of environmental secondary institutions. These changes in state behaviour and diplomatic practices also implied a gradual greening in state identity, though the traces of this deeper transformation are more difficult to discern. Domestically, virtually all states have adopted environmental protection as part of their core responsibility but the shift towards a fully fledged 'green state' or 'eco-state' is incomplete. Internationally, the shift in state identity is most clearly evident in the normative power of environmental multilateralism, which has drawn ever more states into the practice of multilateral policy-making on transboundary environmental issues. Whether they support or reject international regulatory proposals, most states have accepted the need to participate in multilateral environmental processes. However, environmental multilateralism is mainly a procedural, not substantive norm. States can ill afford to disengage from the multilateral environmental agenda, though their rightful membership in GIS is not under threat if they fail to act on it.

Globalising Environmental Stewardship

As we have seen, environmental stewardship emerged as a primary institution in the early 1970s, but its subsequent social consolidation occurred unevenly across the world, following a distinctive spatial pattern of globalisation, contestation and accommodation. Leading industrialised countries were the first to establish domestic institutions and laws in support of more systematic environmental protection. Their experience with institutionalising environmental policy provided important lessons for other countries seeking to go down this path. While environmentalism had many different local roots, horizontal policy diffusion played an important role in the global environmental policy revolution of the late twentieth century, and it also supported the social consolidation of environmental stewardship as a primary institution.

However, the global spread of the new environmental norm was also contested, especially but not exclusively in the developing world, and for environmental stewardship to become accepted in *global* international society, it had to accommodate the expectations and concerns of the Global South. This section traces the globalisation of environmental stewardship up to and beyond the 1992 Rio Earth Summit, with a particular focus on how the North–South divide was addressed in the evolution of GEP.

It is worth noting that not all newly emergent primary institutions become fully universalised. As we know from the standard ES expansion story, the classical primary institutions (sovereignty, territoriality, diplomacy and international law) went global as European international society spread to the rest of the world on the back of colonial expansion and the newly independent countries of the Global South achieved equal membership in GIS during decolonisation (Bull, 1977: 33–40; Bull and Watson, 1984). The globalisation success of more recent primary institutions is more mixed, however. Nationalism, which first appeared in the nineteenth century, quickly became a globally applicable primary institution (Mayall, 1990), as did racial equality and development, and to a lesser extent the market. Other candidates for primary institution status have not achieved universal recognition. Western powers have promoted human rights and democracy but these have come up against resistance elsewhere, and their status as primary institutions of GIS 'remains an open question' (Buzan and Schouenborg, 2018: 77). When environmental stewardship emerged in the 1970s, it was thus entirely conceivable for it to go the way of human rights and democracy and remain part of a sub-global international society. After all, post-Second World War environmentalism was a predominantly Western creation (Dauvergne, 2016: 75), and it was the United States and a handful of European powers that played a dominant role in shaping the new international environmental agenda. What, then, facilitated the growth of environmentalism into a universally accepted primary institution?

Decolonizing Global Environmentalism

As discussed in the preceding chapter, the biggest barrier to the successful globalisation of environmental stewardship was to be found in the Global South. In many ways, the 1972 Stockholm conference came at an inauspicious time for a grand environmental bargain between the North and South. The early 1970s marked the beginning of what Bull (1984) and other ES authors described as the 'revolt against the West' – a major campaign for global justice and institutional reform in global international society. Indeed, UNCHE was one of the first major global summits at which the developing world acted 'as a unified collective' and adopted many of the arguments and strategies that would resurface in its later calls for a New International Economic Order (Najam, 2005: 307). Unsurprisingly, therefore, calls for a globally shared responsibility for the environment fell on deaf ears in the developing world. Leaders from the Global South questioned the very legitimacy of a Northern environmental agenda that was infused with neo-Malthusian fears about overpopulation and economic growth. Fearing that it would turn into an 'effort to sabotage the South's developmental aspirations' (Najam, 2005: 308), they portrayed global pollution as a problem caused by industrialised countries. Poverty and underdevelopment, not industrialism, posed the biggest environmental threat to developing countries. At Stockholm, G77 leaders demanded recognition of their countries' sovereign right to use their natural resources for economic development. They also insisted that additional international assistance would be needed to pay for environmental protection efforts in the developing world (Bernstein, 2001: 35).

Woven into the South's sceptical stance was a long-standing perception of environmentalism as being deeply implicated in the long history of colonial rule and post-colonial exploitation by Western powers. In a line of argumentation that stretches from the 1970s 'revolt against the West' to the 1990s anti-globalisation movement, Southern leaders rejected the creation of an *international* environmental agenda not because of inherent anti-environmentalism but out of fear that it would erode developing countries' autonomy in shaping their own socio-economic future. The 'global' in global environmental politics, as Indian environmental activist Vandana Shiva (1993: 56) famously argued, 'creates the moral base for green imperialism'. Given the strength of Southern contestation of global environmentalism, at Stockholm and in later years, the environmental stewardship norm thus had to be re-framed in terms that included developing country concerns if it was to become universally applicable. Environmentalism had to step out of the colonial shadows of the past and acknowledge the importance of sovereignty and development not only as primary institutions of GIS but also as cornerstones of global environmental management. In other words, environmentalism first had to be decolonised in order to be globalised.

The key political and intellectual shift in the globalisation of environmental stewardship came with the merging of environmental and developmental agendas. This shift occurred in several stages. At the 1972 Stockholm conference, where the underlying tensions between the Global North and South came out into the open, both global environmental responsibility and the right to development were formally recognised, but the tension between them was not successfully addressed. Novel concepts such as 'eco-development' (Sachs, 1974) gained some traction in the preparatory talks for Stockholm (Macekura, 2015: 223–6), but it was not until the 1980s that a new discourse focused on making development sustainable emerged. The concept of 'sustainable development' has since become the 'dominant conceptual framework' (Bernstein, 2002: 3) in GEP. A first formulation of this approach can be found in the 1980 World Conservation Strategy (WCS), which IUCN had developed in collaboration with WWF and UNEP. To win over developing countries to a long-term strategy of protecting biological diversity, WCS's central premise was that developing countries could pursue economic growth while protecting ecosystems and preventing species extinction as part of a long-term environmental and developmental strategy. This shift was not uncontroversial within environmental circles, and it took most environmental groups until the end of the 1970s to embrace developmentalism (Macekura, 2015: 221–2).

The international breakthrough for the sustainable development concept came with the 1987 Brundtland Commission report *Our Common Future*. The report, which famously defined sustainable development as 'development that meets the needs of the present without compromising the ability of future generations to meet their own needs' (World Commission on Environment and Development, 1987: 8), laid the intellectual foundations for the 1992 Rio Earth Summit and sought to keep environmentalism and developmentalism in a delicate balance. This new compromise helped pave the way for the CBD in 1992, which reflected the original conservationist interests of Northern NGOs to protect wildlife and its habitat while also reaffirming developing countries' right to the sustainable use of natural resources and the equitable sharing of benefits arising from genetic resources (Harrop, 2013). The main achievement of the new conceptual compromise was to secure buy-in from competing stakeholders by blending their different preferences without resolving any inherent tensions. Environmentalists had established states' fundamental duty to protect global and local ecosystems and won a concession by developing countries that environmental sustainability considerations would be integrated into their developmental plans. At the same time, Southern leaders were able to reaffirm national sovereignty and developmentalism as core international norms, though the latter was now qualified with reference to its sustainability. Of course, as Steven Bernstein (2001: 97) points out, it would be wrong to overstate the North–South conflict when it comes to the central role of the 'right to development'. At UNCED, both sides came to accept a fundamental commitment to the centrality of the underlying market norm in guiding the path towards a more sustainable global economy. This liberal environmental compromise – a far cry from 1970s style calls for a New International Economic Order – provided a more durable basis on which to build a global consensus around environmental stewardship that supported Southern developmental growth aspirations.

Despite subsequent debates about how to operationalise this vague concept, sustainable development has continued to provide the intellectual glue for keeping diverging interests and constituencies in GEP united. One measure of its success as a unifying political principle lies in the fact that GIS has repeatedly invoked it as a central paradigm for developing new global policy tools. The 2012 Rio+20 conference, officially known as the UN Conference on Sustainable Development (UNCSD), provided a renewed impetus towards integrating environmental and development agendas. One of the outcomes of Rio+20 was the creation of a High-level Political Forum on Sustainable Development, which led the effort to agree the UN Sustainable Development Goals (SDGs) as a global governance tool. Formally adopted by the UN in 2015, the seventeen goals of the SDGs replaced the more narrowly defined Millennium Development Goals of 2000 and re-established sustainable development as an overarching policy framework for alleviating poverty while reducing environmental burdens worldwide. The SDGs have been hailed as marking a 'historic shift' towards one integrated sustainable development agenda after the 'long history of trying to integrate economic and social development with environmental sustainability' (Biermann, Kanie and Kim, 2017: 26).

The globalisation of environmental stewardship cannot be explained with normative accommodation at the international level alone; its success is also down to normative and political shifts in the Global South, and in particular a parallel uptick in domestic environmental concern and mobilisation since decolonisation. This process has been highly uneven and is often difficult to pin down, largely due to the absence of detailed empirical studies of the rise of environmentalism for many developing countries. But the growing shift of attention to 'poverty ecology' and Southern environmental movements provides at least a few pointers to key domestic developments and external influences.

In Brazil, the first environmental conservation groups originate from the 1950s and environmental activism became more widespread from the 1970s onwards. The country was one of the first worldwide to establish a national environmental agency (Special Secretariat of the Environment) in 1973, at a time when such bodies existed in only ten other states (Hochstetler and Keck, 2007: 7, 27). Brazil created a National

Environmental Law in 1981, though it was not until after the 1992 Rio Summit that the Brazilian state began to develop more effective regulatory capacity (ibid.: 31–2). Neighbouring Argentina embarked on creating a domestic environmental policy competence somewhat later than Brazil. The position of Undersecretary of the Environment was introduced in 1985, which provided the necessary administrative structure for developing national environmental policy in subsequent years (Avalle, 1994: 136).

Environmental concerns were largely absent from political debate and national policy in China during Mao's reign. Participation in the 1972 Stockholm conference prompted the Chinese government to initiate the first steps towards developing a domestic focus on environmental policy, and during the 1980s and 1990s the country slowly built the legal and institutional infrastructure for environmental protection (Economy, 2004: 94–5). UNCED in 1992 provided a second external impetus for China's gradual embrace of environmental ideas and policies (ibid.: 98–9). Inevitably, given the severe restrictions on civil society mobilisation, environmental campaign groups have played only a minor role in the greening of Chinese national policy. Still, with Chinese economic growth taking off since the 1990s and placing a rapidly expanding environmental burden on society, food safety and air pollution concerns especially in urban areas have assumed much greater political salience in Chinese politics (Shapiro, 2012).

Civil society mobilisation around environmental threats has played a prominent role in other Asian countries. By the late 1970s, numerous NGOs had emerged throughout South East and East Asia, serving as a vehicle for local communities to express their local grievances and demand governmental action. Especially in the rapidly developing Asian economies (e.g. Taiwan, South Korea, Hong Kong, Singapore), the deterioration in environmental conditions during the 1980s led to 'a gradual emergence of environmental consciousness among the citizenry' as well as organised environmental campaigns (Lee and So, 1999: 5). In India, too, civil society mobilisation around environmental and social justice issues emerged during the 1970s, and by the 1980s the Indian state began to introduce the first comprehensive environmental laws and institutions (Peritore, 1999: 64–9).

The surge in domestic environmental concern, which became a more widespread phenomenon across the Global South from the 1980s onwards, slowly began to make itself felt in international relations. It helped to moderate the resistance that some developing countries still showed to a strengthening of GEP and also led to the first Southern-led efforts to create international environmental rules. Concern over toxic waste imports from industrialised countries was one of the first issues that brought developing countries together behind an environmental regime-building effort. In the 1980s, developing countries emerged as the key *demandeurs* for an international ban on toxic waste trade, a key factor behind the creation of the 1989 Basel Convention (Miller, 1995: 87–107). Similarly, emerging concern in the developing world about the safety risks of novel biotechnology products led to Southern demands for the inclusion of a regulatory instrument in the CBD. Industrialised countries opposed this demand but eventually agreed to launch negotiations on a protocol to the CBD on the question of biosafety (La Vina, 2002). After increasingly acrimonious negotiations that pitted the North against the South as well as the EU against the United States, international rules on trade in GMOs were finally agreed in 2000 in the form of the Cartagena Protocol on Biosafety (Falkner, 2000). Once again, Southern environmental concerns had become a key driver in the growth of international environmental regulation, suggesting a much wider shift in their engagement with GEP (Najam, 2005).

The environmental movement faced its own decolonisation challenge after the Stockholm conference. By the early 1970s, with most colonial empires dismantled, international environmental NGOs had to re-evaluate their strategy for engagement with newly independent developing countries. For some, this involved a difficult process of soul-searching to overcome the long-standing association of nature conservation with colonial rule. As discussed in chapter 3, the colonial system had provided an important context for the formation of environmental knowledge and concern. Europe's colonial experience was an important source of ecological expertise especially in forestry and agriculture, and many innovative conservation practices were first instituted in colonial contexts. Colonial empires also became the focus of the first transnational environmental campaigns in the late nineteenth century, which sought to export a conservation ethic – often seen as part of the West's 'civilisational' standard (Rollins, 1999) – to Africa and Asia. Right until the demise of colonial rule, many international environmental NGOs had benefitted from working in close cooperation with colonial authorities, often promoting conservation practices against a local population that they felt lacked an interest in wildlife protection (Guha, 1997). Russell E. Train, founder of the African Wildlife Leadership Foundation (later to become the African Wildlife Foundation) and influential adviser to US President Nixon in run-up to the 1972 Stockholm conference, spoke for many Northern conservationists when in 1959 he blamed '[t]he ever-increasing native population, always hungry for more land and seemingly indifferent to the fate of the wildlife' for the decline in big mammals in East Africa (quoted in Macekura, 2015: 43). For local communities, the boundaries between environmental paternalism and colonial chauvinism were often difficult to draw.

The International Institute for Environment and Development (IIED), founded in 1971 by Barbara Ward, was one of the first to bring together conservation and development experts in an effort to develop a more holistic approach to nature protection. Influenced by Ward and Maurice Strong, the chair of the Stockholm Conference, IUCN also began to recast conservation policies as a tool for long-term economic development. Throughout the 1970s, various IUCN officials castigated conservationists for failing to engage indigenous communities in the management of protected areas and for neglecting the need for compromise with local decision-makers. By the mid-1970s, many IUCN, WWF and IIED officials had come to endorse a vision of environmental management that promoted the basic needs of local populations and also engaged them in decision-making processes. International NGOs also began to broaden the range of issues that they addressed in their campaigns and included issues of greater concern to developing countries: not just nature conservation but also transboundary issues of deforestation and desertification and the emerging problems of urban pollution in the Global South. There was some resistance among donors and activists in the international conservation movement against this new focus on 'conservation-for-development', but by the late 1970s both IUCN and WWF had completed the strategic shift to bring environmentalism in line with the developmentalist agenda of the Global South (Macekura, 2015: 229–39).

Bridging the North–South divide in global environmental campaigning has remained an ongoing challenge for many international environmental NGOs and activist networks. Despite the progress made in creating a more inclusive global agenda, environmental world society is still characterised by distinct fault-lines between Northern and Southern perspectives. Several international environmental NGOs have faced internal and external criticism over their perceived preference for Northern, and neglect of Southern, environmental concerns (Doherty and Doyle, 2013). While some Northern NGOs (e.g. WWF, Friends of the Earth) have moved more decisively to broaden their agenda to the concerns of the Global South, others (e.g. Greenpeace) have retained a more traditional focus on narrowly defined environmental issues (Rootes, 2006). Tensions between post-materialist values in the North and post-colonial justice framings in the South still surface in GEP, in debates about the right balance between biodiversity protection and use, and between climate change mitigation and adaptation. In climate politics, the rise of the global justice movement has reignited the debate on the marginalisation of Southern perspectives. In terms of representation in international climate forums, environmental campaigners from the North still outnumber those from the South, and their perspectives and strategies continue to diverge (Gereke and Brühl, 2019). Whereas Northern campaigners tend to focus on technocratic solutions to the mitigation problem, global justice campaigners seek to shift the debate to the unequal ecological development that has benefitted industrialised countries in the past (Goodman, 2009). Their efforts to 'decolonize the atmosphere' have directed attention to the North's 'ecological debt' (Warlenius, 2018) that needs to be repaid to late industrialisers. Making global environmentalism more inclusive thus remains a key political challenge in global politics – for both international and world society.

Regional Dimensions of Global Environmentalism

The discussion so far has focused on the universal reach of environmental stewardship and its globalisation from the Stockholm conference to the Rio conference and beyond. As I have argued, environmentalism has indeed become a primary institution for *global* international society. But this is not to suggest that its normative strengthening and institutionalisation have progressed equally in all parts of the world. As the ES literature on regional international societies has shown (for an overview, see Buzan and Schouenborg, 2018: 30–2), GIS is made up of different regional subsystems that either complement and strengthen the normative order of GIS, as is the case with the EU (Diez and Whitman, 2002), or threaten to undermine it and pull international order into a more region-centric future (Acharya, 2014; Buzan and Lawson, 2015). Existing regional subsystems may differ from each other in terms of their level of integration and cultural homogeneity, the extent to which they advance a solidarist or pluralist normative order, and the degree to which the fundamental norms of GIS are established within them. Applied to the environmental field, it should not come as a surprise if we find that the globalisation of environmental stewardship has proceeded at different speeds and reached different depths across the regional subsystems of GIS.

The EU is widely acknowledged to have emerged as an international leader in environmental protection (Wurzel and Connelly, 2010; Vogler and Stephan, 2007; Kelemen and Vogel, 2010; Zito, 2005). Not only has the EU developed some of the most advanced environmental standards and regulations, in areas such as chemicals, recycling, renewable energy and biosafety (Selin and VanDeveer, 2006; Falkner, 2007), it has also done the most to establish environmental responsibility as a fundamental norm in its constitutional order. The founding treaties of the then EEC did not identify the environment as a separate policy area. Until the Single European Act of 1986 elevated environmental protection to a Community responsibility, the environment was a 'thoroughly, thoroughly marginal issue' in the European Commission, as Tony Brenton (2010), the first head of the UK FCO's Environment, Science and Energy Department, recalls. By 1992, the Maastricht Treaty had introduced the precautionary principle as a basis for European environmental policy and later treaty changes (Amsterdam Treaty, 1997, and Lisbon Treaty, 2007) made sustainable development an explicit objective of the Union (Selin and VanDeveer, 2015: 42–8). Environmental stewardship has thus hardened into a constitutional principle for the EU. Furthermore, the EU's commitment to environmental sustainability in Europe is matched with a similar pledge to promote its global diffusion. The EU has identified international environmental leadership as an integral part of its foreign policy identity (Manners, 2008: 53–4) and has promoted its own environmental policy approach through international policy diffusion, both in its neighbourhood and globally (Busch and Jörgens, 2012). It is thus fair to conclude that the EU has developed a particularly strong commitment to environmental stewardship, anchoring the primary institution in its distinctive regional identity and thereby contributing to the solidarisation of environmentalism in GIS (Ahrens, 2017).

Regional systems for environmental governance exist in other parts of the world, too, though they tend to be driven more by the desire to manage transboundary environmental issues than deep political integration as in the EU's case. In North America, where the first bilateral or plurilateral forms of environmental cooperation go back to the early twentieth century, regional integration only picked up momentum in the 1990s. In 1994, when the North American Free Trade Agreement (NAFTA) became the main framework for regional economic cooperation, an environmental side agreement (North American Agreement on Environmental Cooperation, NAAEC) was also established to promote better coordination of environmental standards. Unlike the EU's treaty system, NAAEC offers only a soft instrument for strengthening regional environmental policy. It neither aims at the harmonisation of environmental policies nor does it create supranational institutions with policy-making authority (Schreurs, 2013: 367–8). More recent developments include the creation of transnational emissions trading schemes between the United States and Canada (e.g. Regional Greenhouse Gas Initiative), though these have remained fairly isolated efforts to integrate North American environmental policy across boundaries (Biedenkopf, 2017).

Institutionalised environmental cooperation has emerged in other regions, too, but largely remains in its infancy. The Association of Southeast Asian Nations (ASEAN) has issued various Declarations on climate change, energy and environmental sustainability, while the South Asian Association for Regional Cooperation, the African Union and the Union of South American Nations all serve as regional policy forums to discuss environmental policy objectives. Most, however, shy away from taking any further steps towards integrating environmental policies (Schreurs, 2013: 365–70). In Asia, Africa and Latin America, membership in international environmental regimes rather than regional organisations remains the main anchor for the international norm of environmental stewardship.

As we have seen, the extent to which environmental stewardship has become embedded in regional international societies differs considerably, but this variation is largely in line with the different levels of regional integration that can be found worldwide. Unsurprisingly, the EU represents an advanced form of state-centric solidarism in environmental politics that has found few regional followers in the world. Regional commitment to environmental stewardship thus varies across the world, but the environmental norm is sufficiently globalised to make it a primary institution of *global* international society.

Conclusions

When states established environmental stewardship as a fundamental international norm at UNCHE in 1972, it could not be taken for granted that the environmental norm would become recognised as a universal norm for *global* international society. Not all new norms that are introduced to international society survive and become globally accepted. After all, other solidarist norms that had been championed after the Second World War (e.g. human rights, democracy) have struggled to gain universal acceptance. Environmentalism, however, underwent a process of social consolidation and globalisation that reached its zenith by the time of the 1992 Rio Earth Summit. As I have shown in this chapter, the social consolidation of environmental stewardship was evident from two major processes in international relations: the creation of a large number of secondary institutions that translated the fundamental commitment to environmental responsibility into international regulatory mechanisms; and the adjustments in state behaviour and identity that the deepening commitment to global environmental protection brought with it.

The first measure of global environmentalism's social consolidation can be found in the dramatic increase in MEAs that were negotiated after Stockholm. Few other areas of global policy coordination have witnessed such a dramatic expansion in the number of secondary institutions. Over time, environmental treaty-making has become more systematic in its coverage of global environmental issues. It has also involved an ever-larger number of actors, from states to international organisations, scientists, environmental campaigners and businesses, in the process of global rule-making. A second measure can be found in the gradual greening of state identity and practices. For one, states had to equip themselves with the necessary tools – from scientific expertise to environmental laws, regulations and agencies – to deliver on their global environmental commitments. Starting in the late 1960s, major industrialised countries put in place domestic environmental policy frameworks, which gradually diffused to other parts of the world in subsequent decades. This also included the creation of foreign environmental policy capacity and the integration of environmental and technical expertise into the diplomatic machinery. The proliferation of international sustainability summits and environmental negotiation forums put increasing strain on states' ability to fully engage in environmental diplomacy, especially in the developing world, but the normative pull of environmental multilateralism has become inescapable.

By the time of the 1992 Rio Summit, environmental stewardship was fully recognised as a universal norm. Until then, many developing countries had been sceptical about the new international environmental agenda, largely because of its inbuilt Northern bias and perceived opposition to the South's developmental needs. But the shift in emphasis towards sustainable development as the new leitmotiv in GEP facilitated the normative accommodation between environmentalism and developmentalism and helped to rally developing countries behind the idea of global environmental responsibility. This realignment in GEP was also down to the strengthening of domestic environmental concerns and norms across the developing world, which not only boosted the globalisation of environmental stewardship but also created demand for new international rules and regulations in line with Southern environmental interests. In this way, environmentalism first had to undergo its own process of decolonisation before it could become a universally accepted fundamental norm of GIS.

Ultimately, global environmentalism proved to be a successful normative import into GIS because it could be made to fit into the existing normative order without requiring a major reinterpretation of existing primary institutions, be it sovereignty, territoriality, the market or developmentalism. Although arising in a specific historical context and championed by leading Western powers, it has never followed a narrow Western liberal agenda but appealed to different societies' interest in managing shared ecological threats. Following a 'common fate' logic (Falkner and Buzan, 2019: 132) and resonating with a diversity of environmental values and ideas that are rooted in different societal contexts, it was able to acquire the status of an universal norm. Environmentalism proved to be a malleable global norm, endlessly adaptable to specific regional and national conditions. Inevitably, this flexibility has also limited the scope for turning environmental stewardship into a harder global norm that would require a greater degree of political convergence and harmonisation of environmental practices. Despite being promoted as a solidarist norm that speaks to humanity's shared interest in planetary health, it has been well suited to operating in a pluralist world of normative and ideological diversity.

7 Environmental Stewardship between Consolidation and Contestation

How has environmental stewardship fared as a fundamental international norm once it had become universally accepted in the early 1990s? Has it led to a further expansion of the global environmental agenda? Has GIS built the global governance infrastructure that can deliver on the many internationally agreed environmental goals? And how has the rise of environmental stewardship played out in the context of the existing normative structure of GIS? Not all primary institutions necessarily support a strengthened system of global environmental governance. As many environmentalists have long complained, sovereignty and territoriality prevent the creation of a central international authority that could act in the interest of planetary health. The market norm and developmentalism tend to both support unbridled economic growth and limit regulatory interference with global economic activity. Norm contestation around some, if not all, normative elements of global environmentalism can thus be expected to continue even as the commitment to environmental stewardship strengthens.

In this chapter, I explore both the further strengthening and consolidation of the international environmental norm as well as contestation around it, from the 1990s until the early twenty-first century. In the first section, I examine how global environmental protection has been further embedded in international policy-making and cooperation. This process involved the inclusion of ever more issues in the international environmental agenda; the insertion of green norms into other international policy areas and regimes, most notably those concerned with economic affairs; and a strengthening and broadening of the institutional architecture for global environmental protection. None of these shifts in international practices and norms were uncontroversial or easy to achieve, they were the result of ongoing battles over how to put environmental stewardship into practice. Indeed, the second section of this chapter will discuss in more detail the ongoing normative contestation over global environmentalism. I conclude the chapter with a review of how environmental stewardship fits in with other primary institutions of GIS.

Strengthening Environmental Stewardship: Expansion, Mainstreaming, Institutional Innovation

Expansion of the International Environmental Agenda

The continuous expansion of international environmental policy-making provides an important indicator of GIS's deepening commitment to environmental stewardship. By putting ever more environmental issues on the international agenda, states confirmed one of the key insights of ecological thinking: that the increasingly global character of environmental problems undermines the state's ability to deal with these problems at the national level and that global ecological interdependence requires states to cooperate at the planetary scale (Hurrell, 2007: 218–22). The internationalisation of environmental protection, already in evidence after Stockholm but accelerating in subsequent decades, was driven by at least three distinct forces.

First, it was the result of improved scientific identification and understanding of global ecological problems and interdependencies. From the 1970s onwards, Earth System Science thinking became an influential framework for integrating research into biophysical systems on a planetary scale and went on to inform major international research programmes (e.g. International Geosphere-Biosphere Programme, World Climate Research Programme, Future Earth) (Warde, Robin and Sörlin, 2018: 152–4). Scientists from around the world not only promoted the idea of globally interconnected ecological systems but also made it the basis for advocating internationally coordinated responses to environmental problems (Kohler, 2019).

Norm entrepreneurship by world society actors proved to be a *second* influential force behind the globalisation of environmental politics. In the post-1945 era, environmental NGOs had themselves adopted an increasingly global outlook in their campaigns, tackling transboundary issues such as whaling, air pollution and radioactive waste. With the founding of new environmental organisations in the 1960s and 1970s (especially WWF, Greenpeace, Friends of the Earth), a new phase of global green activism had arrived that was 'unconstrained by national boundaries' (McCormick, 1989: 147). Through their global campaigns, NGOs not only created awareness for transboundary environmental issues (e.g. biodiversity, ozone layer depletion, climate change) but also put pressure on states to tackle them via international cooperation (Betsill and Correll, 2008).

A *third* factor can be found in the distinctive political economy that underpins the internationalisation of environmental policies. As leading industrialised economies developed more demanding environmental standards and rules at the domestic level, business sectors that faced international competition argued – often in alliance with environmental NGOs – for the internationalisation of such regulations so as to create a level playing field in the global economy (Falkner, 2008). Thanks to such 'Baptist-bootlegger' coalitions of business and environmental interest groups (DeSombre, 2000), the expansion of domestic environmental regulation soon created its own dynamic of regulatory export to the international level. Leading economic powers, in particular, have thus driven the internationalisation of environmental regulation, whether through direct forms of international environmental leadership or indirect processes of 'trading up' (Vogel, 1995).

Chemicals regulation is a good example of an environmental policy field that was gradually transformed from an exclusively domestic to an increasingly international domain. The environmental and health risks of modern chemicals were one of the key concerns that sparked the rise of environmental concern after the Second World War. In her 1962 book *Silent Spring*, which is widely credited with having launched the modern environmental movement, Rachel Carson documented the numerous environmental effects that indiscriminate use of pesticides has had, including the decline in bird populations resulting from the aerial spraying of DDT. Initially, chemicals safety was defined as a matter for local and national environmental authorities. However, as became apparent in the late twentieth century, the production of chemicals is a highly globalised business, and emissions of persistent organic pollutants (POPs) and other toxic substances are transported over long distances and across national borders (Selin, 2013: 110). In response to the globalisation of chemicals risk, governments increasingly worked together to coordinate their regulatory approaches. One of the first international actions against a specific chemical risk was the 1973 OECD decision that its member states should restrict the production and use of polychlorinated biphenyls (PCBs). Subsequently, the OECD adopted further decisions against dangerous substances and became a major forum for coordinating and strengthening industrialised countries' chemical safety rules (Eckley and Selin, 2004: 88–9). The chemical industry itself came to recognise the need for greater international coordination of safety approaches. Following a series of damaging industrial accidents in the 1980s, North American chemical firms created private governance mechanisms that promoted stricter chemical safety standards across the world (Garcia-Johnson, 2000).

The international regulatory response to the global risks of chemicals was piecemeal and incomplete, however. Instead of dealing with all chemicals under one overarching framework convention (as is the case with climate change and biodiversity), GIS adopted a series of independent treaties that covered only certain narrowly defined aspects of the global regulatory challenge. Four international treaties and several regional agreements form the heart of this regime complex for chemicals: the 1989 Basel Convention on the Control of Transboundary Movements of Hazardous Wastes and Their Disposal; the 1998 Rotterdam Convention on the Prior Informed Consent Procedure for Certain Hazardous Chemicals and Pesticides in International Trade; the 2001 Stockholm Convention on Persistent Organic Pollutants; and the 2013 Minamata Convention on Mercury. Cognisant of the fragmented nature of global chemicals governance, the UN has sought to promote better coordination between the different elements of the regime complex. UNEP, for example, sponsored the creation of the 2006 Strategic Approach to International Chemicals Management (SAICM), which provides an umbrella mechanism for promoting environmentally safe chemical management but stops short of establishing a common international legal framework (Selin, 2013: 108). Despite the gradual strengthening and expansion of international chemicals regulation, serious gaps in coverage and effectiveness persist. Plastics pollution of marine environments, for example, has recently emerged as an urgent problem, with the amount of plastic flowing into the ocean set to double between 2010 and 2025 (Dauvergne, 2018). Despite growing alarm about the problem, no functioning global governance framework has been put in place to stem the global tide of plastic waste.

Biodiversity protection, long considered a local or national concern, similarly underwent a process of internationalisation in the late twentieth century. Although some of the first international environmental treaties were aimed at the protection of individual species, such as birds, seals and whales (see Chapter 4), it was only from the 1970s onwards that a more systematic international approach to preserving animal and plant species and their natural habitats began to emerge. Following on from the first two modern nature conservation treaties, the 1971 Ramsar Wetlands Convention and the 1973 CITES, the 1992 CBD became the first treaty to put in place a comprehensive governance framework. Environmental NGOs had long been campaigning for a more holistic approach that focused not on saving individual species but on protecting entire ecosystems. The CBD, which has its origins in the IUCN's 1980 World Conservation Strategy (McCormick, 1989: 162–79), responded to these demands but also sought to address developing countries' concerns about balancing conservation and sustainable use of natural resources. The convention became an important tool for globalising biodiversity protection and led to the creation of national biodiversity strategies in 191 out of its 196 parties. However, it imposes only 'soft' obligations on states (Harrop, 2013) and has largely failed to strengthen international oversight over the underlying drivers of species extinction and ecosystem destruction.

More recently, global biodiversity governance has switched to a targets-based strategy in order to stem the tide of biodiversity loss. Many of the Aichi Biodiversity Targets from 2010 have not been met, however. Some have since been incorporated into the 2015 SDGs and others were renewed for 2020. As with other environmental policy areas, states have continued to develop new biodiversity initiatives even when they have failed to deliver on past ones. For instance, the adoption of a target for marine protected areas (MPAs) to cover 10 per cent of the world's oceans is behind a significant upswing in the creation of new MPAs, which increased from ca 2 million km^2 in 2006 to over 14 million km^2 in 2014 (Chan, 2018: 539). At the same time, many states have been accused of 'target gaming' by shifting the focus away from coastal areas to larger but often uninhabited ocean spaces (Humphreys and Clark, 2020; Chan, 2018). As states struggle to implement internationally agreed policies and targets, biodiversity governance has broadened to include an ever larger range of non-state actors at the transnational and subnational level. Empirical analysis of this expanding regime complex shows how the majority of the 56 transnational institutions involving over 10,000 state, non-state and subnational actors were created from 2000 onwards (Pattberg, Widerberg and Kok, 2019: 386), suggesting a dramatic expansion of the actors involved in global biodiversity governance.

International environmental policy-making has also been extended to deal with novel technologies and their associated risks, an area that is characterised by both high levels of scientific uncertainty and deep political contestation. The initial impetus for this shift came in the context of the CBD negotiations in 1992, when developing countries demanded the inclusion of modern biotechnology in the remit for the biodiversity treaty (Zedan, 2002). Developing countries were concerned that the international movement of GMOs would expose them to a wide range of risks – environmental, health, as well as socio-economic – that they would find difficult to manage without legally binding international rules. After some initial resistance, industrialised countries acceded to these demands in 1996. The biosafety negotiations opened in 1996 and led to the adoption of the Cartagena Protocol on Biosafety to the CBD in 2000 (Falkner, 2000). The biosafety treaty broke new ground in GEP in that it dealt with an anticipatory regulatory challenge, 'one where the very existence and nature of risk and harm remains scientifically and normatively contested' (Gupta, 2013: 89). Instead of trying to deal with past or current forms of industrial pollution, it empowers importers of GMOs to adopt a precautionary approach when deciding whether to allow such imports to happen. For the first time, the precautionary principle, one of the notable innovations of the 1992 Rio Declaration, was included in the operational part of the Cartagena Protocol (Graff, 2002). The Cartagena Protocol thereby pushed international environmental policy-making in a direction that positively acknowledges scientific uncertainty and diverging domestic choices in risk regulation (Gupta, 2013), though this approach remains controversial in GEP (see subsequent discussion). Subsequent efforts to extend precautionary global governance to other novel technologies were less successful. The question of how to deal with the risks of newly emerging nanotechnologies, for example, led to some regulatory coordination and harmonisation efforts at the OECD and International Organization for Standardization (ISO), but the leading nanotech countries have not deemed it necessary to launch negotiations on a formal international regime. Different national preferences with regard to balancing technology promotion and precautionary regulation have so far held back a shift towards the global governance of nanotechnologies (Falkner and Jaspers, 2012).

The one issue that has come to dominate international environmental policy-making in recent years is climate change. After a long agenda-setting phase – the UN initiated the first debate on the threat from global warming in the late 1970s – GIS embarked on creating an international climate regime in the early 1990s, when climate change was still seen as just one among many global environmental challenges. After the UNFCCC (1992) and its Kyoto Protocol (1997) had been established, climate change steadily moved up the international agenda. By the early twenty-first century, it had 'grown so big that other dimensions of environment became marginalized in the popular media' (Warde, Robin and Sörlin, 2018: 119). Non-state actors have played a major role in the growth of climate change into an all-encompassing global threat. For one, scientists have been able to demonstrate with ever greater precision and certainty the unprecedented scale of the past global warming trend and how unmitigated climate change would wreak ecological and economic havoc on the world. The IPCC became the authoritative source of regular reviews of scientific knowledge that were fed into the international climate negotiations, while social scientists provided evidence of the extraordinary economic damage that global warming would produce if left unchecked (Stern, 2007). Many international NGOs also came to focus their environmental campaigns on the threat from climate change and put pressure on states to agree more ambitious international action. With the annual Conferences of the Parties (COPs) to the UNFCCC attracting ever larger numbers of delegates and observers, the UN climate regime became the single most important international forum in which states and non-state actors interacted to shape the global environmental response. Through a process of 'climate change bandwagoning' (Jinnah, 2011), the climate regime has also become enmeshed with other international secondary institutions (e.g. forest protection, energy governance), with the latter taking on climate objectives partly in order to support climate change mitigation and adaptation, partly to derive benefits for themselves from being associated with the high-profile climate agenda. Climate change also came to play an ever more high-profile role in other non-environmental forums, from the G7/8 and G20 summits and World Bank/IMF annual meetings to the Davos gathering of the World Economic Forum (Kirton and Kokotsis, 2016). Thus, the all-encompassing nature of climate change, the magnitude of the global threat and the political mobilisation around it have turned climate change into the defining global challenge of the twenty-first century. Few other environmental issues have come to demonstrate just how critically important GIS's collective environmental responsibility is for the future of the planet.

The Greening of International Economic Regimes

A second indicator of the strengthening of environmental stewardship in GIS can be found in the gradual acceptance of environmental values and concerns by other, non-environmental, secondary institutions, especially those governing the global economy. Without such mainstreaming of environmentalism in global governance, environmental protection would be restricted to playing a fairly marginal role in the governance of global economic activity. Indeed, we can find considerable evidence of such mainstreaming across a wide range of global policy areas, from forests and agriculture to education, health and population policy, as well as in the core economic areas of international trade, finance and investment. However, this is not a straightforward story of the gradual 'greening' of secondary institutions. The degree to which environmental principles have become embedded in other international regimes depends on the congruence or conflict between environmental and other international norms. In some areas (e.g. forests, agriculture), environmentalism offers a good fit with established ideas of resource conservation and sustainability. In other areas, most notably economic governance, certain environmental values and principles pose a more profound challenge to dominant liberal principles of market-based exchange, the free movement of goods and services, and economic growth. The market norm, in itself a primary institution that initially depended on support by Western powers but has become universally accepted after the end of the Cold War (Buzan, 2004: 235), serves as a strong barrier to the adoption of more radical environmental principles. The greening of the international economic order that can be observed from the 1990s onwards has therefore involved a considerable degree of normative accommodation on the environmental side, leading to what Bernstein (2001) has described as the compromise of 'liberal environmentalism'.

In the early days of modern environmentalism, the idea of introducing environmental principles into the global economic order still seemed to be a somewhat fanciful idea. Throughout the 1970s and 1980s, environmental protection and a liberal economy were often discussed as mutually exclusive ideals. Even in the late 1990s, it was still common to describe attempts 'to reconcile environmentalism and economic freedom' as a 'seemingly quixotic endeavor' (Block, 1998: 1887). Given the long history of antagonism between environmentalism and economic liberalism, it is unsurprising that some international economic regimes were much slower to respond to the rise of global environmentalism than other parts of the international governance architecture. However, with the expansion of environmental multilateralism the growing number of international environmental rules inevitably encroached on, and clashed with, international economic regimes. Bit by bit, the environmental impacts of international trade, finance and investment flows came under closer scrutiny, and one economic regime after another had to confront the global ecological challenge. Some were prompted to reconcile environmental and economic imperatives as early as the 1980s, as was the case with the World Bank that established its first Environmental Department in 1987. The international trade regime formally adopted sustainable development as a core principle only in 1995, when the newly created World Trade Organization (WTO) commenced its work. Other institutions of global economic governance took much longer to respond to rising environmental concern. It was only in 2003 that the first internationally agreed environmental rules were created for export credit agencies (ECAs), in the form of the voluntary OECD Common Approaches on Environmental and Officially Supported Export Credits. This process of normative adaptation is far from complete and remains deeply contested, as will be discussed further subsequently.

The international trading system has come a long way in its engagement with the international environmental agenda. Back in 1947, when the General Agreement on Tariffs and Trade (GATT) was adopted, the treaty that came to underpin the post-war trading system was not given an explicit environmental mandate. Based on the core norms of non-discrimination and reciprocity, its main objective was to reduce the overall level of trade tariffs and other trade barriers through multilateral negotiations. Trade restrictions for environmental reasons were allowed only in exceptional circumstances, as defined in Article XX of the GATT (Jaspers and Falkner, 2013: 414–15). The GATT system was thus not blind to the need to deviate from free trade in certain circumstances, but on the whole the restrictive nature of the GATT rules and jurisprudence ended up having a 'chilling' effect on environmental regulations and multilateral environmental agreements (Eckersley, 2004b). As the environmental stewardship norm strengthened in GIS, the international trade regime came under growing pressure to recognise the legitimacy of environmental policies even if they interfered with international trade. The key turning point came with the GATT's Uruguay Round (1986–94), which extended its disciplines to new trade areas and led to the creation of the WTO. In the preamble to the Marrakesh Agreement Establishing the WTO, the parties for the first time included a reference to sustainable development and environmental protection as explicit objectives of the reformed trading system. The Uruguay Round also added new agreements (on Technical Barriers to Trade, and on Sanitary and Phytosanitary Measures) that recognised member states' right to impose trade-restrictive measures in order to protect human health and the environment (Charnovitz, 2007). The strengthening of environmental norms within the WTO is most clearly visible from the way that its trade-related jurisprudence has since evolved. Whereas in the pre-WTO era, the dispute settlement mechanism declined to consider environmental justifications in deciding cases (e.g. 1991 Tuna-Dolphin case), rulings in the post-1995 WTO era have been able to strike a better balance between trade concerns and sustainable development objectives (e.g. 1998 Shrimp-Turtle case; 2003 EC Biotech Products case) (Trachtman, 2017).

However, despite the considerable mainstreaming of the environment into the international trading system, fundamental tensions between WTO rules and environmental policies remain unresolved. These tensions centre on the use of precaution in environmental regulation and the scientific justification for trade measures, the WTO's stance against environmental trade restrictions that target pollution in industrial processes (so-called process and production measures) and the thorny issue of whether non-parties to MEAs have recourse to the WTO's dispute settlement and whether MEAs are subordinate to WTO rules in such cases (Jaspers and Falkner, 2013; see also the subsequent discussion). The search for a negotiated solution to these latent conflicts, whether through the WTO Committee on Trade and the Environment or as part of the Doha Round, has been an elusive quest so far.

In the debate on how to reconcile environment and trade, the multilateral trade regime of the GATT/WTO has received the lion's share of attention. However, with the multilateral Doha trade round stalled, plurilateral or regional trade agreements (RTAs) have gained in significance as they offer a more promising route for ambitious trade liberalisation deals. Many RTAs have also generated far more detailed provisions on the trade–environment link, usually with stronger support for environmental measures than can be found in WTO agreements. Taken together, the number of environmental provisions in multilateral and regional trade agreements has risen significantly since the early 1990s. A recent study listed a total of 288 such references to environmental norms by 2016 (Morin, Pauwelyn and Hollway, 2017: 373), and another study identified North–South RTAs as showing the strongest growth trend in this regard (Morin, Dür and Lechner, 2018). Industrialised countries have been the main driver behind this partial greening of RTAs, using bilateral and regional trade agreements as a lever to push for stronger environmental action worldwide. This has been the case with the United States ever since it agreed with Canada and Mexico to include environmental provisions in a side agreement to NAFTA. The European Union has similarly used trade agreements with other nations to demand stronger environmental protection measures and promote environmental multilateralism more generally (Jinnah and Morgera, 2013). As can be expected, the inclusion of environmental provisions in RTAs has not gone uncontested. Many developing world representatives have criticised the imposition of environmental rules via RTA negotiations, even though they have been largely unable to reverse the trend towards greater enmeshment of trade and environmental policy-making.

Environmental norms have also made some inroads, albeit slowly, into the trade-promoting practices of ECAs. In the late 1990s, the OECD responded to growing pressure by transnational activist groups and agreed the 2003 Common Approaches for Officially Supported Export Credits and Environmental and Social Due Diligence, a set of non-binding rules for harmonising environmental and social standards in the provision of export credits. The Common Approaches represent only a small, tentative, step towards greener ECA practices. They have been criticised for lacking enforceability and failing to cover the growing activities of ECAs from emerging economies (Hopewell, 2019). But they suggest that, shortcomings in coverage and effectiveness notwithstanding, the momentum for integrating environmental objectives into international trade rules and practices continues in line with the broader social consolidation of environmental stewardship.

The gradual greening of the trade regime has been complemented and reinforced by the emergence of an increasingly dense network of private environmental standards and regulations in the global economy. The origins of this trend can be traced back to the early 1990s, when NGOs stepped up their campaigns against multinational corporations' (MNCs) environmental impact in the developing world (Newell, 2001). A growing number of multinationals responded to civil society pressure by establishing sustainability standards for internationally traded commodities (e.g. timber, coffee), whether for their subsidiaries or global supply chains (Pattberg, 2007; Gulbrandsen, 2010). For the most part, these initiatives have developed in a relatively uncoordinated fashion, with private governance approaches often diffusing from one sector to another. Taken together, they can be seen as part of the wider trend towards a strengthening of environmental norms in both international and world society.

Most private standard setting has occurred in contexts of transnational social contestation, usually in response to NGO campaigns or consumer pressure, and more recently states and international organisations have also begun to steer corporate self-regulation in an effort to complement international governance efforts. While some private sustainability standards provide a general framework for improving the environmental performance of companies or transparency around sustainability (e.g. Global Reporting Initiative), most have emerged within the context of global supply chains and are focused on specific commodities or sectors. The Forest Stewardship Council, Marine Stewardship Council, Roundtable on Sustainable Palm Oil and Global Sustainable Tourism Council are just a few prominent examples of a complex and diverse web of transnational governance initiatives (Derkx and Glasbergen, 2014; Ponte, 2019). In order to facilitate greater coherence between state-centric and transnational governance, international organisations have more recently developed meta-frameworks for private standards, and some actively seek to encourage business actors to engage in private standard setting. The UN Global Compact and the UN Forum on Sustainability Standards are two such international initiatives that provide a certain degree of 'orchestration' (Abbott, et al., 2015a) of private environmental governance. In this way, the spread of environmental norms throughout international and world society is creating 'complex governance beyond the state' (Hurrell, 2007: 95), an ever more interconnected web of authority relations involving state and non-state actors.

Apart from international trade, environmental stewardship has also affected the global governance of international finance. The greening of multilateral development banks (MDBs) started as early as the 1980s, when the World Bank came under fire for financing infrastructure projects in the developing world that caused widespread environmental damage and social displacement (Nielsen and Tierney, 2003). A transnational NGO campaign against high-profile investment projects (e.g. Polonoroeste road project in Brazil; Narmada Dam in India), together with unilateral pressure by the United States, the largest shareholder of the World Bank, prompted the Bank to take environmental concerns more seriously (Keck and Sikkink, 1998: 135–7; Park, 2007). The World Bank was the first MDB to establish a central environment department in 1987, and it also began to change its practices by applying formal environmental assessment criteria across an ever larger share of its lending portfolio. As critics of the World Bank's record have pointed out, bureaucratic inertia and conflicting interests among donor and recipient countries have held back its environmental reforms (Hicks et al., 2008; Gutner, 2005: 778–9). However, several scholars have shown that the policy shift went beyond a tactical response to external criticism, that it amounted to a deeper form of institutional greening, with the Bank treating environmental objectives no longer as an exception but as routine (Wade, 1997; Park, 2005). Indeed, from the 1990s onwards, the World Bank increased environmental lending and reduced its support for environmentally destructive mega-projects. It went some way towards internalising environmental sustainability as part of its core mission even if that journey is as yet incomplete (Park, 2005, 2007).

Significantly for this discussion, the World Bank also became a source of environmental norm diffusion to other MDBs and global financial institutions (Wade, 1997; Wright, 2013). Its model of environmental impact assessment provided the basis for those developed by regional development banks, from the Asian Development Bank and Inter-American Development Bank to the African Development Bank (Hunter, 2007); and by establishing the World Commission on Dams and Extractive Industries Review, it introduced international expert panels as a mechanism for developing lending guidelines for environmentally sensitive hydroelectric dams and mining projects (Schaper, 2007). Furthermore, working through the International Finance Corporation (IFC), its private sector financing arm, the World Bank also managed to orchestrate some degree of environmental change in the private banking sector. Responding to NGO criticism of their project finance lending to developing countries (Park, 2007: 537), a group of investment banks came together in the early 2000s and sought the IFC's guidance on how to establish environmental principles for their sector. The resulting Equator Principles, which were agreed by ten banks in 2003 and have since been adopted by over 100 financial institutions, established a common set of environmental principles based on the IFC's environmental performance standards (Wright, 2012). In this way, environmental ideas and norms that originated in a 1980s NGO campaign aimed at greening World Bank lending diffused over time via the IFC and into the commercial banking sector.

One important mechanism in the greening of international finance has been the move towards greater transparency around companies' environmental performance. Enhanced transparency does not in itself drive improvements in the environmental behaviour of business, instead it acts as a lever in the hands of key stakeholders that seek to exert pressure on the corporate sector: it allows environmental NGOs to target those companies that fail to reduce their environmental burden; it enables institutional investors to discriminate between companies according to environmental performance criteria and thereby channel capital towards greener businesses; and it assists efforts by regulatory authorities to better assess the environmental risk that exists in the financial sector.

Overall, the trend towards greater disclosure of environmental, social and governance issues is well established by now (Vukić, Vuković and Calace, 2018). According to consultancy firm KPMG (2017: 4), corporate responsibility reporting is now 'standard practice for large and mid-cap companies around the world', but it has taken at least two decades to get to this point. Transparency regimes have become firmly embedded in the global economy not least due to a long history of transnational initiatives that have sought to create globally harmonised systems for environmental accounting and disclosure (Pattberg, 2017). Some of the earliest forms of environmental disclosure go back to the Coalition for Environmentally Responsible Economies (CERES), an NGO created in 1989 by a group of institutional investors and environmental campaigners. Starting with the Valdez Principles, CERES developed a number of tools to promote corporate transparency on environmental practices. The Greenhouse Gas Protocol grew out of a collaboration between the World Resources Institute and the World Business Council on Sustainable Development. It provides a widely used standardised approach to accounting greenhouse gas emissions from the corporate and public sector. The Global Reporting Initiative (GRI) was created by CERES and UNEP in 1997. GRI grew into a more comprehensive global reporting framework to cover both environmental and social criteria, with around two-thirds of corporate responsibility reports covered in the 2017 KPMG survey applying GRI Guidelines or Standards. Finally, the independent non-profit organisation CDP (formerly known as Carbon Disclosure Project) has established itself as the leading investor-driven register for carbon emissions, initially from companies but increasingly also from cities, states and regions. It issued its first questionnaire on corporate carbon emissions in 2003 and has since grown into the world's largest database on corporate environmental performance (Pattberg, 2017: 1446).

Environmental transparency is one area where public regulatory requirements have lagged behind private initiatives but where private and public governance increasingly overlap and interact. Major industrialised countries have incorporated provisions for environmental disclosure into national laws (e.g. UK Companies Act, Sarbanes–Oxley Act in the United States, EU Directive 2014/95/EU). In addition, international organisations have tried to orchestrate more harmonised reporting practices at the international level. Among the most prominent initiatives are the Principles for Responsible Investment Initiative (PRI), an international network of investors that operates in partnership with the UNEP Finance Initiative and UN Global Compact, and ISO, which based its 14064-I standards on the GHG Protocol's corporate standard (Pattberg, 2017: 1449).

Climate change has more recently provided a dramatic boost to the growth of environmental disclosure schemes in the global economy. This trend reflects growing recognition of the systemic risk that climate change poses to the stability of the financial sector. This financial risk is now well understood and documented by academic researchers, who have identified the impact of climate-related risks on economic stability and prosperity (Dafermos, Nikolaidi and Galanis, 2018) as well as the regulatory responses available to financial authorities (Campiglio et al., 2018). Some central banks and financial regulators have accordingly begun to put in place the first policy frameworks that require companies to assess and disclose climate-related risks more extensively and systematically than ever before. The first wave of such regulatory initiatives started in the 2000s, mainly in leading industrialised countries. They were focused on measures that introduced public insurance schemes for natural disasters as well as disclosure requirements for pension funds and listed companies (McDaniels and Robins, 2018). The momentum for this policy shift accelerated in the 2010s, with new international bodies emerging that seek to stimulate the diffusion of national regulations for the financial sector and to coordinate them internationally (Ameli et al., 2020; Keenan, 2019). In 2015, following an initiative of the G20, the Financial Stability Board set up the Task Force on Climate-related Financial Disclosures (TCFD) to develop recommendations on climate-related financial disclosures in line with the objectives of the Paris Agreement. The Task Force has since become a major vehicle for collaboration among national financial authorities from G20 countries.

Two years later, a group of eight central banks and financial supervisors from Europe, China, Mexico and Singapore established the Network of Central Banks and Supervisors for Greening the Financial System (NGFS), to serve as a platform for sharing best practices and develop climate risk management approaches in the financial sector. By 2020, the NGFS had attracted sixty-three members and twelve observer organisations. The growing attention that central banks and financial regulators are paying to climate-related financial risks thus adds another layer to the ongoing greening of international finance, further reinforcing the complex intertwining of public and private environmental governance.

Institutional Strengthening and Innovation

Apart from establishing environmental stewardship as a fundamental norm in GIS, the Stockholm conference's 'most visible and lasting achievement' (Bauer, 2013: 320) was the creation of UNEP. For the first time, environmental protection had been given its own institutional home within the family of UN organisations. But UNEP's purpose and capabilities were kept comparatively small. Rather than create a free-standing specialised agency or organisation, UNEP was designed as a programme under the UN's ECOSOC. Operating with a limited budget and located in Nairobi, far from the main UN sites in New York and Geneva, UNEP was unlikely to do more than seek to shape the international agenda and facilitate international rule-making in a supportive role (Ivanova, 2012). As international society's commitment to environmental stewardship hardened after 1972 and its ambition for global environmental protection expanded, the question soon arose as to what institutional framework states would build to support this global policy area. One option would have been to turn UNEP into a fully fledged international environmental organisation with greater policy-making authority and a dispute settlement mechanism to strengthen compliance with international environmental rules. This was the model that many environmentalists had demanded in the 1970s – the creation of global environmental governance along state-centric solidarist lines (see chapter 3). Proposals for a World Environment Organization have surfaced repeatedly since (e.g. Charnovitz, 2002; Whalley and Zissimos, 2002; Biermann, 2014) but have largely fallen on deaf ears. The path that GIS chose instead followed a more modest model of strengthening the institutional architecture of GEP through incremental UN reform and institutional innovation. The steps that have been taken in this regard provide further evidence of the social consolidation of environmental stewardship, though they also demonstrate states' unwillingness to radically reimagine and upgrade global environmental governance.

As the international environmental agenda expanded after Stockholm, institutional strengthening followed two principal paths. The first was centred on the environmental treaties that a growing number of states negotiated. MEAs were supported by their own organisational structures: annual meetings of the COP took on the role of chief decision-making body that further developed a treaty's rule book and soft law (Brunnée, 2002), while treaty secretariats provided the necessary support functions for the COP and for the implementation of MEAs (Jinnah, 2014). The second path involved the creation of new international organisations in and around the UN that complemented UNEP's role. Major UN environmental summits often galvanised states to embark on such international institution-building. The 1992 Rio Earth Summit thus led to the creation of the UN Commission on Sustainable Development (CSD) as a functional commission of ECOSOC, with a mandate to oversee the outcomes of the 1992 summit. The Rio Summit also saw the transformation of the World Bank's provisional Global Environment Facility (GEF) into a permanent institution to deliver international environmental aid in support of action on climate change, ozone layer depletion, biodiversity, waters, land degradation and POPs. Other MEAs with major aid commitments established their own financial instruments, such as the Multilateral Fund for the Implementation of the Montreal Protocol and the Green Climate Fund (GCF) to support the UNFCCC's mission (Streck, 2001; Bowman and Minas, 2019). One benefit of this proliferation of environmental organisations and treaty secretariats was that these environmental bureaucracies came to perform an ever larger set of functions, from distributing environmental knowledge to reporting and monitoring, initiating negotiations, developing soft law and capacity building, thereby building up layer upon layer of environmental governance (Biermann and Siebenhüner, 2009). At the same time, this strategy risked creating an overly complex institutional architecture, with a proliferation of weak and underfunded bodies that serve poorly coordinated and overlapping mandates.

Recognising the threat of fragmentation, GIS considered on various occasions whether and how to strengthen the institutional architecture for environmental protection. To date, incremental reform rather than radical overhaul has been the dominant response. After the disappointing outcomes of the 2002 World Summit on Sustainable Development (WSSD) and an unsuccessful proposal in 2005 to push for a UN Environment Organization that would replace UNEP, momentum was building in the late 2000s behind a major institutional reform debate.

The issue was taken up in the preparatory process for the 'Rio+20' summit in 2012, formally known as the UN Conference on Sustainable Development. Proponents of a strengthening of the UN's environmental role were arguing for a 'double upgrade' – turning UNEP into a UNEO specialised agency and the CSD into a Sustainable Development Council (Bauer, 2013: 329). In the end, resistance to such a profound institutional transformation proved too strong. Rio+20 ended with a compromise agreement to upgrade UNEP, mainly by providing it with more secure and stable financial resources and adopting universal membership for its governing council. This outcome was much closer to an earlier, more modest, proposal that had emerged from WSSD in 2002. Countries sceptical of the need for a strengthened international environmental authority (the United States, Canada, Japan, some G77 members) had thus prevailed over the alliance of more ambitious countries (mainly from Europe and Africa) that wanted to turn UNEP into a specialised UN agency. Rio+20 also decided to abolish rather than strengthen the CSD – long seen as ineffective and increasingly irrelevant – replacing it instead with a high-level intergovernmental forum on sustainable development (Ivanova, 2013). In many ways, the 2012 summit seemed to provide the best chance yet to empower the UN with environmental policy-making authority. Instead, it brought to an end the long-standing debate about the future of the state-centric solidarist vision for environmental governance. Incremental adjustment has since been the order of the day in GEP.

While the UN reform debate may have run out of steam, this has not stopped the creation of ever more transnational initiatives to advance environmental protection. If anything, it redirected attention to the potential for creating environmental governance mechanisms outside the states system. The 1992 Rio Summit provided an important first impetus for the emergence of transnational environmental governance initiatives. The Summit both acknowledged the importance of civil society engagement in delivering sustainable development solutions and managed to galvanise a record number of non-state actors to participate in the conference. Public–private partnerships (PPPs) became one of the early forms of governance innovation. Partnerships are deliberately designed to bring together state and non-state actors in network-type arrangements to combine their respective problem-solving capabilities in pursuit of global public goals (Andonova, 2017: 7–10; Pattberg et al., 2012: 3). PPPs became a particularly popular instrument for delivering the global sustainability agenda after Rio. They were actively promoted at the 2002 WSSD, and by the time of the 2012 Rio+20 Summit, a total of 349 partnerships working on environmental issues were registered by the UN CSD (Beisheim, 2012).

Since the beginning of the twenty-first century, a variety of non-state actors – NGOs, businesses and also subnational actors – have dramatically stepped up their engagement with GEP and also developed new and innovative forms of environmental governance. Such environmental governance experimentation (Hoffmann, 2011) was driven by a combination of supply and demand factors. On the supply side, the deepening commitment to environmental values in world society has pushed non-state actors to seek new solutions to environmental problems that are not dependent on formal state-level or international regulation. On the demand side, states and international organisations themselves have increasingly encouraged and invited non-state actors to take on environmental governance roles, whether in support of international treaties or to fill existing governance gaps. As discussed earlier, this has had a profound effect on the rules and standards governing the global economy, with private environmental authority (Green, 2013) on the rise across a wide range of sectors. Environmental self-regulation by business actors has become a particularly prominent feature of global climate politics, where environmental activism and public pressure have prompted ever more companies to take action for themselves (Bulkeley et al., 2014). A '"Cambrian explosion" of transnational institutions, standards, financing arrangements, and programs' has thus led to the emergence of a large-scale 'transnational regime complex' in the climate area (Abbott, 2012: 571). Other environmental areas (e.g. forests, fisheries, maritime shipping) have similarly witnessed a dramatic increase in the number and scope of transnational governance initiatives (Overdevest and Zeitlin, 2014; Gulbrandsen, 2010; Lister, Poulsen and Ponte, 2015).

Thus, even if state-centric solidarism may have reached a temporary peak in GEP, solidarist developments in world society have strengthened and accelerated. Transnational initiatives involving non-state actors are now one of the most dynamic areas of governance innovation and growth in GEP. Their significance lies in the contribution they can make to global capacities for environmental protection, but also in what they signal about the social consolidation of environmental stewardship. Environmental values and norms, originating in world society, have been adopted by international society. They have also broadened their appeal within world society as it has expanded and deepened in recent decades. Global environmentalism has thus gone through sustained social consolidation in both international and world society. In this sense, the rise of transnational environmental governance speaks to the important agency of non-state actors, acting as norm entrepreneurs within international society and as agents of political change outside it. At the same time, it is worth noting that many transnational governance initiatives have sprung up in a context that is heavily structured by state-centric international governance. From the UNFCCC to the SDGs, international treaties and targets often serve as focal points for non-state actor initiatives that seek to strengthen or implement internationally agreed policy goals. Furthermore, states and international organisations have increasingly sought to stimulate and orchestrate transnational governance by non-state actors, in an effort to extend the means for delivering global governance on complex issues (Andonova, 2017; Abbott et al., 2015a). GEP has, therefore, undergone a profound process of 're-scaling', whereby '[a]n international sphere dominated by interactions among nation-states has been replaced by one in which international organizations, substate governments, scientists, nongovernmental organizations (NGOs) and multinational corporations play major roles' (Andonova and Mitchell, 2010: 256). In this way, the rise of global environmentalism has turned into an important stimulus for the integration of international and world society, a topic I shall return to in chapter 10.

Normative Contestation in Global Environmental Politics

The discussion so far has focused on the question of whether the primary institution of environmental stewardship went through a process of social consolidation since the 1990s. It identified several empirical indicators that suggest a distinctive strengthening of environmentalism in both international and world society. This is not to suggest, however, that the journey towards a greener international system has been a smooth one. Far from it, as I shall discuss in this section, it has been challenged repeatedly over the last few decades. Global environmental politics remains an issue area characterised by deep international conflicts and normative contestation. If the above account appears to the reader to be infused by a teleological belief then this is far from the author's intention. There is no preordained path towards ever greater sustainability in international politics, in fact many global ecological trends, from species extinction to global warming and plastics pollution, suggest the contrary. In many important ways, the greening of international and world society remains incomplete and environmental values and norms continue to be resisted. This raises the question of what continued environmental norm contestation means for the validity of environmental stewardship as a primary institution. Does the simple fact of such contestation weaken the argument that environmentalism has emerged as part of international society's normative structure?

That international norms – whether fundamental norms (e.g. primary institutions) or issue-specific ones (e.g. norms contained within secondary institutions) – are dynamic and not static is now widely accepted in the IR norms literature. Norms have a distinctive life cycle that includes their emergence, acceptance and internationalisation (Finnemore and Sikkink, 1998) but also their potential decay (Panke and Petersohn, 2016). As norms develop and diffuse, they assume a greater regulative and constitutive role in shaping actors' identities and interactions. However, even established norms are always at risk of being challenged by states or other actors. While the early norms literature had a tendency to 'freeze' norms (Hoffmann, 2010), more recent theoretical work has moved latent and actual contestation of international norms centre stage in its analysis (Lantis, 2017). Both the content and the scope of an international norm can change as norm setters and followers engage in norm contestation. Critically, such contestation does not necessarily undermine the social validity of a norm, it may in fact strengthen its legitimacy (Wiener, 2004). What matters, therefore, is not whether norms are being challenged as such, but what aspects of norms (or norm complexes) are being contested. Contestation may challenge a given norm overall, or specific meanings that norm entrepreneurs attach to it, or the way normative principles are applied in historically bounded contexts (Wiener and Puetter, 2009). Depending on the mode of norm contestation, it can either lead to the demise of a norm or it may help entrench its legitimacy. Where it concerns the appropriateness of a norm for specific situations and determining required actions, it often clarifies and thereby strengthens a norm; where it is focused on the justification or righteousness of a norm's core, it is more likely to lead to its weakening (Deitelhoff and Zimmermann, 2020).

The question that arises in the context of this chapter is thus two-fold: whether international contestation of environmental stewardship has focused on the primary institution's normative core or its specific meaning in given historical contexts; and whether existing modes of contestation have strengthened or weakened its status as an element of GIS's normative order. As I shall argue subsequently, global environmentalism has come up against repeated, and occasionally fundamental, opposition. After Stockholm and especially in the wake of the Rio Summit in 1992, however, most international contestation has concerned the meaning of environmental stewardship and its application to specific problems. The ideological battle between environmentalists and anti-environmental forces has become less pronounced at the international level. Instead, it now plays out mostly within, rather than between, individual societies. As I argue in this section, ongoing norm contestation does not call into question the long-term social consolidation of environmentalism as a primary institution.

The Shifting Ground of Anti-Environmentalism

The world may have entered the 'age of ecology' (Radkau, 2011), but environmentalism's core values and norms continue to be questioned and challenged. There is nothing new or surprising in this. Anti-environmentalism, in the sense of a fundamental and often organised opposition to environmental ideas, values and politics, is as old as the environmental movement itself. Rachel Carson's book *Silent Spring* (1962), widely viewed as the foundational text of modern environmentalism in North America, provoked an immediate and hostile response from representatives of the chemical industry and science. In one of the first responses entitled 'Silence, Miss Carson!', published in *Chemical and Engineering News*, the reviewer lumps Carson together with other anti-modern and anti-rationalist forces in society and warns the reader that her perspective would lead to 'the end of all human progress' (quoted in Smith, 2001: 738). Ever since, renewed warnings about novel environmental threats, whether toxic chemicals, acid rain or genetically modified organisms, have similarly encountered stiff and often shrill resistance. The discovery of a link between anthropogenic chlorofluorocarbons (CFCs) and ozone layer depletion in 1974 initially led to a brief 'ozone war' (Dotto and Schiff, 1978) in the United States, involving direct attacks from industry aimed at the scientists that had identified the environmental threat. Of all environmental issues, climate change has arguably attracted the strongest anti-environmental backlash in recent years. Drawing on financial and organisational support from major corporations, conservative think tanks and wealthy individuals, critics of climate policy have sown doubt about the underlying science of global warming and attacked what they regard as intolerable regulatory restrictions on economic liberty and growth (Weart, 2011; Oreskes and Conway, 2011).

Anti-environmentalism has thus been a constant feature of environmental debates from its origins until today, but the transformation that it has undergone in the last fifty years supports the argument that environmental stewardship has strengthened in global politics. In the early phase of modern environmentalism, it was still common for critics to dismiss environmental concerns out of hand and to treat environmental protection as incompatible with a free and prosperous economy. The 1970s debate on the limits to growth argument, for example, pitted neo-Malthusians against cornucopians, with the former predicting ecological catastrophe if the world continued on its current path of unlimited economic growth and the latter expecting technological progress and the market to overcome any temporary resource scarcities (Robertson, 2012: 180–1; Beckerman, 1995). Since the 1990s, with sustainable development now serving as a commonly accepted reference point, the debate has narrowed to questions of the appropriateness and affordability of environmental policy measures. Anti-environmentalists carry on to attack some of the core premises of environmental policy, at both the domestic and international levels, but on the whole fundamentalist opposition has given way to critical engagement with the environmental agenda. In a kind of role reversal, anti-environmentalists now portray themselves as David fighting Goliath. The new anti-environmentalists that came to the fore since the 1990s routinely describe themselves as 'contrarian' (Ridley, 1995) or 'skeptical' (Lomborg, 2003), arguing against what they see as a globally dominant norm. They still draw on ideological resources that are deeply critical of environmentalist beliefs, particularly libertarian beliefs in economic freedom and the market and opposition to an overbearing regulatory state (White, Rudy and Wilbert, 2007). However, the focus of contention has moved to a narrower range of policy questions that accept the need for environmental protection but challenge specific policy designs or the costs of regulation. In a sense, the shifting ground of anti-environmentalism thus confirms just how successfully environmentalism has come to establish itself in global politics.

If anti-environmentalism continues to challenge many key premises of the environmental movement today, the main divisions in the environmental debate now run through, rather than between, societies. Early on, at the time of the 1972 Stockholm conference, international society was still deeply divided over whether to accept a general responsibility for the global environment. While a small group of leading industrialised countries were pushing for the creation of an international environmental agenda, many developing countries and members of the Soviet bloc oscillated between scepticism and outright opposition (see Chapter 5). By the time of the 1992 Rio conference, a broad international consensus had formed on the need to integrate sustainability into economic development. In a sign of the global acceptance of environmental stewardship, many MEAs managed to reach nearly universal membership, and even where individual states objected to specific regulatory proposals they broadly accepted the legitimacy of international environmental policy-making. In the case of climate change, which has gradually moved centre stage in GEP, both the 1992 UNFCCC and the 2015 Paris Agreement attracted the support of all major greenhouse gas emitters, whatever their remaining reservations about specific policy instruments.

Thus, while fundamental contestation over environmentalism carries on, it does so mainly at the domestic level within states and in transnational spaces across national boundaries, but less so within GIS.

Contestation over Specific Environmental Principles

The intellectual and political foundations for outright anti-environmentalism may have weakened, but this does not mean that contestation has disappeared from GEP. Far from it, states continue to clash over key environmental policy objectives and principles, and many fail to live up even to the policies and targets they have agreed internationally. The perilous state of many global ecosystems speaks to the weakness of existing approaches to global environmental protection, and international conflicts over environmental policy continue to hold back the implementation of existing, and creation of new, international commitments. What has changed in GEP, however, is the nature and focus of contestation – away from fundamental objections to the idea of environmental stewardship as such and towards competing interpretations of its meaning and application in specific contexts. Environmental stewardship, just like other primary institutions, needs to be translated into more specific principles and approaches that can guide policy-making and resolve conflicts and tensions with other policy objectives. It can lead to stronger or weaker, radical or reformist, ideas about how to protect ecosystems and species against the demands of economic development and growth, and about how to distribute the responsibility and financial burden of global environmental protection. In this section, I review normative contestation in GEP over some of the most prominent and controversial principles and norms: the polluter pays principle (PPP), the precautionary principle, the common heritage of humankind idea, conservationism, and distributional justice and equity norms.

The PPP has become an integral part of domestic environmental regulation around the world. First introduced in a 1972 OECD document outlining *Guiding Principles Concerning the International Economic Aspects of Environmental Policies*, it states that polluters should bear the cost of the environmental damage they cause, including clean-up and remediation. By forcing polluters to internalise environmental costs, the PPP has become the basis for a range of market-friendly approaches that use the price mechanism to reduce pollution (e.g. emissions trading systems). It therefore plays a key role in aligning environmentalism with the market norm (Bernstein, 2013: 133) and informs leading industrialised countries' domestic regulations (Larson, 2005).

Translating it to the international level has proved more problematic, however. Although the 1992 Rio Declaration formally acknowledged the PPP, the wording of its Principle 16 ('the polluter should, in principle, bear the costs of pollution') is comparatively weak and ambiguous. In fact, the PPP does not feature widely in international environmental law, and where it exists it is generally couched in soft language, reflecting strong international opposition against the principle's application to interstate relations (Sands and Peel, 2012: 229). Apart from being reflected in a few treaties dealing with civil liability for damage from hazardous activities, the PPP has not gained wider significance in GEP. Much like the 'no harm' principle that has its roots in nineteenth-century international law (see Chapter 3), the PPP has failed to develop into an effective regulative norm in interstate relations. One important reason for this is that industrialised countries themselves have resisted its internationalisation. Despite having championed it domestically and within the OECD, they have long harboured concerns that it might give rise to developing country claims for international environmental compensation.

The question of historical responsibility for transboundary environmental damage has taken on particular political significance in the climate negotiations. The Global South and many environmental NGOs have persistently argued that those countries that have historically contributed most to the global warming problem should provide financial compensation to those that have a low emissions profile but are likely to be the main victims of rising temperatures. In fact, the question of historical responsibility and compensation has proved to be one of the thorniest issues in the over two decades-long climate negotiations. Only in 2013 did developing countries and their allies in global civil society succeed in establishing the Warsaw International Mechanism for Loss and Damage associated with Climate Change Impacts (Vanhala and Hestbaek, 2016: 112). However, industrialised countries resisted any moves to establish a formal commitment to compensation payments based on their historical responsibility. At their insistence, the decision adopting the 2015 Paris Agreement incorporated a provision that states unambiguously that the climate treaty 'does not involve or provide a basis for any liability or compensation' (United Nations Framework Convention on Climate Change, 2016: para 52).

The *precautionary principle* is a relatively late addition to the growing complex of international environmental norms. It was first explicitly applied in international law in the mid-1980s (e.g. 1985 Vienna Convention for the Protection of the Ozone Layer, 1985; Ministerial Declaration of the International Conference on the Protection of the North Sea, 1984) and was formally acknowledged in the 1992 Rio

Declaration. Principle 15 refers to the 'precautionary approach' and states: 'Where there are threats of serious or irreversible damage, lack of full scientific certainty shall not be used as a reason for postponing cost-effective measures to prevent environmental degradation' (Sands and Peel, 2012: 218–19). Environmentalists consider precaution an important principle as many environmental threats are long-term and shrouded in scientific uncertainty. Applying a 'better safe than sorry' approach thus legitimises taking early and drastic action rather than waiting until full scientific proof of harm has been established, particularly where an environmental threat can cause serious and/or irreversible damage.

Much like the PPP, the precautionary principle has found widespread application in domestic contexts, especially in Europe (Tosun, 2013), but remains contested internationally. During the negotiations on the Cartagena Protocol on Biosafety, for example, a transatlantic conflict emerged over the extent to which the treaty should allow precautionary restrictions on international trade in GMOs. While the European Union came out in support of applying the precautionary principle in the biosafety regime, not least because it mirrored its own domestic approach to GMO regulation (Falkner, 2007), the United States and other leading agricultural exporter nations strongly objected to it on the grounds that it clashed with the WTO's demands for science-based decision-making (Isaac and Kerr, 2003). References to precaution have since been included in other international treaties (e.g. Stockholm Convention on POPs, 2001), but international contention over its use in MEAs and trade-related environmental measures has not abated. Most international lawyers acknowledge that the precautionary principle has become a norm of customary international law, but normative contention has limited the impact it has had on the development of international environmental law (Wiener, 2007). If anything, normative conflicts around precaution are set to intensify in future years as novel but risky technologies are introduced to deal with environmental problems, as is the case with proposal for geo-engineering to combat global warming (Talberg et al., 2018).

The concept of *common heritage of humankind*, one of the guiding principles of global environmentalism at the time of the 1972 Stockholm conference (see chapter 3), has similarly failed to harden into a strong international environmental norm. Introduced in the context of the Law of the Sea negotiations and in the regimes to protect outer space and Antarctica (Wolfrum, 1983), it has also informed proposals to internationalise the protection of endangered species and ecosystems. Although expressing a core element of environmental stewardship, namely that states have a collective duty to preserve the global environment for future generations, it has come up against strong opposition by states that resist its application to domestic environmental resources (e.g. forests, fossil fuels). The principle thus points to a fundamental tension between the primary institutions of environmental stewardship and sovereignty. Whereas the former points to the need for a truly collective response to global environmental threats, the latter protects the nation-state and its exclusive authority over the domestic environment. The conflict over the 'common heritage' idea thus reflects not only competing national interests but a deeper tension over how to balance national and global environmental responsibilities.

Forest protection is one prominent area where this conflict came out into the open and prevented the creation of an international environmental regime. The destruction of tropical rainforests had become a global concern by the mid-1980s. Scientists pointed to the global significance of forests as a source of biodiversity and as a carbon sink, and environmentalists and governments in the North called for an international forest treaty to protect what was increasingly being framed as a global ecological asset – the 'common heritage of humankind' (Chasek, Downie and Brown, 2017: 217). Proposals for a legally binding instrument to protect forests were introduced in the preparatory process for the 1992 UNCED but were soon dropped due to irreconcilable differences over whether forests should be defined as a global responsibility (Dimitrov, 2005: 7). Subsequent efforts to establish alternative mechanisms for international forest governance didn't get very far, they merely led to the creation of a series of UN policy forums without decision-making authority (Humphreys, 2013: 80). Since the creation of the Forest Stewardship Council in 1993, a transnational initiative by NGOs and business, policymakers' attention was also beginning to shift to non-state market-based certification schemes that promote sustainable forestry practices and target international timber trade flows (Gulbrandsen, 2010).

The deadlock over whether to internationalise forest protection based on common heritage ideas has never been broken. Developing countries' resolve to defend their sovereign rights over tropical forests never weakened, particularly as their demands for substantial international compensation for forest protection measures have not been met. Only in recent years has the linking of deforestation with the global climate regime brought some movement into this policy area. Proposals to compensate developing countries for the 'opportunity costs forgone' if they prevent deforestation has informed G77 proposals for international financing instruments such as REDD+ (Humphreys, 2013: 81). The hope is that such linkage politics can help recast the ecology–sovereignty conflict into a scenario where developing countries retain sovereign control over forests but are compensated for preventing their destruction as providers of global environmental services (Carvalho, 2012).

Nature conservation, and in particular the preservation of endangered species, has similarly pitted Northern and Southern countries against each other, with the latter often resisting efforts to internationalise animal protection efforts. In the case of elephant poaching for ivory, which caused a significant decline in elephant populations during the 1970s and 1980s (Stiles, 2004), environmental campaigners pushed for a more hard-line approach to enforce nature conservation principles, rejecting notions of the sustainable use of animal species. This campaign paid off by the late 1980s, when CITES introduced a complete ban on international ivory trade in 1989 to reduce the demand for the precious commodity. The ban, which had been proposed and supported by a broad coalition of Northern governments and NGOs, was contested by some developing countries that argued for continued but controlled ivory trade. This was not simply a conflict over the sovereign control of national resources – all CITES members have come to support the idea of a global regime for species protection – but involved normative contestation over whether preservation or conservation principles should guide the protection of endangered species (see chapter 3 on the preservation–conservation divide). Northern countries favoured the preservation of elephant populations by banning their commercial use, while some African countries argued for a conservation approach that allowed them to derive commercial benefits from ivory as a renewable resource (Duffy, 2013). Thus, norm contestation in this case did not challenge environmental stewardship and the idea of global governance per se, it was focused on the particular meaning that individual countries attached to it and how they proposed to balance environmentalism with developmentalism.

The question of how to protect endangered whale species provoked a similar conflict between preservationist and conservationist perspectives, although in this case the lines of conflict ran within the group of industrialised countries. Following a NGO campaign against whaling that started in the 1970s and developed global momentum in the 1980s, the International Whaling Commission underwent a gradual normative shift, away from its original support for the whaling industry that allowed but limited global catches and towards the banning of commercial whaling. As one whaling nation after another started to succumb to the naming and shaming tactics of environmentalists, the IWC eventually enacted a global whaling ban in 1986. By this time, only a few countries were still carrying on with limited whale catches in defiance of a global NGO campaign that had given rise to an international prohibition norm against whaling (Bailey, 2008). Japan, despite being recognised as an international leader on other environmental issues (Maddock, 1994; Schreurs, 2004), continues to contest the global whaling ban within the IWC. The country has consistently argued for a limited form of whale hunting not because it rejects the objective of species protection but because it believes that it can be achieved while allowing whaling operations that are strictly regulated and limited to those species no longer under threat of extinction. Similar to African arguments for regulated ivory trade, Japan favours a conservationist approach that allows 'sustainable whaling' as an alternative to the global whaling ban (Scott and Oriana, 2017).

The conflict over whaling is often portrayed as involving diverging national interests, but at its heart are different versions of environmentalism, preservationism versus conservationism, which divide not just international society but also world society (Epstein, 2006; Epstein and Barclay, 2013). Just as the anti-whaling lobby has built a transnational coalition of NGOs and states that have targeted the IWC, so have pro-whaling interests formed a transnational network that links Japanese whalers with aboriginal communities in other parts of the world that defend whaling on cultural grounds (Bailey, 2008: 292). The struggle over how to protect nature, be it endangered elephants or whales or entire ecosystems, thus reflects a more fundamental divide within global environmentalism, and this divide runs between states as well as between different parts of world society (on divisions within world society, see the discussion in chapter 10).

Questions of *distributional justice and equity* have provoked some of the most heated debates in GEP. These debates have traditionally pitted countries in the Global North against those in the Global South, although the rise of emerging economies in the early twenty-first century has complicated and blurred this line of conflict. In the run-up to the 1972 Stockholm conference, it was still common for developing countries to question the framing of environmental protection as a collective international responsibility, mainly on the grounds that economic development ought to take precedence (see chapter 5). As more and more countries came to accept environmental stewardship after Stockholm, the North–South conflict came to focus more narrowly on how to distribute the responsibility for global environmental protection (Najam, 2005). Global justice and equity thus became the main focal point in developing countries' contestation of global environmentalism (Okereke, 2008). Once sustainable development had been accepted as the basis for international environmental action, North–South struggles in international negotiations focused increasingly on questions of burden sharing, differential treaty commitments and international environmental aid.

In this regard, the inclusion of the principle of differential treatment in the Rio Declaration and several key MEAs was a major success for the developing world. Principle 7 of the Declaration stated unambiguously that '[i]n view of the different contributions to global environmental degradation, States have common but differentiated responsibilities' (CBDR). Together with the inclusion of the similarly phrased principle of 'common but differentiated responsibilities and respective capabilities' (CBDR-RC) in the UNFCCC (Article 3.1), this marked a high point in developing countries' fight for strong equity norms in GEP. Significantly, the CBDR norm was operationalised and further strengthened in the 1997 Kyoto Protocol, which excluded developing countries from legally binding climate mitigation measures. However, this proved to be a short-lived victory. The G77 countries may have 'managed to secure a good deal in 1992 when they were materially a good deal weaker' as Hurrell and Sengupta (2012: 467) point out, but these achievements have since come under attack from some industrialised countries. Their contestation of strong equity norms succeeded in reversing some of the gains made at UNCED, with the climate change regime, in particular, witnessing a 'retreat from [...] differential treatment in central obligations' (Rajamani, 2012: 622).

The United States led the international pushback against strong differentiation in the climate regime and other environmental issue areas. When US President George W. Bush withdrew the country's signature from the Kyoto Protocol in 2001, he did so not least because of longstanding opposition to the rigid distinction between Annex I (developed) and non-Annex I (developing) countries, which exempted the latter from legally binding emission reductions (Bang et al., 2007). Other industrialised countries (e.g. Japan, Canada) gradually swung behind the United States in its opposition to Kyoto-style differentiation, and by the time of the 2009 Copenhagen conference virtually all Northern countries had demanded a more balanced distribution of the climate mitigation burden that would include the rapidly rising emitters in the Global South (Hurrell and Sengupta, 2012: 471–2; Falkner, Stephan and Vogler, 2010). Northern contestation of strict differentiation has thus been helped by the accelerating shift in the global distribution of GHG emissions, with China becoming the world's largest emitter in 2006 and other emerging economies (India, Brazil) rising fast in the GHG emissions league table. Faced with a dramatically expanding ecological footprint, emerging powers have come under pressure, including from other

G77 countries, to redefine their position vis-à-vis differentiation and accept greater environmental responsibility. The first signs of this shift were clearly visible at the 2009 Copenhagen conference, when five emerging economies formed the new negotiation group of the BASIC countries and made the first promise to accept some limitation on their rising emissions (Hochstetler and Milkoreit, 2015). By the time of the 2015 Paris climate conference, the success of Northern contestation of strict differentiation was on full display. Based on the model of the 2009 Copenhagen Accord, the Paris Agreement abandoned Kyoto-style differentiation and instead adopted a much weaker form of voluntary climate mitigation pledges, with countries determining their own level of ambition. Most references to differentiation in the agreement suggest a more ambiguous approach – they are either recommendations or create expectations rather than binding obligations (Rajamani, 2016: 510). With the CBDR norm thus diluted in the Paris Agreement and compensation claims based on historical responsibility rejected, distributive justice claims had finally been relegated to the margins of the international climate agenda (Falkner, 2019).

Arguably the strongest form of contestation in GEP – and possibly the toughest test for the environmental stewardship norm – has arisen in the context of the recent rise of anti-environmental populism and authoritarianism in several major powers, most notably in the United States but also in other countries (e.g. Brazil, Russia). The election of US President Trump in 2016 is but the most prominent example of this trend. Although the United States has a long history of environmental exceptionalism – it has failed to ratify most new MEAs since the early 1990s (Brunnée, 2004) – Trump's nationalist protectionism and anti-regulatory rollback has taken US unilateralism to a new level. His decision to withdraw from the Paris Agreement, despite the fact that the treaty reflects many core US demands (Milkoreit, 2019), merely reinforces a growing international perception that the US challenge to the climate regime signals broader discontent with environmental multilateralism and global environmental responsibilities. Similar resistance against the Paris Agreement has surfaced in Brazil, where President Jair Bolsonaro embarked on his own programme of rolling back domestic environmental regulations, and the fear persists that it will lead to a domino effect of wavering countries withdrawing their support for the climate regime and other MEAs. However, the strong international opprobrium that Trump and Bolsonaro have received suggests that the environmental multilateralism norm has so far withstood America's challenge. Furthermore, as can be seen from the strength of domestic US support for the Paris climate accord (Bomberg, 2017; Gies, 2017), commitment to environmental stewardship runs much deeper in American society than the recent turn towards populist politics would suggest. The anti-environmental backlash, whether in the United States or elsewhere, has certainly shown the fragility of the global consensus on environmentalism. However, the domestic divisions over environmental policy have yet to spill over into serious environmental contestation *between* states over the future of global environmental protection. The status of environmental stewardship as a fundamental norm in international society remains, as yet, secure.

Conclusions

Once environmental stewardship had emerged as a fundamental international norm at the 1972 Stockholm conference, it underwent a gradual process of globalisation. By the early 1990s, it had grown into a universally accepted primary institution of *global* international society. For many diplomats and analysts, the 1992 Rio Earth Summit represented the high point of global environmentalism, the 'zenith of commitment on environment and development' as Malaysia's Ambassador Razali Ismail, the President of the UN General Assembly in 1997, put it (quoted in MacDonald, 1998: 3). Coming just after the dissolution of the Soviet Union and the end of the Cold War, UNCED encapsulated the solidarist hopes for a new international order, one marked by deeper interstate cooperation and non-state actor participation in global governance. Yet, the subsequent path towards implementing this global environmental agenda was far from smooth. As I have argued in this chapter, environmental stewardship after Rio experienced further consolidation but also renewed contestation in international relations. The international commitment to the environmental norm strengthened at least in three ways: international policy-making expanded to ever more global environmental issues; environmental principles and objectives were integrated into non-environmental secondary institutions (so-called mainstreaming), especially in the economic field; and existing international institutions were (partially) strengthened while new transnational governance initiatives emerged. All three trends suggest continued normative entrenchment in GIS.

At the same time, just as GEP became more ambitious, so did contestation of specific environmental principles and norms increase, for example with regard to the PPP, precaution, common heritage and international equity. Despite gaining universal acceptance, environmental stewardship is still an evolving norm, with competing understandings of its core meaning vying for influence in international politics. Many normative fault-lines continue to run between developed and developing countries, as they did in the early 1970s. Yet, shifts in the global distribution of economic strength and power have complicated the picture, and new normative conflicts within the Global North and Global South have come to the fore. The question of how to deliver on GIS's commitment to environmental stewardship is anything but resolved. Importantly, however, ongoing contestation has not significantly weakened this commitment. Fundamental contestation of environmentalism is now mostly found within rather than between states.

As discussed in this chapter, much of the international contestation around environmental stewardship concerns its relationship with other primary institutions, many of which predate the arrival of modern environmentalism and are central to the functioning of GIS. This is most clearly the case with sovereignty and territoriality. Many proposals for global environmental governance call into question classical understandings of these Westphalian norms. Moves towards internationalising the policy response to some environmental problems (e.g. deforestation) have been resisted by states keen to protect their national sovereignty, while many states are generally unwilling to support deeper levels of international institution-building (e.g. climate change). I will discuss the limits to the greening of sovereignty in more detail in the next chapter.

In a similar vein, environmental stewardship has come into conflict with two of the newer primary institutions, the market and development. Where transboundary forms of environmental degradation require international restrictions on the free movement of goods and services and economic growth, existing international economic regimes (e.g. WTO) still limit the scope for such environmental action, despite adopting some environmental mandates. Likewise, most states and international organisations remain wedded to a developmentalist logic in their economic policy-making, which curtails more radical proposals for setting ecological boundaries for future economic growth. At the same time, mainstreaming of environmentalism into economic regimes has produced some normative shifts in the interpretation of market and development norms, with sustainable development serving as a widely accepted leitmotif. By implication, environmental stewardship has itself gone through normative adaptation in order to become established within the dominant political–economic order. This has led to a marked preference for market-friendly instruments within GEP and a general reluctance to fundamentally challenge the global economic growth imperative. The 'compromise of liberal environmentalism' (Bernstein, 2001) that emerged in the 1990s still stands in sharp contrast to radical ecologists' demands for a global economic restructuring (Newell, 2020).

When it comes to other primary institutions of the international normative order, environmental stewardship offers a much smoother fit. This is the case with diplomacy and international law, as well as with human rights. The rise of GEP has opened up a new agenda for international diplomacy and has generally led to a vast expansion of multilateral forums and diplomatic interactions focused on environmental matters. Supporting international environmental processes is now a firmly established goal in most states' foreign policy machinery. International law has similarly incorporated environmental principles and norms, with environmental law becoming one of the major growth areas in international legal development. However, normative contestation around some of the more progressive legal principles (e.g. precaution) continues to slow down the strengthening of international environmental law (discussed in more detail in chapter 8).

Finally, international society remains divided over the degree to which environmental stewardship gives rise to a global justice agenda involving strict differentiation of responsibilities and international redistribution of wealth. Initial advances in establishing strong equity principles within GEP have been challenged more recently, and to some extent reversed. With normative contestation focusing on questions of the historical responsibility of established industrialised countries versus new responsibilities of emerging economies, North–South differentiation has been weakened, especially in the climate regime. As more and more emerging powers are faced with demands for shouldering a greater burden for global environmental protection, environmental stewardship has moved closer to engaging existing notions of great power responsibility alongside the push for strengthening global governance. As Shunji Cui and Buzan (2016) argue, the rise of the non-traditional security agenda has already brought about some convergence between the agenda and actors of global governance and great power management. Whether increasing environmental securitisation will further drive this merging of agendas remains unclear at this point. I explore this question in more detail in Chapter 9.

Part III

Analytical Perspectives

8 Solidarist Ambition

In this chapter and the next two (Chapters 9 and 10), the focus shifts from a mostly historical account of the rise of environmental stewardship towards an analytical inquiry into the underlying logics that drive, but also restrict, the greening of global international society. I am interested in understanding how far we have come in transforming the normative structure of international relations and embedding within it a strong, if not dominant, environmental dimension. I engage with this question by employing the theoretical framework of the ES as laid out in Chapter 2 and its application to global environmental politics (see Chapter 3). In this chapter, I explore how a solidarist logic of cooperation has pushed states to expand the scope for global environmental action via international institution-building and international law. I am interested in understanding the extent to which this emerging system of global environmental governance has brought about a gradual greening of the sovereignty norm – a key marker of environmental solidarisation. Chapter 9 will explore whether a pluralist logic of coexistence alone can form the basis for effective environmental action within a Green Westphalian order, and what role, if any, great power management and responsibility might play as part of global environmental protection. Chapter 10 broadens the perspective to consider the contribution that non-state actors have made to the growth in global environmental action, the normative tensions that exist between eco-globalist and eco-localist elements of world society and the degree to which global environmentalism has promoted an integration between international and world society.

The conceptual dyad of solidarism and pluralism captures the two principal normative positions in the ES's 'great conversation'. It also provides a conceptual lens for a more analytical assessment of the development of international society (Jackson, 2000; Mayall, 2000; Hurrell, 2007) or specific international institutions (Buzan, 2014). It is in the latter sense that I use these two concepts, as analytical ideal types (Weinert, 2011) that provide a yardstick for evaluating normative development in international relations. I am interested in understanding how far GIS has travelled down the path towards creating global environmental governance based on solidarist principles, and why this development may still be incomplete. This analytical use of solidarism and pluralism does not, however, exclude normative concerns altogether. Indeed, rather than separating out analytical and normative inquiry, the ES has made a virtue of keeping both in play. As Reus-Smit (2002: 501) observes, ES scholars (unlike some constructivists) have not shied away from 'systematic reflection on the nature of international social and political life, what constitutes ethical conduct in such a world, and how this might be realised'. It is in this spirit that an ES approach allows us to explore what kind of global environmental action is possible as well as desirable, and how much further the international normative order needs to change if we are to advance a green agenda in international relations.

In discussing the solidarist perspective on GEP, I follow the common distinction in the ES literature between state-centric solidarism, which is focused on shared values and norms and the logic of cooperation in interstate relations, and cosmopolitan solidarism, which speaks to the notion of the 'great society of humankind' based on universal rights (Buzan, 2014: 114–16). Because this distinction has not always been made explicit in the ES literature, it has led to some confusion over whether the solidarist project in international society necessarily involves a transcendence of the state system as envisioned by cosmopolitan thinkers. To avoid this confusion, this chapter focuses on state-centric solidarism in international society, while the cosmopolitan dimension of solidarism will be explored as part of the discussion of world society in Chapter 10.

A solidarist international society is generally characterised by a deeper level of international cooperation, with states sharing values that go beyond a narrow concern for survival and coexistence. It seeks to realise these shared values by promoting international institution-building, stronger international law and a move towards global rule-making. Solidarism may rely on some degree of enforcement of international norms, but it should be noted that the extent to which coercion, calculation or belief are the foundation for compliance with international norms remains an open question (Hurrell, 2007: 58–9; Buzan, 2004: 152–7). Based on this understanding of state-centric solidarism, it is clear why environmentalism has been widely portrayed in the ES literature as a solidarist value (Reus-Smit, 2002: 499; Wheeler and Dunne, 1996: 106; Diez, 2017; Jackson, 2000: 176–7; Buzan, 2004: 150; Hurrell, 2007: 224–8). Much like other global issues such as human rights, distributive justice and arms control, GEP is closely linked to normative and political projects that seek to create an international order based on universal values, deep international cooperation and strong international institutions. As discussed in Chapter 3, state-centric solidarism envisages the creation of a system of global environmental governance, which accepts the reality of a world divided into sovereign nation-states but seeks to redefine what it means to be a legitimate member of international society. In this view, sovereignty is contingent on the state and international society accepting a normative commitment to environmental stewardship and acting in the interest of planetary health.

Much of the history of GEP since the 1972 Stockholm conference can indeed be interpreted as a series of efforts to build a green international order based on solidarist principles. As environmental stewardship became universally accepted (see chapters 6 and 7), global environmental policy-making expanded to include ever more environmental issues. States negotiated a myriad of international environmental treaties and established specialist international organisations for political coordination, scientific information exchange and the distribution of environmental aid. Environmental objectives have been integrated into the mandates of non-environmental international organisations, and on numerous occasions GIS reaffirmed its commitment to work together to tackle some of the biggest environmental threats. In a sense, the story of GEP fits neatly with other trends – the rise of humanitarian intervention, the deepening of economic globalisation and the expansion of international law – that suggest a solidarisation in international relations, especially after the end of the Cold War (Hurrell, 2007: 58–65; Wheeler, 2003).

Yet, the demand for stronger global environmental governance has only partially been met, and attempts to expand the scope for international environmental action have come up against considerable resistance in recent years. In some cases, the very idea of a solidarist solution to global environmental problems – based on internationally agreed policies and targets that are enshrined in international law and supported by strong institutions – has been rejected in favour of a more decentralised, voluntary, response by individual states. As I have argued in chapter 7, the deepening consensus behind the environmental stewardship norm has not lessened international conflicts over the question of how to promote global environmental protection.

Pursuing a solidarist environmental solution thus raises a number of important questions that are at the heart of this chapter. To what extent has the consolidation of GIS's commitment to environmental stewardship gone hand in hand with a strengthening of the solidarist agenda in international relations? Has the rise of global environmentalism served to redefine the meaning of sovereignty in an age of global ecological interdependencies? Has it promoted a deeper sense of common values and interests that transcend national boundaries, thereby helping GIS to evolve from a 'thin' to a 'thick' sense of community and common purpose? I seek to answer these questions first by exploring the extent to which environmental policy-making has become internationalised and institutionalised, while the second section examines the role that international law has played in driving the global environmental agenda forward. Both serve as a measure of GIS's commitment to collective action and common values, and the extent to which environmentalism has succeeded in transforming core foundations of international society, most notably the sovereignty norm.

Internationalising the Green State: The Expansion of Global Environmental Rule-Making and Institution-Building

The expansion of GEP after Stockholm has benefitted from, but also contributed to, the post-1945 boom in global governance. Indeed, the story of global environmentalism's rise is closely interwoven with the broader solidarisation trend that has led to an expansion of global policy-making and the creation of strong international institutions after the Second World War (Hurrell, 2007: 59–60). In order to deal with the multiple forms of complex interdependence – economic, political and social, but also ecological – that globalisation and increasing levels of interconnectedness produced, international society began to reconfigure political authority at the international level. This involved a certain degree of concentrating authority at the highest level (e.g. UN's collective security system), but GIS also began to diffuse governance functions 'downwards and outwards' (Buzan 2014: 147), away from great powers and towards secondary institutions and hybrid governance mechanisms. This trend has been particularly pronounced in the field of environmental protection, where the complexity of tackling environmental degradation requires new levels of cooperation not just between states, at the intergovernmental level, but also between a wide range of state and non-state actors that are operating at different scales, all the way from international organisations down to subnational and local actors. The move towards global environmental governance has involved increasingly intrusive international rules and norms that affect domestic politics as much as interstate relations, leading to profound adjustments in the way domestic social and economic actors behave (Hurrell, 2007: 220–1). In this sense, state-centric solidarism represents both an increase in the moral ambition of GIS and a functional response to the ever more complex coordination and adjustment needs of an interdependent world.

From the Green State to Green International Society

Given the pressures that the ecological challenge creates for domestic societies and the states system, there are good reasons to expect this functional need to translate into a continuous greening of not only the state but also international society. As is widely recognised, addressing environmental problems at the domestic level increasingly requires concomitant advances in global environmental protection. This intricate domestic–international nexus in environmental politics is at the heart of the literature on the 'green state', 'ecostate' or 'environmental state' (Dryzek et al., 2003; Eckersley, 2004a; Barry and Eckersley, 2005; Meadowcroft, 2012; Bäckstrand and Kronsell, 2015; Duit, Feindt and Meadowcroft, 2016), which combines a normative argument about the need for a deep greening of the state with the empirical observation that this transformation is already underway.

At the domestic level, the contours of the emergence of the 'environmental state' are well established. The rise of modern environmentalism in the 1960s helped establish the creation and expansion of the environmental policy domain, with environmental policy constantly extending its 'social reach'. The state's engagement with the environmental agenda has deepened over time in that environmental policy has become more systematic and focused on broader structural transformations. The state has become more preoccupied with environmental steering at the domestic level and environmental negotiations in international contexts, thereby increasing its footprint in environmental politics. And with the environment turning into a 'permanent area of political contention', environmental issues have become a regular fixture in daily politics (Meadowcroft, 2012: 65–7). When applied to interstate relations, green state arguments tend to be more tentative and qualified but nevertheless point to a substantial international greening process. Robyn Eckersley (2004a: 203) notes that global environmental discourses have 'begun to shift shared understanding of legitimate state conduct in a greener direction', while James Meadowcroft (2012: 82) similarly detects a shift in the basis for international legitimacy: 'International environmental policymaking [...] provides support for the idea that environmental protection is an essential part of what it means to be a modern state'. Importantly, international actors have also become a source of demand for domestic environmental action, thereby reinforcing the initial political impetus that moves from domestic to global environmental debates (Meadowcroft, 2012: 82).

The green state literature and ES solidarist arguments overlap and reinforce each other to a large extent, not least because they interpret the rise of global environmentalism as more than just the arrival of a new policy domain in a crowded marketplace of political issues. Their argument speaks to the transformative potential of environmental ideas, for the nation-state as much as for the international normative order. They share an interest in understanding how the sovereignty norm, the cornerstone of the Westphalian system and a major barrier to deeper international cooperation, is affected by environmentalism. Both argue – contra the pluralist perspective – that sovereignty is a historically contingent norm (Eckersley, 2004a; Wheeler, 2003), and that the history of sovereign statehood is marked by several profound shifts in how sovereignty was interpreted. Originally closely aligned with the primary institutions of territoriality and dynastic succession, sovereignty came to be reinterpreted in the nineteenth century as nationalism established itself in the normative order of international society (Mayall, 1990; Barkin and Cronin, 1994). It retained its close link to territoriality but was now seen to express a different kind of statehood, one that connects authority and political legitimacy with national communities, which are defined by a particular language or culture, or both (popular sovereignty). The new concept of national sovereignty underwent yet further changes in the twentieth century, with a shift from ethnic to civic forms of nationalism normalising the existence of multi-ethnic nations and an expansion of the core purpose of the state to include developmentalism and welfarism (Reus-Smit, 1999). On the basis of this interpretation, it becomes possible to imagine a further transformation of the state's moral purpose to include a more explicit environmental mandate, to protect the ecological integrity of the planet. It is important to note that the greening of the state and international society as the next stage in international normative development does not necessarily amount to the demise of sovereignty, as anticipated by some early environmentalists. Instead, we should expect a reinterpretation of its meaning: a green evolution *in* sovereignty (Eckersley, 2004a: 207) and a shift in what it means to be a sovereign state in an age of global interdependence and ecological values (Reus-Smit, 1996; see also Camilleri and Falk, 1992; Litfin, 1998).

How should we assess the progress that has been made on the journey from Green Westphalia towards global environmental governance? What does the historical record of the rise of global environmental politics (see Chapters 6 and 7) tell us about the depth and strength of the normative changes brought about, in terms of the greening of the core purpose of the state and of international legitimacy? Green state theorists themselves warn against unrealistic expectations when it comes to the malleability of the international system. For Eckersley (2004a: 209, 240), there is 'no inevitability about this possible green trajectory', although an accumulation of 'unit-level transformations' may eventually pave the way for 'more general, system level transformations'. What does the experience with half a century of GEP tell us about this accumulation of small changes?

There can be little doubt that GIS has moved considerably beyond the Green Westphalian model in its attempt to build global environmental governance. As discussed in Chapters 6 and 7, states have advanced global environmental policy-making in a number of ways. They have added new environmental issues to the international agenda, created ever more environmental treaties and organisations, widened participation in international decision-making, integrated scientific advice functions into international processes and developed innovative governance approaches to implement internationally agreed policies. Slowly but steadily, GIS has put together an institutional infrastructure to support environmental governance at all levels, from the global to the regional and national. As regime theorists (e.g. Haas, Keohane and Levy, 1993) have demonstrated, international institution-building is critical to the success of any long-term project of collective environmental management in international relations: secondary institutions increase the flow of information between states, they prescribe roles and incentivise behavioural changes, and they reduce uncertainty about states' future behaviour and shape expectations about their likely adherence to international rules. In short, international institutions create durable institutional and legal structures that allow international society to pursue a more ambitious normative agenda, one that goes well beyond the narrow pluralist logic of coexistence.

One important but easily overlooked indicator of the success of global environmentalism can be found in the fundamental shift in the way states approach environmental problems. Up until the 1970s, environmental protection was a predominantly domestic policy responsibility, often left to local or national authorities. Where states engaged in international environmental initiatives, they would do so on an ad hoc basis and not in recognition of a global responsibility. Following the Stockholm and Rio conferences, states began to systematically build domestic environmental policy capacity and also accepted the need to engage in multilateral environmental cooperation. The resulting expansion of global policy-making can be seen in the dramatically increased number of international environmental treaties that had been signed by the early twenty-first century, involving more and more states. In the

1970s, the then 140 members of the UN had on average ratified only ten MEAs. By the early 1990s, UN membership had gone up to ca. 180 and the number of MEAs that they had each joined on average shot up to 50 (Mitchell et al., 2020: 8).

The rise of environmental stewardship had thus produced a remarkable socialisation effect, though it is felt more in its procedural than substantive dimension. As more and more states took part in the routines of environmental multilateralism, with its ever-expanding set of international negotiation rounds, conferences of the parties, technical and legal subsidiary bodies, reporting schemes and compliance committees, they have gradually become accustomed to choosing collective over unilateral environmental action. To participate effectively in environmental multilateralism, states had to train a cadre of environmental diplomats and regulators that serve on the many forums, committees and agencies that have mushroomed in GEP. And by drawing a wide range of non-state actors with relevant expertise into global governance – from the scientists, environmental campaigners and business leaders that participate in international negotiations to the technical, legal and regulatory experts that serve on MEA's subsidiary bodies – states have also nurtured and empowered new generations of environmental actors in world society.

Environmental issues also gained in salience outside environmental organisations. After Stockholm, the UN inserted an environmental mandate into virtually all of its programmes, funds and initiatives (Conca, 2015: 3), and other global forums also began to address environmental problems as a matter of routine, from the G7 and G20 summits to major economic institutions (WTO, World Bank, IMF) and transnational organisations (World Economic Forum). By the early twenty-first century, most states appeared to have accepted environmentalists' longstanding claim that, to be effective, environmental protection needed to be organised on a planetary scale. International environmental cooperation, whether on the preservation of endangered species or the protection of the ozone layer, was no longer a question of national choice but one of global necessity.

Growing participation in MEAs has also increased the vertical linkages in environmental governance, by tying domestic policy-making and international regimes together. This enmeshment of domestic and international processes has allowed environmental norms to circulate more widely across the world, with some states shaping international regulations and global principles (e.g. sustainable development, environmental precaution) finding their way into other countries' domestic laws and regulations. It is important to note that in this 'cross-fertilization between national and international environmental law and policy' (Eckersley, 2004a: 215), policy ideas and approaches flow in both directions, bottom-up and top-down. Countries with strong domestic regulations often try to export them to the international level, and many international regimes do indeed reflect regulatory choices that had previously been made at the domestic level (Economy and Schreurs, 1997; DeSombre, 2000). Once internationally agreed, such norms and rules are then transposed into domestic law among a wider range of countries. One of the most successful examples of such domestic–international linkages in GEP is the Montreal Protocol of 1987. The treaty came about as a compromise among a small group of leading industrialised countries that had each developed their own domestic regulations for ozone-depleting substances (ODS). Once an ODS phase-out schedule had been agreed after protracted negotiations, it quickly became a globally applicable framework for the design of national ODS restrictions, and all major ODS producers moved in a coordinated fashion to bring down ODS production and use (Falkner, 2008: chapter 3). The Montreal Protocol, arguably the most successful MEA ever created, is a classic example of how international cooperation along solidarist lines allowed GIS to solve a global environmental problem that none of the leading powers could have solved through unilateral action.

The internationalisation of environmental policy has helped to strengthen national capacity for environmental protection in other ways, too. Away from the most visible manifestations of international governance – UN summits, international organisations and COPs – various transboundary networks have emerged that connect national regulators and experts from different countries. Such transgovernmental networks have grown up to support regulatory cooperation among government officials in a wide range of issue areas, from international finance to environmental protection (Raustiala, 2002; Slaughter, 2004). By facilitating information exchange and promoting harmonisation of regulatory approaches, they act as key sites of environmental policy learning and diffusion across borders. Transgovernmental networks tend to be informal but are often clustered around, or supported by, formal international organisations (e.g. OECD, UNEP). For example, the OECD has established two expert working groups to address the environmental, health and safety risks of nanotechnologies. Although no international treaty on nanotechnology safety exists, the OECD working groups are coordinating national safety approaches with a view to facilitating regulatory harmonisation and diffusing best practice (Falkner and Jaspers, 2012: 43–4; Marchant and Abbott, 2012: 400–1). The International Network for Environmental Compliance and Enforcement was founded in

1989 with the purpose of strengthening compliance with environmental law. The network consists of some 4,000 environmental regulators, investigators, prosecutors, judges and other experts in the field of environmental protection and maintains regional networks in all major regions. It has grown into a major vehicle for the export of regulatory expertise and capacity-building from the United States to Latin America (Raustiala, 2002: 43–9). These transgovernmental networks do not perform global governance functions in the narrow sense – they usually have little authority to create or implement new rules – but they facilitate regulatory harmonisation and diffusion and thereby help to strengthen global environmental governance. However, because of their often-informal nature, they reproduce unequal power structures in GEP. Most transgovernmental networks tend to be dominated by representatives of Northern countries where regulatory capacity is strongest, which is why they have come in for criticism for lacking broader international legitimacy.

Finally, in another sign of the solidarisation of GEP, some states have strengthened the rights of citizens to gain access to policy-relevant information and enlarged their ability to participate in environmental decision-making. The most prominent example of this is the United Nations Economic Commission for Europe (UNECE) Convention on Access to Information, Public Participation in Decision-Making and Access to Justice in Environmental Matters, commonly known as the Aarhus Convention. Signed in 1998 and in force since 2001, the Aarhus Convention has forty-seven parties from Europe and Central Asia. Widely seen as an ambitious attempt to advance principles of environmental democracy in international governance, it establishes that citizens in member states have rights to environmental information, public participation and access to justice. Eckersley hails Aarhus as a 'significant step toward transnationalizing ecological citizenship' (Eckersley, 2004a: 194) while Duncan Weaver (2018: 209) sees in it 'clear solidarist potential' as it advances the cosmopolitan rights of individuals vis-à-vis the state. However, the broader impact that the Aarhus Convention has had on GEP is fairly limited. The rights that Aarhus establishes are restricted to procedural matters of participation and information access, not substantive questions of environmental governance and the role of non-state actors in it. Furthermore, Aarhus does not represent a universally accepted model for citizen participation, given that it applies only to European and Central Asian countries. Even if it were to become globally established, its impact on the normative basis of GEP can best be summed up as 'evolutionary and incremental', not revolutionary (Weaver, 2018: 210). In the end, Aarhus reinforces the state-centric nature of policy-making in GEP, it confines itself 'within the safety and familiarity of sovereign statehood' (ibid.).

The Vexed Question of Compliance and Enforcement

However impressive the expansion of global environmental policy-making since the 1970s, GEP is still far from reaching the level of deep international cooperation that the solidarist vision for a green international order entails. While elements of global environmental governance have been established, the underlying institutional architecture is fragmented and incomplete, and the international authority invested in international environmental institutions remains severely circumscribed. If, as Bull (1966: 52) notes, it is states' solidarity 'with respect to the enforcement of the law' that marks out solidarist from pluralist international society, then GIS is still far from creating a solidarist green order.

To be sure, environmental secondary institutions provide a range of governance mechanisms that support their implementation and seek to strengthen states' compliance with internationally agreed rules. Once in force, MEAs are reviewed, revised and further developed in annual COPs, which provide a regular forum for international environmental decision-making. They typically include provisions for national reporting, which require parties to disclose relevant environmental indicators as well as how they are implementing international obligations. Many MEAs also include a financial mechanism which administers international environmental aid in support of the treaty's objectives, with the majority of funds usually targeted at developing countries. Support mechanisms may also include funding for capacity-building to boost countries' scientific, administrative or regulatory expertise, with a view to strengthening domestic implementation. In a few cases, financial support also extends to technical cooperation and the facilitation of technology transfer. Finally, many MEAs include compliance and enforcement mechanisms that allow parties to identify, and where necessary address, cases of non-compliance. Viewed through a solidarist lens, the hope is that such governance mechanisms help states meet their treaty obligations by strengthening domestic capacity for implementation and information exchange. Where financial and compliance mechanisms are included, MEAs also seek to incentivise or induce states to change their environmental behaviour. In this way, the proliferation of MEAs is meant to gradually enmesh states in an ever-tighter web of international environmental rules and regulations that eventually add up to a coherent system of global environmental governance.

Impressive as the continuous institutionalisation of global environmental policy has been, it has not lived up to solidarist hopes for a deeper 'greening of sovereignty' (Litfin, 1998). Despite agreeing a myriad of international environmental targets and commitments, states have been careful throughout to retain a large degree of national autonomy when it comes to the implementation of MEA obligations. Proposals to move towards more comprehensive monitoring of state behaviour and sanctioning of non-compliance have been repeatedly resisted, by developing countries as much as by leading industrialised countries. The strength of international environmental law certainly cannot be reduced to the question of its enforceability (Brunnée, 2006), but weak national implementation of international agreement remains the Achilles heel of the UN system of global environmental governance.

The aspiration to develop stronger provisions for monitoring compliance and addressing non-compliance is evident from some of the major environmental treaties negotiated after Stockholm. The standard model widely used in modern MEAs involves the COP, serving as the Meeting of the Parties in the case of protocols to conventions, taking appropriate measures against non-compliant states. COPs usually establish subsidiary bodies to monitor levels of compliance and recommend action against non-compliant states (Goeteyn and Maes, 2011). In the case of the CBD, for example, the treaty creates a relatively weak basis for a compliance procedure. Article 23.4(i) merely authorises the COP to 'consider and undertake any additional action that may be required for the achievement of the purposes of this Convention in the light of experience gained in its operation', while Article 27 refers to the International Court of Justice for purposes of dispute resolution. Similar provisions exist in the Kyoto Protocol (Art. 18), Cartagena Protocol on Biosafety (Art. 29.4(f)) and the Ramsar Convention (Art. 6.2(f)). The Kyoto Protocol's Compliance Committee stands out as a noteworthy and innovative example as it has operated in a quasi-judicial manner with limited political interference (Oberthür and Lefeber, 2010). Established as an independent body that reviews how states implement the climate treaty, the Kyoto Protocol's compliance mechanism allows not only states but also expert review teams to trigger a compliance procedure. It has at its disposal various 'carrots and sticks' to deal with cases of concompliance. At a minimum, the Committee can issue a simple request to a party to submit a plan for returning to compliance. If the Committee's Enforcement Branch finds that a party has failed to meet its emissions targets in the first commitment period, it may impose sanctions, either by demanding that the party meet more stringent emissions targets in the second period or by suspending it from participation in international emissions trading (Oberthür and Lefeber, 2010: 148). In this way, the Kyoto Protocol has played an important role in developing and expanding the scope for compliance measures in support of internationally agreed emission reduction targets.

But despite its innovative nature, Kyoto's compliance mechanism could not deal with the root causes behind the failure of major emitters to reduce their greenhouse gas (GHG) emissions. The United States, the world's largest GHG emitter at the time of the 1997 Kyoto conference, simply refused to ratify the treaty. In the case of Canada, which had acceded to the treaty but failed to meet its emissions target, the Facilitative Branch of the Compliance Committee initiated a non-compliance procedure in 2012 and asked the Canadian government to clarify its position (Zahar, 2015: 79). The intervention came far too late, however, as Canada had already announced its departure from the Kyoto Protocol in late 2011 (United Nations, 2011). Faced with international and domestic criticism for non-compliance, the Canadian government availed of its right to withdraw from the treaty altogether. The experience with the United States and Canada in the Kyoto Protocol thus illustrates a core dilemma in the solidarist ambition to strengthen secondary institutions in GEP. Environmental stewardship may have gained universal acceptance as a primary institution of GIS, but this came at the cost of restricting the international authority invested in its institutional architecture. As long as powerful states resist any moves to mandate and enforce environmental action against their will, global environmental governance can only be built on a voluntarist basis, based on belief and calculation, not coercion (Falkner and Buzan, 2019: 150).

Given the persistent constraints on enforcing international environmental rules, it is more appropriate to think of compliance mechanisms in MEAs as essentially political or managerial, not judicial, instruments (Bodansky, 2010: 235–8; Goeteyn and Maes, 2011: 797). For one, even where the parties to an environmental treaty have agreed to create special bodies that administer compliance procedures, it is usually the COP in its entirety, not an independent judicial body, that takes decisions on how to respond to cases of non-compliance. Furthermore, most MEA compliance procedures tend to be facilitative rather than punitive in nature. They are focused on supporting rather than sanctioning states that fail to live up to their international obligations (Bodansky, 2010: 243–5). Even those MEAs that include punitive measures (e.g. Montreal Protocol) have shied away from using them and have chosen to provide technical or financial assistance instead. To some extent, this reflects the reality of environmental management on the ground, especially in the developing world. Many states lack the capacity, rather than the will, to comply with international obligations. As in other global policy areas, non-compliance with MEAs is a complex phenomenon that reflects a variety of underlying causes: from ignorance about treaty obligations to insufficient monitoring; lack of technical, regulatory or administrative capacity; legal ambiguity; or simply economic hardship (Goeteyn and Maes, 2011: 798). But the general weakness of compliance mechanisms and the reluctance to use enforcement tools also reflects active resistance by sovereignty-conscious members of GIS. States may profess their support for planetary sustainability and international environmental cooperation, but they also carefully guard their sovereign rights, including when it comes to shirking global environmental responsibilities.

The limitations of state-centric solidarism in GEP can also be seen in the debate on creating stronger international organisations for the environment. As discussed in Chapter 3, proposals for creating a strong international authority that can rein in global environmental degradation featured prominently in early designs for a green global order. In the run-up to the 1972 Stockholm Conference, several diplomats and authors (e.g. Kennan, 1970; Falk, 1971) called for a new international body that would have the power to safeguard the planetary interest against narrow national interests. For some, this was to be the first step on the path towards a fully fledged green world government. But even if such a degree of centralisation in international society remained a utopian dream, many environmentalists have continued to call for a strengthening of the authority of the UN and its associated international organisations as a way out of the fragmented and inefficient system of environmental governance that was emerging after Stockholm. The Declaration of The Hague in 1989, for example, called for a new international body to deal with the threat of global warming (Biermann, Davies and van der Grijp, 2009: 356). Debates about creating a new UN environment organisation flared up repeatedly in the last three decades, especially in the run-up to the UN sustainability summits in Johannesburg (2002) and Rio de Janeiro (2012), both of which were meant to address the lack of implementation of international agreements and systemic weaknesses in the existing institutional architecture. Encouraged by the UN's desire to put institutional reform on the international agenda, several expert groups developed three principal options: the strengthening of UNEP, the transfer of environmental responsibilities to the UN Trusteeship Council and the creation of a new World Environment Organization (Brack and Hyvarinen, 2002; Charnovitz, 2002; Esty and Ivanova, 2002; Runge, 2001; Newell, 2002; Najam, 2003; for an overview of this debate, see Biermann and Bauer, 2005).

State-centric ideas about strengthening global environmental governance are never too far from the surface in GEP debates.

In the end, however, the enthusiasm for a beefed-up international governance system fizzled out when it met the reality of international politics. The 2012 'Rio+20' summit concluded the reform debate with a modest agreement to boost UNEP's role by enlarging membership of its Governing Council to all UN members and putting its finances on a more stable and secure footing (Bauer, 2013: 329). However, none of these changes amounted to the kind of radical centralisation of decision-making authority in a single international organisation that environmentalists have routinely demanded. The outcome of the UNEP reform debate is thus symptomatic of the deeper malaise in environmental multilateralism and international environmental governance more generally (Bernstein, 2013). State-centric solidarism has put in place a comprehensive institutional architecture for dealing with environmental problems, but has failed to translate this into integrated and authoritative decision-making at the international level. Even though states' commitment to the environmental stewardship norm has hardened over the last fifty years, their resistance to creating stronger governance institutions has not waned.

If anything, the direction of travel has shifted in recent years, reducing the solidarist impulse for stronger legal commitments in favour of voluntary targets or non-binding goals (on the popularity of goal-setting in GEP, see Biermann, Kanie and Kim, 2017). This is particularly the case in areas where environmental issues have gained in political salience and thus increased the sovereignty costs of strong international action. In this respect, the evolution of the climate regime provides a good example of the rise, and eventual decline, of state-centric solidarism in GEP. In many ways, the 1997 Kyoto Protocol came to epitomise the success of solidarist policy-making after the Rio Earth Summit, signalling major powers' willingness to act in accordance with notions of collective environmental responsibility and climate justice: the Kyoto compromise included a multilaterally agreed distribution of the climate mitigation burden, enshrined national emission reductions in legally binding international commitments, applied a strong international equity norm ('common but differentiated responsibilities') and established a compliance mechanism to oversee the implementation of national commitments. This success, however impressive, was not to last. It began to unravel when the 2009 Copenhagen conference failed to agree a successor agreement to Kyoto and instead paved the way for a different, voluntarist, logic of climate action (Falkner, Stephan and Vogler, 2010). Although Kyoto entered into a second commitment period after

2012, it has since seen the departure of major emitters (Canada, Russia, Japan) and was superseded by the 2015 Paris Agreement with its focus on nationally determined emission pledges and an international review mechanism (Falkner, 2016b). To get major emitters (mainly the United States but also China and India) on board, the new climate treaty had to avoid any legally binding environmental action. By moving international climate policy onto a different trajectory, the Paris Agreement came to symbolise the slow but continuous retrenchment of solidarist ambition in GEP (Falkner, 2017a).

Resistance to legally binding environmental targets and measures is also evident from the broader shift towards informality in global governance, including in GEP (Stone, 2013; Roger, 2020). In response to increasing gridlock in multilateral institutions (Hale, Held and Young, 2013), states have come to rely on innovative processes for agreeing common objectives but without creating legally binding obligations, formal organisational structures and resource-intensive bureaucracies. Several such informal environmental institutions have come into existence in the last two decades, including the Major Economies Forum, Carbon Sequestration Leadership Forum, the Powering Past Coal Alliance and the Coalition of Finance Ministers for Climate Action. They all combine the benefits of uniting like-minded actors behind a common purpose while keeping the level of legal commitment low, thereby lowering the barriers to join such initiatives. Several factors lie behind this trend: informal governance reflects the growing demand for international regulatory capacity at a time when domestic constraints in powerful states limit their ability to commit to formal international rules (Roger, 2020); it points to deep-seated distributional conflicts and power asymmetries among international actors (Reinsberg and Westerwinter, 2021); and it suggests a growing willingness to experiment with new initiatives that draw on the governance capacities of other, usually non-state, actors (Abbott et al., 2015a). Informal governance is certainly a multifaceted phenomenon, and much like the use of minilateral forums it can support more traditional multilateral cooperation (Falkner, 2016a). However, its growing popularity in GEP underlines the difficulties that the state-centric solidarist project faces at a time of fractured great power relations and a resurgence of sovereignty concerns.

To sum up the discussion so far, GIS has undoubtedly taken a series of important steps beyond a minimalist Green Westphalian order, but remains far from realising the state-centric solidarist ideal of global environmental governance. The greening of the state's core purpose is well underway at the domestic level, but a corresponding greening of sovereignty in international relations has only just begun. The large number of MEAs and environmental organisations that have been established since Stockholm have created an increasingly complex web of international environmental rules and obligations. As such, the expanding scope for global environmental policy-making has been impressive. At the same time, however, states have closely guarded their autonomy in deciding on how to translate international commitments into environmental action, including when to ignore such commitments. If anything, many states have pushed back against a further erosion of their sovereign rights, preferring voluntarist language and practice. Sovereignty in the age of environmentalism has been reinterpreted, but not transformed. The contours of the 'green state' (Eckersley, 2004a) may be becoming visible at the national level, but 'green sovereignty' remains an elusive ambition in GIS.

Restraining Sovereignty: The Prospects and Limitations of International Environmental Law

One of the lasting consequences of the 1972 Stockholm conference has been the dramatic expansion of international environmental law. Still considered to be a 'relatively new field' (Bodansky, Brunnée and Hey, 2007: 2), the international law of the environment has become a major growth area for new international legal rules and norms. The dramatic speed and scope of its development in the modern era can only be grasped if we compare it to its early roots in traditional international law. Until the early twentieth century, environmental issues were dealt with on an ad hoc basis, as part of international claims over infringements of states' sovereign rights. In the classic Westphalian tradition, international law comprised only minimal rules that sought to regulate interstate relations. Their purpose was to guarantee 'good neighbourship' (Sand, 2007: 31) based on the state's universally acknowledged right to territorial sovereignty and integrity. This right ensured that states were free to act within their own territory as they saw fit but also implied an entitlement to freedom from interference by neighbouring states. As a result, a state's main environmental responsibility was not to inflict significant environmental damage on other states, but this had to be balanced with its sovereign right to pursue its own developmental path. Peaceful coexistence, not the protection of planetary health, was at the heart of traditional international law.

In its approach to transboundary environmental issues, traditional international law was 'inherently bilateral and confrontational in character' (Birnie and Boyle, 2002: 178). To manage disputes arising from transboundary environmental externalities, states could either negotiate bilateral agreements or seek judicial settlements – provided institutional arrangements for arbitration existed. One of the first and still prominent cases of transnational harm that resulted in international arbitration was the *Trail Smelter* dispute, which concerned smoke from a Canadian lead and zinc smelter blowing over the border and damaging forests and farmlands in the US state of Washington. When the case was finally settled in 1941, it produced the first international ruling on transboundary air pollution that helped to establish the harm principle in international environmental law (Wirth, 1996). According to the ruling, sovereign equality requires states to refrain from activities that cause significant transboundary environmental harm – a principle with potentially far-reaching consequences for other environmental matters, such as climate change (Simlinger and Mayer, 2019: 186–7). But far from setting a precedent for further international enforcement action and a strengthening of the harm principle, the *Trail Smelter* ruling remained the exception to the rule.

Despite its potential to become a regulative instrument for global environmental management, states have tended to avoid using the law of state responsibility and international adjudication of transboundary pollution disputes. Applying the no-harm principle to specific cases has proved difficult. Most forms of international pollution happen over a long time, often at a low level of intensity, and usually as a result of ordinary economic activities that are spread over multiple countries. The few cases that have ended up in front of international courts addressed individual acts with significant impacts that could be traced back to a single state, and these have rarely involved significant reparation payments to the injured party (Simlinger and Mayer, 2019: 188).

Nevertheless, it is not too far-fetched to imagine states devising governance mechanisms that connect the general responsibility not to cause transboundary environmental harm with specific liability regimes (Wapner, 1998: 280–2). This could be in line with established customary law principles that ensure 'good neighbourship' and would lead to claims for reparation and compensation in cases of transboundary environmental harm. International society came close to going down this path at UNCED in 1992. Principle 13 of the Rio Declaration includes the commitment that states shall 'cooperate in an expeditious and more determined manner to develop further international law regarding liability and compensation for adverse effects of environmental damage caused by activities within their jurisdiction or control to areas beyond their jurisdiction' (United Nations General Assembly, 1992). Indeed, many subsequently negotiated MEAs contain provisions on liability and redress. However, these tend not to go beyond stating the need to develop principles, rules and operational guidelines in this area. In reality, virtually all efforts to move beyond declarations of intent have proved futile. For example, the inclusion of substantive provisions on liability in the 2000 Cartagena Protocol on Biosafety as demanded by developing countries proved highly contentious. The issue was only resolved when the parties agreed in Article 27 of the treaty to postpone work on liability rules until after the treaty had entered into force (Cook, 2002). The inclusion of liability and compensation in the climate regime proved similarly divisive during the negotiations on the Paris Agreement. Developing countries' demands that the Warsaw Mechanism for Loss and Damage should include liability and compensation provisions were explicitly ruled out at the insistence of Northern countries (Falkner, 2019: 275).

As we have seen, efforts to create a working liability regime based on state responsibility have been unsuccessful. To some extent, such arrangements are 'largely impractical' (Wapner, 1998: 281), mainly due to the difficulty to create precise rules on how to establish causality, apportion responsibility and determine compensation in specific cases. Many environmental problems are multi-causal in nature, making it difficult to trace them back to a single act by a specific actor. But even if such legal complexities could be overcome, major industrialised countries would still resist demands for binding rules on liability and redress, reflecting a straightforward self-interest in preventing multi-billion dollar claims being brought against them. Connecting the no-harm principle with specific liability rules and compensation payments has thus never progressed beyond the level of customary principles in international law. Because it is closely tied to the protection of state sovereignty, traditional international law proved inadequate as the basis for an expanding system of global environmental law, governance and enforcement. Rooted in a pluralist logic of coexistence, it failed to evolve into what might otherwise have been a powerful tool for changing the environmental behaviour of states.

Progressive Legal Developments

The main shift in the evolution of international environmental law came in the 1960s and early 1970s, at a time when states were beginning to accept a fundamental responsibility for the global environment. As discussed earlier, the rise of environmental multilateralism turned environmental protection into one of the most dynamic areas of international law. This legal dynamism expressed itself in the accumulation of ever more rules and obligations resulting from internationally agreed treaties, but also in a shift away from the traditional focus on interstate dispute resolution and towards the use of law as a tool for managing global ecosystems in a proactive and forward-looking manner. In this way, international environmental law became one of the central elements of a solidarist project of building global environmental governance.

Five trends, in particular, stand out as markers of international society's progressive approach to international environmental law: the dynamic quality of norm-creation; the inclusion of common environmental areas and concerns; the focus on the interests of future generations; the linking of environmentalism and human rights; and the emergence of an eco-centric conception of environmental responsibility. I shall explore each of them briefly with a view to assessing the extent to which they have taken international environmental law in a more solidarist direction.

The first trend has to do with the process of international norm creation itself. Before the Stockholm conference, some European and North American states had already agreed a few international environmental treaties but these were isolated attempts to manage transboundary pollution or wildlife protection issues. The situation changed after Stockholm as a larger number of states embarked on a near-continuous process of negotiating environmental agreements. Many such treaties not only became the source of novel international environmental rules and norms, they also generated a routine process of further decision-making by the parties to the agreement, through the annual COPs. Modern environmental treaties therefore 'no longer reflect a static set of rules' (Bodansky, Brunnée and Hey, 2007: 21) but have become sites of an iterative process that allows legal norms to be adapted to new circumstances. Such ongoing processes of norm development and evolution have created opportunities for a wide range of actors (regulatory, scientific and technical experts, as well as civil society and business representatives) to seek to influence global rule-making and regulatory processes. They have also created a closer link and cross-fertilisation between international and national environmental policy processes. It is not least in this sense that international environmental law is increasingly being talked about as a form of 'global administrative law', with international regimes reaching directly into domestic regulatory systems and relying on hybrid public–private rule-making (Kingsbury, 2007). The growing use of informal approaches to environmental standard-setting has also broadened the range of norm-creation mechanisms at the international level, with intergovernmental regimes giving rise to and interacting with transgovernmental regulatory networks (Raustiala, 2002). This expansion of participation in environmental rule-making has gone some way towards fulfilling state-centric solidarism's promise of more systematic global decision-making. At the same time, however, this development has led to concerns about the legitimacy of the international process, particularly where the push for collective action is seen to have eroded the requirement of state consent as the main form of legitimation in the creation of international law (Bodansky, 2007: 714).

A second area in which international environmental law has progressed well beyond its narrow Westphalian origins is the growing attention paid to 'common areas' and 'common concerns' (Brunnée, 2007). Traditional international law, with its focus on the balancing of competing sovereign state interests, found it difficult to deal with environmental problems that lie outside the jurisdiction of sovereign states. No single state can comprehensively regulate the use of common areas, such as Antarctica and the high seas, and in the case of common concerns, such as the protection of global biodiversity and the global atmosphere, nearly all states need to contribute to their environmental management. As international society started to tackle such global problems, international environmental law underwent a slow but ongoing transition towards greater recognition of collective, alongside national, environmental concerns. Legal scholars have identified this move towards managing global public goods in a number of areas: in the shift from bilateralism to community interests (Simma, 1994), in the emergence of 'international public law' (Hey, 2003) and in the weakening of the state-consent norm in international law (Krisch, 2014). It is most evident in the growing use of multilateral environmental treaties that establish a common global purpose and institute processes for administering internationally agreed policies. As ever more MEA obligations expressed the fundamentally shared nature of the global environment, international law came to express general obligations that a state owes towards the community of states, and that can be enforced by all. In its decision in the Barcelona Traction case, the International Court of Justice confirmed the existence of such obligations *erga omnes*, to the 'international community as a whole' (quoted in Tams and Tzanakopoulos, 2010: 782). The application of this principle to the environment remains contested, but some legal experts express the expectation that 'we shall witness in coming years an obligation *erga omnes* to safeguard the Earth's environment' (Robinson, 2018). Despite the fact that humanity's interest in planetary protection is now reflected in international norm development, the legal standing of humanity is still incomplete, and it remains unclear how violators of such emerging international obligations can be held responsible based on customary international law alone (Brunnée, 2007: 556). What enforcement of international environmental law there is, it remains limited to existing compliance procedures in MEAs.

A third trend that suggests a move towards solidarist international order can be found in international law's growing recognition of the environmental interests of future generations. Intergenerational equity found its clearest expression in the Brundtland Commission report of 1987, which defined sustainable development as 'development that meets the needs of the present without compromising the ability of future generations to meet their own needs' (World Commission on Environment and Development, 1987). The report provided an influential framing of the international environmental agenda and is widely recognised for having helped to bridge the North–South divide in the run-up to the 1992 Earth Summit (see Chapter 6). It also played an important role in shifting the focus to more long-term environmental threats that affect the interests of future generations. A growing number of MEAs have endorsed this emphasis on intergenerational equity, seeking to promote environmental governance that serves the interest of both current and future generations. Efforts to protect natural habitats and species (e.g. CITES and CBD) have traditionally been predicated on the idea of protecting biological diversity in the long run. The fight against climate change (UNFCCC) is similarly driven by a concern that future generations' well-being will be compromised if the global warming trend is not slowed down. Future-oriented environmental concerns also inform the precautionary principle, which has gained in importance as an environmental norm since its inclusion in the Rio Declaration (Principle 15) and later international treaties. The Cartagena Protocol on Biosafety (2000) stands out as a notable achievement in this regard as it includes the precautionary principle in its operative part (Graff, 2002). As discussed in Chapter 7, it has proved more difficult to extend precaution to other areas of global regulation of risky technologies, not least due to contestation by some of the leading industrialised countries that seek to protect their high-tech industrial sectors against intrusive international regulation (Falkner and Jaspers, 2012). Again, these normative developments suggest tentative moves in the direction of a more cosmopolitan viewpoint in international law, although the legal standing of future generations remains uncertain and ill-defined.

Establishing environmental protection as a basic human right has become a further important avenue for the solidarist push to strengthen international environmental law. Anchoring environmental protection in human rights offers at least three advantages over the traditional route of legal norm creation by intergovernmental treaties: it provides a basis for addressing the impacts of environmental degradation on individuals' life, health and property rather than on states; it creates a comprehensive obligation for states to take action against such environmental impacts on individual humans; and it opens the path towards legal action by individuals against states that fail to regulate environmental pollution (Boyle, 2012: 613–14). In short, framing environmental protection as a human right is a strategic move in the solidarisation of international environmental law. It empowers individuals and world society actors to demand environmental action from the state and international society. GIS has taken the first steps towards establishing a stronger link between environmental rights and human rights, though this remains work in progress. The 1972 Stockholm Declaration includes an early statement of a human right in the environment (Principle 1), as does the 1992 Rio Declaration in a more indirect way (Principle 1). Other international declarations have similarly stated a close link between human rights and the environment, while a growing number of countries have incorporated a right to a healthy environment in their constitutions (Weiss, 2011: 16; Boyd, 2011). Finally, human rights courts have developed case law around environmental issues (Boyle, 2012), and the idea of constitutionally anchoring environmental rights, whether at the state or international level, has developed widespread interest in legal scholarship (Hayward, 2005; Humphreys, 2010; May and Daly, 2015).

The human rights-based solidarisation of international law has made considerable progress in the area of procedural environmental rights – 'the most important environmental addition to human rights law since the 1992 Rio Declaration on Environment and Development' according to Boyle (2012: 616). MEAs (e.g. CBD, UNFCCC, Paris Agreement, Minamata Convention on Mercury) now routinely include references to the rights of the public, with regard to access to information, participation in decision-making processes or access to education and training (Peters, 2017: 1–2). And as mentioned earlier, the 1998 Aarhus Convention took an important step towards codifying such procedural rights. It is the first convention to put Principle 10 of the Rio Declaration into legally binding form, establishing a formal right for the public to information, participation in decision-making and access to justice in environmental matters (Weaver, 2018).

While the link between environment and human rights is thus generally thought to be strengthening (Weiss, 2011: 16–17), GIS has so far refrained from formally declaring a human right to environment. Several initiatives to bring about this shift in the human rights regime have been launched. An early attempt, the politically ambitious Draft Principles on Human Rights and the Environment from 1994 (United Nations Economic and Social Council, 1994), failed to win enough support among states to be adopted (Popovic, 1995). By 2011, enough momentum had built up for the UN Human Rights Council to begin examining the relationship between human rights and the environment. A year later, the Council decided to establish a mandate on human rights and the environment and proceeded to create the position of Special Rapporteur on human rights and the environment (Knox, 2018). One area where the push for greater human rights recognition in GEP came to the fore is climate change (Humphreys, 2010). With the UN Human Rights Council having adopted several resolutions on climate change and human rights, parties to the UNFCCC came under pressure from campaigners to reflect this development in a post-Kyoto Protocol agreement. The 2015 Paris Agreement thus became the first climate treaty to contain a reference to human rights, though this was included in the legally non-binding preamble and is balanced with references to a plethora of other rights, including those of 'indigenous peoples, local communities, children, persons with disabilities and people in vulnerable situations' and, critically, 'the right to development' (United Nations Framework Convention on Climate Change, 2016: recital 12). It is fair to conclude that the environment–human rights link in international law has hardened, but it would take a leap of faith to argue that it has also begun to curtail states' sovereign right to determine what international environmental obligations they accept and implement.

The fifth area in which international environmental law is being pushed beyond its state-centric tradition is the emerging concept of ecocentric rights. Whereas traditional international environmental law is rooted in a strictly anthropocentric approach, some recent developments in GEP and legal scholarship point to a more radical extension of the rights concept to the non-human world. Should it gain ground, this move would represent a radical departure from the established legal approach in GEP. As mentioned in Chapter 3, the primary institution of environmental stewardship follows a predominantly anthropocentric logic of protecting the global environment to safeguard human progress today and for future generations. Both the 1972 Stockholm Declaration and the 1992 Rio Declaration take the human interest in a healthy environment as the paradigmatic starting point for developing international environmental law, a position that most MEAs have adopted, too. Only in some cases did international society hint at the possibility of nature protection for its own sake. The Antarctic Treaty of 1959 is the best example of what an ecocentric approach would look like in GEP. It was the first international treaty to protect a vulnerable ecological area by removing it completely from human use. In the 1980s, a global NGO campaign helped shift the IWC to adopt a preservationist approach. Instead of regulating the commercial exploitation of whales, as it had done since it was established in 1946, the IWC switched to a near-complete ban of whaling. At around the same time, environmental organisations also lobbied for a global biodiversity regime that reflected ecocentric values. The 1992 CBD incorporated at least some of these ideas, though with a strongly anthropocentric tinge. The treaty opens with recognition of 'the intrinsic value of biological diversity and of the ecological, genetic, social, economic, scientific, educational, cultural, recreational and aesthetic values of biological diversity and its components', but switches quickly to speak more specifically of 'biological resources' and the right of states to exploit those resources. To some extent, these references to protecting nature for its own sake hark back to older preservationist motives in nineteenth-century environmentalism, but they also speak to the growing resonance of ecocentric ideas in the late twentieth century.

This ecocentric turn is only slowly gaining ground in legal thinking, however, and it has yet to have a lasting impact on the way that legal rights and responsibilities are defined in an international context. One of the more radical shifts in jurisprudence has been the proposal that animals, and nature more generally, should be given their own legal status. The anti-whaling campaign was one of the first transnational campaigns that relied in part on arguments about whales, which are among the most intelligent animals on the planet, deserving protection for their own sake, and not simply as part of an effort to safeguard long-term commercial whaling interests (Epstein, 2008: 206–7). Some lawyers went as far as to argue for 'an emergent entitlement of whales – not just "on behalf of" whales – to a life of their own' (D'Amato and Chopra, 1991: 23). Proposals to recognise animals (or trees, or rivers) as bearers of rights first surfaced in the early 1970s (Stone, 1972). Such a move towards acceptance of ecocentric rights would go beyond political symbolism. It would allow anyone to bring a lawsuit on behalf of animals or ecosystems, not just the landowners to which these 'natural resources' belong. It also reflects a profound shift in ethical orientation, away from a purely human-centred towards a broader, more inclusive, humanist consciousness, with humans having duties and obligations towards nature (Nash, 1989: 10). Originally languishing on the fringes of legal theory, the argument for ecocentric legal rights has slowly gained momentum among environmental legal scholars and campaigners, though mostly within a domestic context. In 2008, Ecuador became the first country to recognise the rights of nature in its constitution (Tanasescu, 2013), and a few other nations (Bolivia, New Zealand, India) have since taken tentative steps in the same direction (Gordon, 2018). An ecocentric vision for international law has now been articulated more clearly, but anthropocentric environmentalism continues to be the most powerful rationale for expanding international environmental action. The solidarisation of international environmental law still follows a more classically liberal, human-centred, script.

The question of criminal responsibility for acts of environmental degradation is another area where progressive ideas have entered international legal discourses, though without corresponding shifts in international law and its implementation. Proposals to categorise some forms of environmental harm as international crimes – referred to as 'ecocide' in particularly egregious cases – have been made at least since the 1972 Stockholm conference, when Swedish Prime Minister Olaf Palme controversially referred to America's military campaign in Vietnam as an act of 'ecocide, which requires urgent international attention' (Flippen 2008, 631). In subsequent years legal experts made several attempts to establish an international statute for crimes against the environment, from Richard Falk's (1973) draft 'Ecocide Convention' to the deliberations of the International Law Commission on the 1998 Rome Statute of the International Criminal Court (ICC) (Malhotra, 2017: 52). In the end, the Commission decided not to include a generic reference to ecocide in the ICC's mandate, following objections by the United States, the United Kingdom and the Netherlands, with only acts of war that cause 'widespread, long-term and severe' environmental damage included in Article 8(2)(b) (Gauger et al., 2012: 2–3). The idea of criminal responsibility for environmental destruction has not disappeared, however. Several international lawyers have continued to push ecocide to be recognised as one of the crimes against peace that fall under the ICC's jurisdiction (Higgins, Short and South, 2013; Mehta and Merz, 2015). In 2016 the ICC adopted a policy paper on case selection and prioritisation in which it declared to 'give particular consideration to prosecuting Rome statute crimes that are committed by means of, or that result in, inter alia, the destruction of the environment, the illegal exploitation of natural resources or the illegal dispossession of land' (Vidal and Bowcott, 2016). Although the ICC's move provides a glimmer of hope for the criminalisation of transnational environmental harm, this area remains one of the many as yet unfulfilled hopes for a solidarist future for international environmental law.

To conclude, the international law of the environment has witnessed an impressive expansion of international rules and obligations since Stockholm. It has long left behind its Westphalian origins in customary norms of good neighbourship that seek to resolve transboundary disputes and has successfully established more comprehensive norms that aim to protect global ecosystems in a more preventive fashion. In line with the solidarist ambition of codifying in law the obligations that states have towards each other and the planet, international environmental law has shown ingenuity in developing novel environmental norms and innovative regulatory mechanisms. Yet, time and again, such developments have come up against sovereigntist resistance in GIS. Much of the solidarist edifice of international environmental law remains an unfulfilled ambition.

Conclusions

Environmental stewardship, the international norm that has emerged as a primary institution of international society since the 1970s, is predicated on the idea of states' shared responsibility for the global environment. To act on it in a way that goes beyond mere diplomatic rhetoric and symbolic politics, states would need to share a deeper level of common purpose and values and embark on a serious effort to create the institutional framework for sustained international cooperation. Unsurprisingly, environmental stewardship is, therefore, frequently cited as a solidarist international norm, and the greening of GIS is usually portrayed as a process of state-centric solidarisation in international relations. However, unlike some other solidarist norms (e.g. human rights, democracy) that have come up against resistance and failed to establish themselves as universally accepted norms, environmental stewardship has acquired the status of a fundamental norm for *global* international society.

Indeed, virtually all states have taken part in the post-Stockholm effort to build a comprehensive web of international treaties, norms and regulatory mechanisms that seek to put states' environmental commitments into operation. Environmental stewardship has found its most potent expression in the rise of environmental multilateralism, a procedural norm that expects states to participate in global rule-making on global environmental issues. The result has been a dramatic increase in the number of environmental treaties, many of which have reached near-universal levels of ratification. In this respect environmental sustainability has come to redefine good citizenship expectations for most members of GIS. Participation in multilateral environmental processes is now a routine practice that even countries that oppose specific regulatory proposals find difficult to elude. However, environmental multilateralism remains a mostly procedural norm. The sanctions for failing to live up to substantive obligations are relatively weak. Past proposals to strengthen compliance mechanisms in MEAs and invest international organisations with international authority have come up against stiff resistance by sovereignty-conscious states. The existing international environmental order may have left behind the narrow strictures of Green Westphalia, but is still far from the solidarist ambition of a fully fledged system of global environmental governance.

The rise of environmental stewardship has also gone hand in hand with a substantial expansion of international environmental law. Much of the legal norm creation has occurred through the negotiation of MEAs. As such, the increase in international legal norms concerning the environment speaks to the solidarist ambition behind GEP. International environmental law is also characterised by a certain degree of legal dynamism that has pushed it beyond traditional Westphalian boundaries. Innovative environmental norms, concerning common environmental areas, the interests of future generations, environmental human rights, precaution, criminal responsibility and ecocentric rights, suggest an increasingly progressive and cosmopolitan interpretation of the nature and purpose of international environmental law. Yet, none of these normative developments has successfully called into question sovereignty and territoriality, or the right of non-intervention, even if it is becoming less difficult to imagine calls for international intervention against states that commit egregious acts of environmental harm.

Overall, environmental stewardship has changed understandings of what it means to be a sovereign state, but without redefining the criteria for rightful membership of GIS. States have taken on ever more environmental commitments and have agreed to be tied into a dense network of international treaties and governance mechanisms, but this has not come at a significant cost to their sovereign rights. In fact, most environmental secondary institutions operate within a sovereigntist straight-jacket, which limits the authority they have over states and thereby undercuts solidarist environmental ambition. The many small steps that have been taken towards global environmental governance do not amount to a radical transformation in international legitimacy. As yet, the environmental norm is not strong enough to lead to a state's expulsion from GIS or demotion within its hierarchy. States accept a moral responsibility to safeguard the planet's health and have taken some measures to that end, at both domestic and international levels. The question remains whether the accumulation of these small evolutionary changes amount to a more fundamental, transformative, change when viewed over the long term, an effective greening of the moral purpose of the state and of international society.

9 Pluralist Constraints

Environmental stewardship is a distinctly solidarist norm in international relations. It enlarges the moral ambition of international society and seeks a fundamental reorientation of key elements of the international normative order, most notably sovereignty and territoriality. Yet, its realisation as a guiding principle for interstate relations has come up against the pluralist constraints of the existing international order: the predominance of power politics and a deeply unequal distribution of power, profound value diversity and ideological conflict, and a weak institutional basis for international cooperation. It is thus tempting to conclude that the rise of global environmentalism in international relations is a story of solidarist ambition against pluralist resistance. But to what extent can environmental stewardship arise in a pluralist context? Could international recognition of the need to protect the global environment emerge from a pluralist logic of coexistence, rather than a solidarist logic of cooperation? And if pluralist international society and environmental stewardship are in some sense compatible, how would the pluralist logic translate into international environmental action?

As mentioned earlier, in Chapter 2 and at the beginning of chapter 8, solidarism and pluralism are used in ES theory for both normative and analytical purposes. My use of the concepts is predominantly analytical in nature, though I do not exclude a more normative assessment of societal development in international relations. I employ solidarism and pluralism as analytical ideal types (Weinert, 2011), not as fixed descriptions of an actual empirical state of affairs. In reality, most international societies blend solidarist and pluralist elements: 'world order is and always has been both pluralist and solidarist' (De Almeida, 2006: 68). For the purposes of this analysis, however, I focus on the pluralist elements of international society and isolate the pluralist logic of international action in relation to environmental issues and values, what I have referred to as Green Westphalia in Chapter 3. In a first step, I explore this pluralist logic and the extent to which it can give rise of sustained international environmental cooperation. I then move on to consider whether the emergence of the environment as a security threat (e.g. climate change) could translate into a distinctly pluralist response to global environmental problems. In the final section of this chapter, I take this argument one step further and explore whether environmental securitisation could engage great power management, a classic pluralist institution in international society, as the source for international environmental action and leadership.

Green Westphalia: The Pluralist Logic of Environmental Action

The ES's original theoretical concern was with the question of order, and specifically with the extent to which 'the inherited political framework provided by the international society of states continue[s] to provide an adequate basis for world order' (Hurrell, 2014: 143). For the first generation of ES scholars writing during the Cold War, international society's political framework was predominantly pluralist in nature, and the question of how to establish and maintain order was to be resolved within the constraints of this framework. Pluralist international society is based on a limited range of shared norms and institutions. Because states have profoundly differing values and interests, they are unlikely to cooperate on more than a narrow range of issues. Where cooperation is possible, it is mostly concerned with maintaining coexistence or managing shared fate concerns. International order tends to be minimalist in nature, and demands for international justice and deeper levels of cooperation are likely to go unmet in a pluralist context (Bull, 1977; Hurrell, 2007: chapter 2; Buzan, 2014: chapter 6).

Unsurprisingly, therefore, pluralist ES theory injects a strong dose of scepticism into the debate about the chances for a successful greening of international society. For one, if state behaviour is governed by a logic of coexistence, not cooperation, and states are primarily concerned with protecting their survival and independence as sovereign members of international society, the scope for sustained international environmental cooperation is likely to be severely limited. A pluralist international society should still be able to achieve some level of international coordination of environmental action, mainly to deal with shared threats that are relatively easy to resolve and do not involve distributional conflicts. However, the norms and rules underpinning such international efforts are bound to be narrow. Furthermore, as pluralist international society is characterised by deep cultural and ideological differences, states will find it difficult to establish a universally agreed understanding of the kind of environmentalism that international cooperation needs to be based on.

Value diversity characterises not only the society of states but also global environmentalism. As discussed in Chapters 3 and 10, environmental discourses have produced different conceptions of nature and what it means to protect the natural environment. This inescapable reality of value pluralism in both international and world society thus makes it difficult for states to advance a more solidarist agenda in global environmental cooperation.

Hedley Bull provides a classic statement of this pluralist perspective on global environmental politics. In a short passage of *The Anarchical Society* (1977), Bull discusses the rise of global environmentalism against the backdrop of the perceived legitimacy crisis of the state-centric international order. Bull's intention is to repudiate environmentalists' claims that the Westphalian nature of international society is itself partly to blame for the global environmental crisis. At the time when Bull was writing his book, it was not uncommon for environmentalists to argue for the creation of a strong international authority that would overrule narrow national interests in order to safeguard the planet's future. Richard Falk's *This Endangered Planet* (1971), in particular, served as a foil for Bull's defence of state-centricity and pluralist order against solidarist ambition. Bull (1977: 293–4) makes three interconnected points: First, international society is inherently pluralist, in terms of the diverse societal values and interests that it represents, and this prevents sustained international cooperation for the global common good, including environmental protection. Second, effective environmental solutions cannot be found without the order and stability that the state-centric international society provides. Third, even if the environmental crisis were to exceed international society's problem-solving capacity and a solidarist response was needed, it is only through the states system that 'a greater sense of human solidarity in relation to environmental threats may emerge' (Bull, 1977: 294; see Falkner, 2017b, for a more comprehensive discussion of Bull's argument).

It is clear from Bull's discussion that he views pluralist international society as a necessary, if not sufficient, condition for tackling global environmental problems. But how far does the pluralist logic take global environmental action? How strongly does it embed environmental stewardship in the normative fabric of international society? Is it compatible with strong versions of interstate environmental cooperation? To answer these questions, we need to examine in more detail how the pluralist logic of coexistence motivates states to pursue environmental cooperation alongside other foreign policy objectives. The answer is twofold: without a universally shared vision of global environmental sustainability, states act on environmental threats for two main reasons: either out of *raison d'état*, that is subject to individual states' calculation of their national interest in an anarchic environment, or through a shared sense of *raison de système*, when the leading members of international society decide to act together 'to make the system work' (Watson, 1992: 14; see also Buzan and Schouenborg, 2018: 35).

The *raison d'état* perspective defines environmental objectives as matters of national, not planetary, interest. Simply put, where a state perceives environmental degradation to affect its core interests, it is likely to act on it provided other vital national interests don't stand in the way of environmental measures. In a pluralist context, we should therefore expect states to undertake environmental action where the costs of such action are either minimal or clearly outweighed by the benefits of environmental protection. States may choose to coordinate their actions with other states on an ad hoc basis, and occasionally enter into lasting institutional arrangements, such as multilateral environmental agreements. Given the pluralist constraints of international society, however, any such arrangement will be based on a shallow sense of shared purpose. Most importantly, institutionalised forms of environmental cooperation will stop short of infringing on the classical primary institutions, most notably sovereignty and territoriality. A pluralist international society is thus entirely consistent with some degree of collective environmental policy-making at the international level, as Buzan (2004: 145) argues, though this will fall short of the solidarist ambition for globalised decision-making and legalisation. Cooperation between the superpowers during the Cold War provides a blueprint for such international environmental coordination. Despite being locked into a profoundly adversarial relationship, the United States and the Soviet Union were able to establish a range of international norms and rules to manage their relations, including on environmental issues (Brain, 2016). Environmental issues offered an opportunity to strengthen the East–West dialogue during the era of détente (Hünemörder, 2010), not least as they were widely seen as 'low politics' at the time and did not directly play into the core dynamic of Cold War competition.

It is not difficult to find instances of states agreeing to internationally coordinated action out of environmental *raison d'état*. The first environmental treaties signed from the late nineteenth century onwards were bilateral or plurilateral agreements mainly aimed at resolving conflicts over transboundary environmental problems. None of them involved any serious effort at international institution-building or legalisation. The 1909 Great Lakes Water Quality Agreement between the United States and Canada is one of the first examples of two states jointly managing shared natural resources through international diplomacy, and the two countries went on to sign other such bilateral agreements in the early twentieth century, to protect migratory species and to manage the environmental quality of their shared lakes and rivers (Dorsey, 1998). In a similar fashion, European states sought to manage shared ecosystems through limited forms of environmental diplomacy. After the Second World War, the riparian states of the Rhine (Germany, France, Luxembourg, the Netherlands and Switzerland) signed various treaties to limit the pollution of the river. This included the creation of the International Commission for the Protection of the Rhine in 1950. As in North America, these early forms of international environmental management in Europe did not involve a solidarist commitment to deep cooperation. The Rhine Commission served mainly as an advisory body and was never given sufficient authority to function as an international regulatory agency (Bernauer and Moser, 1996).

One of the most ambitious and successful global environmental treaties, the Montreal Protocol on ozone layer depletion, can also be interpreted as following a logic of coexistence, in that it dealt with a 'clear and present global environmental danger for which countermeasures were within reach' and the emphasis was on 'measures necessary to maintain the conditions of existence for the members of the [international] society' (Buzan, 2004: 145). The leading producers of ozone-depleting chemicals (the United States, Japan, Britain, France, Germany) were also those that were faced with the gravest threat from ozone layer depletion, which was at its strongest in the regions closest to the polar regions, so in this sense the rational calculations of self-interested actors and not their desire to promote global solidarity for nature preservation was the driving force behind international ozone cooperation (Falkner, 2008: 51). However, coming up with an effective solution to the ozone problem required more than just a pluralist version of low-level environmental coordination. As discussed in the preceding chapter, the Montreal Protocol stands out among the environmental regimes of its time not least for its innovative provisions on financial aid to developing countries and a strong compliance mechanism, which included the possibility of trade sanctions (Benedick, 1991). In this sense, the growing ecological interdependence and complexity of the world economy pushed states to experiment with new forms of international cooperation that went well beyond what a narrow pluralist logic would dictate.

A *raison de système* perspective takes the argument about self-interested state behaviour one step further and identifies a shared interest in preserving international society among the leading powers as the main motive for international environmental action. As Watson (1992: 14) argues, collective action by the most powerful members of international society is possible where they share a 'belief that it pays to make the system work'. This collective hegemony is usually concerned with matters of international peace and security, as in the case of the nineteenth-century Concert of Europe (Clark, 2011). It is no stretch of the imagination to apply this logic to environmental problems too, particularly where they are so severe that they threaten peaceful coexistence among sovereign nation-states. In such cases, leading powers may wish to work together to avert an ecological crisis from spilling over into a system-wide conflict among states. International environmental cooperation would originate not from a shared ecological vision but out of a concern for survival and preventing international disorder. Given that the majority of environmental problems do not pose such an existential threat to pluralist international society, we should expect a collective environmental hegemony to arise only in a limited number of extreme ecological cases, such as catastrophic climate change.

As the earlier discussion has shown, a pluralist international society is not incompatible with some level of international environmental cooperation. That said, it is also clear that in a pluralist context environmental stewardship is unlikely to harden into a prominent fundamental norm that could challenge or modify the established classical primary institutions. Should certain environmental problems intensify and begin to threaten either the core interests of individual states or the survival of international society as a whole, then the pluralist logic of coexistence could provide the basis for stronger forms of coordinated environmental action at the international level. It would involve states recognising ecological threats as security threats, leading to partial or complete environmental securitisation. Furthermore, a form of collective hegemony could emerge that would develop a coordinated international response to avert catastrophic ecological trends. Both developments, the securitisation of the environment and the emergence of collective environmental hegemony, offer a clear pluralist logic for enhanced international environmental action. In the next section I explore the conditions for successful environmental securitisation, the first signs of which can already be identified. In the final section of this chapter, I consider whether great powers could potentially engage in some form of collective hegemony, or great power management, for environmental reasons.

Securitising the Environment? *Raison d'état* Meets the Environmental Crisis

Under normal circumstances, global environmentalism cannot be expected to play anything but a marginal role in pluralist international society. In an international context where sovereignty is the prime organising principle and states are deeply divided along cultural and ideological lines, maintaining international order is bound to be the predominant concern. This will inevitably come at the cost of other common interests, including environmental protection, which can only be pursued to the extent that they do not interfere with the principles of coexistence, national sovereignty and value pluralism. Deeper forms of international environmental cooperation based on solidarist values are simply out of reach. But what if some environmental problems were to become so calamitous and destabilising that they posed a threat not just to planetary health but also to states' core interest in their independence, integrity and survival? What if *raison d'état* made it necessary for states to participate in internationally coordinated action against an existential ecological threat? In other words, could environmental degradation become a source of insecurity, comparable to traditional security threats emanating from outside a state's borders? If such a scenario were to occur, it could propel environmental protection from the margins to the centre of the foreign policy agenda. In this section, I explore how such environmental securitisation could come about, and whether it would provide a basis for the strengthening of global environmentalism as a fundamental norm in pluralist international society.

Securitisation and the Environment

The concept of securitisation is well established in the IR literature and needs only a brief summary (Wæver, 1995; Buzan, Wæver and De Wilde, 1998). Securitisation is the process through which actors construct an existential threat that mobilises a given political community, usually a state, to generate a corresponding emergency response in order to counter the perceived threat. In contrast to realism, which treats security threats as an objective condition, securitisation theory and the ES view them as socially constructed. It is through a social process – a discursive representation or speech act – that a political issue is defined as an existential threat and is elevated above normal politics: 'If we do not tackle this problem, everything else will be irrelevant (because we will not be here or will not be free to deal with it in our own way)' (Buzan, Wæver and De Wilde, 1998: 24). Securitisation relies on such speech acts to succeed, but actors pursuing securitising moves only produce a successful outcome – full securitisation – if the audience has given its consent. Given its intersubjective nature, with society determining which issues are to be seen as an existential threat, security is never fixed in its meaning. It can encompass a wide range of issues, not only military threats but also migration (Huysmans, 2000), transnational crime (Emmers, 2003), minority rights (Roe, 2004) and infectious diseases (Elbe, 2006).

States are the usual referent object for security, though securitisation can also target individuals, subnational communities, humanity or even other living beings in the non-human environment (McDonald, 2018). Especially in a pluralist international context, security discourses tend to favour the state as the main referent of securitisation. The same can be said for the securitisation of environmental problems, which traditionally occurs at the level of states, particularly when one state's sovereignty is threatened by environmentally damaging activities emanating from other states. In this scenario, environmental securitisation pits one state against another, which is often considered to be the most promising form of securitisation as rivalry helps to rally society behind the state (Buzan and Wæver, 2009: 255). But the environment can also be constructed as an overarching security threat that affects not just a particular state but all states collectively. This higher-level securitisation would treat international society, and potentially also world society, as its referent object. Climate change, insofar as it threatens the well-being and survival of all societies on the planet, is one such environmental problem that is increasingly being talked about as a high-level security threat (Buzan and Wæver, 2009: 254). The question remains, however, whether securitisation moves for such macro-threats can be successful, given that the environmental impacts are often complex, indirect and diffuse. Moreover, world political community may be the ultimate referent in such macro-securitisation but it does not normally command the same political affection and loyalty as the nation-state. In a pluralist international society, environmental securitisation is thus likely to be pushed back to the state level (I discuss higher-level securitisation aimed at international society in the next section).

There are good reasons why global environmental problems have been conceptualised as both a traditional and a non-traditional security concern. Many forms of environmental degradation can dramatically reduce the well-being of individuals, local communities and entire societies, and some even pose an existential threat to them. Air pollution is widely recognised to have a negative effect on human health, causing chronic disease and premature deaths on a large scale; toxic effluents from factories that seep into the soil, rivers and lakes threaten drinking water supplies; and plastic waste can find its way into the human food chain as micro plastics via the world's oceans and marine life. In an enlarged framing of security, such environmental bads can all be conceived of as a threat to human security (Barnett, 2003). Some environmental problems even pose a direct, existential, threat to the survival of individuals, communities and societies, in which case it is also possible to mobilise traditional security frames (Trombetta, 2008). Climate change is by far the most prominent example of such an existential threat. Should global warming reach 3°C or more by the end of this century, the resulting melting of glaciers and polar ice caps is expected to lead to a global sea level rise of 1 m or more. This would not only undermine the livelihoods of millions of people living in densely populated coastal areas, but also threaten the very survival of low-lying island states and might set off migration-induced interstate conflict (Dupont, 2008).

Securitising the environment has proved popular not least because of the political expediency of defining problems as security risks. Actors have applied a security framing to a wide range of non-military threats in the hope of mobilising a more urgent and effective political response. As Copenhagen School theorists argue, the very point of defining an issue as a security threat is to remove it from the 'normal haggling of politics' (Buzan, Wæver and De Wilde, 1998: 29) and to elevate it to the level of an existential threat that justifies extraordinary measures. Securitisation involves a process of depoliticisation that does away with routine politics (debate, contestation, compromise seeking) and enables authorities to centralise political power in order to respond more effectively to an emergency. Establishing the environment as a security threat can thus help sidestep the slow-moving processes of environmental policy-making that usually produce lowest-common-denominator outcomes. It cuts through issue complexity and scientific uncertainty that stand in the way of urgent action, particularly when environmental responses need to be coordinated internationally through regime-building processes (Underdal, 2010; Keohane and Victor, 2011).

There is some debate in IR about what it means for securitisation to be successful. In Copenhagen School terms, successful securitisation occurs when authorised actors take up emergency measures that would not be available (or difficult to achieve) through normal politics. Within democratic states, this usually refers to a concentration of power in the hands of the executive and a dispensation of conventional democratic processes – a form of 'executive unilateralism' as developed in Carl Schmitt's decisionist political theory (Williams, 2003). However, securitisation that leads to extreme measures, such as the breaking or suspension of law, is rarely found in democratic politics. Critics have therefore suggested that this is far too narrow a conceptualisation, and that it cannot be easily applied to non-democracies and international contexts where 'normal politics' are less clearly defined. Rita Floyd (2016: 678), for example, proposes a more meaningful notion of (domestic) exceptional measures, which she extends to the passing of new laws to empower the executive, the granting of emergency powers or the use of a state's existing security apparatus. What matters is what relevant actors agree to constitute as extraordinary measures in dealing with an identified threat. Applied to the international context, we should similarly be able to identify successful securitisation where the agreement to identify a problem as a security threat is followed by a change in the behaviour of relevant agents. As Floyd (2016: 684) notes, 'the action taken is justified by the securitising actor with reference to the threat they identified and declared in the securitizing move'. This could result in the creation of new institutions or the empowerment of existing institutions to act in new ways, or a shift from an institution that is unable to act towards one that has the capacity to organise a global emergency response. Successful securitisation empowers relevant actors to overcome blockages in 'normal politics' that would otherwise persist if securitisation was unsuccessful.

Environmental Securitisation: The Record (So Far)

What does the historical record tell us about past environmental securitisation moves? Even on the basis of this enlarged conceptualisation, we find that most attempts at securitising the environment have had only limited success. Neither have states taken emergency measures to bring the underlying drivers of major pollution trends under control, nor have relevant actors within states been empowered to deliver more effective political solutions. In the field of climate change, which has dominated environmental securitisation moves, the long-standing trend towards ever-greater greenhouse gas concentrations in the atmosphere continues unabated, despite the growing attention paid to the climate-security link. Securitisation remains a popular discursive strategy within environmental policy debates, but has yet to deliver the results its advocates hope to achieve.

It is important to note that there is considerable variation among past and current securitisation moves. The first efforts to define environmental degradation as a security issue can be traced back to the 1970s (Falk, 1971; Brown, 1977), though it was only in the late 1980s that a more comprehensive debate among academics and practitioners about the links between environmental degradation and security emerged (Ullman, 1983; World Commission on Environment and Development, 1987; Myers, 1989; Mathews, 1989; Gleick, 1991). Boosted by the end of the Cold War, which many interpreted as opening up opportunities for redefining the global order, the environmental security concept soon developed resonance within various military and political organisations, from the North Atlantic Treaty Organization (NATO) and the US Department of Defense to the Organization for Security and Co-operation in Europe (OSCE) and the EU. Under pressure to re-legitimate their security role in a dramatically altered international context, these state institutions and international organisations explored different ways in which environmental degradation can act as a (direct) source of domestic and international conflict or as an (indirect) threat multiplier. Unsurprisingly, military organisations focused on the contribution that runaway global warming makes to conflict within or between states. In contrast, international organisations in the development field emphasised a wider, human security-based, notion of environmental security. In its landmark Human Development Report of 1994, the UN Development Program (1994) broadened securitisation to the daily insecurity of individuals caused by both conflict and the erosion of sustainable living conditions. In this version, environmental damage increases individuals' and communities' vulnerability and threatens their 'freedom from harm and fear' (Adger, 2010: 275). There was thus more than one way to attempt securitisation of the environment, with important consequences for how a securitised response to environmental threats would be organised.

The environmental security discourse is well established among leading powers, particularly in the West, but is also recognised in other parts of the world. In fact, some 70 per cent of UN member states have declared climate change to be a national security threat (Scott, 2015: 1330). The push for environmental securitisation is not progressing equally among major powers, however. Comparative analysis of security discourses has shown that they vary considerably across different political, cultural and historical contexts and produce different levels of engagement with the environmental security agenda (Hayes and Knox-Hayes, 2014).

In the United States, securitisation moves first gained momentum during the Clinton Administration in the 1990s. They reached a high point during the first Obama Administration in 2008–9, at a time when the White House and its Congressional allies made the case for federal climate regulation to combat rising greenhouse gas emissions (Floyd, 2010), before declining again with the arrival of the Trump Administration. Security politics has remained an attractive framing in the US context, however, as a way to overcome public reticence to acknowledge and engage with the climate change agenda (Hayes and Knox-Hayes, 2014: 90).

Efforts to establish a climate security link have been more even, and somewhat more successful, in Europe, where human security and human vulnerability tend to prevail over the national security-oriented discourse that is more common in the United States (Diez, Von Lucke and Wellmann, 2016). The German government commissioned a major report by its German Advisory Council on Global Change (2008), which highlighted multiple dimensions of climate security: water and food security, livelihood security and the threat of climate change-induced disasters. Ever since the publication of its first national security strategy in 2008, the United Kingdom has made climate change an integral part of its national security planning (Harris, 2012: iv). The United Kingdom has also sought to establish climate security as a policy priority at the international level: during the UK Presidency in 2006, the G8 for the first time accepted a fundamental link between energy, security and climate change; and in April 2007, the UK chaired the first UN Security Council debate on climate change, which discussed rising temperatures as an amplifier of conflicts within and between states (Scott, 2012: 221). More recently, the securitisation debate has also gained growing popularity in developing countries. Support for climate securitisation in the Global South is more uneven, however, and where environmental/climate security threats are identified they tend to be more closely associated with energy, water and food security concerns (Nyman and Zeng, 2016; Schäfer, Scheffran and Penniket, 2016).

It is fair to say that across the world there is a considerable gap between the diffusion of environmental securitisation discourses and the effects that existing securitisation moves have had on environmental policy. Despite the growing popularity of viewing environmental threats through a security lens, most countries have failed to take the next step towards a full securitisation of environmental response strategies. This is even the case in the most likely candidate for environmental securitisation: climate change. In the case of Australia, for example, the government of Prime Minister Kevin Rudd, despite explicitly framing climate change as a security threat at national and international level, failed to introduce even 'relatively mainstream policy measures' (McDonald, 2012: 580), let alone exceptional measures to avert the climate threat. The Australian public may have been sympathetic to depictions of climate as a security concern, but political contestation of climate policy by the main opposition scuppered efforts to establish a carbon emissions trading system or tax. In a context in which political polarisation around climate change deepened, the Australian government and military were unable to act successfully as a securitising agent (Thomas, 2015).

In other countries, too, the growing popularity of securitising discourses failed to generate the kind of political responses that would suggest successful securitisation of climate change. In the United

States, fluctuating support for climate policies has resulted in an uneven process of attempted but unsuccessful environmental securitisation. After a few failed attempts to establish federal policy frameworks for climate mitigation under Obama, the Trump Administration not only rejected the climate-security link in its national security strategy but also sought to de-legitimise climate policy altogether, ordering the removal of references to climate change from all federal policy documents (Calderwood, 2019). European countries have been far more consistent in their attempt to integrate climate change objectives into core economic policies, and securitisation discourses have if anything intensified. But even in Europe, calls for extraordinary measures to tackle the climate security threat remain the exception, indeed most securitisation moves end up emphasising the importance of established national and multilateral processes such as those of the UNFCCC (Oels, 2012: 191). Far from transforming politics, securitisation is largely used as a tool to reinforce 'normal' climate change politics and bolster European efforts to claim an international climate leadership role within established political channels.

What explains the limited success of environmental securitisation? Despite the growing popularity of securitisation discourses, few if any countries have managed to define a securitised policy response to environmental threats. For most states, except perhaps those low-lying island states threatened by rising sea levels from climate change, the pursuit of *raison d'état* within a pluralist logic of coexistence does not yet include the pursuit of environmental sustainability as an urgent purpose of foreign policy. Two major factors explain the limits to environmental securitisation to date.

One reason lies in the very nature of the environmental policy problem. Environmental degradation is a pervasive phenomenon in modern societies that has multiple and complex causes, which are to be found in a myriad of industrial processes and consumptive patterns. For many environmental problems, and most certainly for complex global ones such as climate change, an effective policy response requires structural changes to industrial processes, energy systems, transport infrastructure and agriculture on a global scale. The scope for emergency responses and quick fixes to deal with these challenges is extremely limited. As long as securitisation favours short-term, territorially defined and even militarised policy responses, it is unlikely to bring about the kind of long-term and internationally coordinated changes to energy systems, industrial processes and cultural norms that are needed to deal with global environmental threats. It is for this reason that some environmentalists have questioned the utility of a security framing as it serves to promote counterproductive responses that end up legitimating state-centric policy approaches (Deudney, 1990). The comparison with the global response to the outbreak of COVID-19 is telling in this regard. Governments around the world were able to introduce draconian restrictions on their citizens' mobility, including the shutdown of factories, schools, shops and restaurants as well as the closing of borders, in an effort to contain the spread of the virus. In this case, emergency measures were possible not least because it was widely understood that they were needed for an effective response to an imminent threat to people's lives.

A second reason lies in the difficulty of mobilising a security framing for threats that emanate from the structural conditions of industrial capitalism rather than human agents (I am grateful to Barry Buzan for this point). The threat of military aggression by other states is the classic case in which most societies are willing to suspend normal politics and mobilise for a military defence. Securitisation around terrorist threats or even migration has similarly proved successful in the past, not least because it involves conflicts that pit one group of humans against another. Such threats create a zero-sum logic that helps mobilise traditional security responses and justifies empowering the executive to take emergency-type action. In contrast, environmental threats, although ultimately caused by human activities, manifest themselves as anonymous forces (e.g. rising sea levels, loss of biodiversity, more severe weather events) that produce human suffering in a more indirect way. They may still trigger traditional threats involving a zero-sum logic of distributional conflict, for example by causing migratory flows or competition for scarce resources, but any securitisation of such issues would address the symptoms, not causes, of the underlying environmental stresses.

If indeed securitisation, as the Copenhagen School suggests, is a form of de-politicisation that removes the environment from normal political processes and activates an emergency response mode, then the empirical record suggests an unambiguous conclusion: existing securitisation moves have been about mobilising support for more ambitious, but ultimately conventional, climate policy. For reasons outlined above, securitisation discourses are strategically employed to create political momentum behind environmental policy but struggle to define how securitised responses would help to deal with the root causes of environmental threats. Applied to the international realm, we find that the securitisation of the environment has similarly ended up mobilising international support for existing multilateral efforts, rather than transforming the international politics of climate change into a securitised field of action. It has resulted in a form of 'macropoliticisation' (Buzan and Wæver, 2009: 271), which helped to get climate change and other environmental threats onto the global agenda. Whether securitisation moves could eventually engage existing mechanisms for international emergency measures (e.g. as part of the UN's collective security system) remains a matter of debate and will be examined more closely in the next section. As far as the major global powers are concerned, however, none have as yet chosen to go down this path out of a conventional logic of *raison d'état*.

Could this change in the future? What would move major environmental threats (e.g. climate change) from politicisation towards full securitisation? Would the escalation of global environmental stresses to an existential ecological threat trigger a corresponding shift in problem perception and response strategies? The current debate about the ecological consequences of runaway global warming offers a glimpse into such a future. Should current efforts to limit global warming to below 2°C by the end of this century fail, then global ecosystems may reach various ecological tipping points that would lead to accelerated and potentially catastrophic climate change. In a 3°C or 4°C warming scenario, the melting of polar ice caps and glaciers would lead to a rise in sea levels of anywhere between 1 and 3 m. Major ecosystems (coral reefs, tropical rainforests) would be lost, and extreme weather events (hurricanes, flash floods), would occur more frequently and with far greater damage to human populations (Lenton et al., 2019). The livelihoods of hundreds of millions of people would be directly affected, leading to greater distributional conflicts over natural resources as well as a dramatic increase in transboundary migration. Under such conditions, it is at least conceivable that states will act in ways that mirror responses to more conventional external security threats. In this perspective, a pluralist logic of coexistence would be consistent with states enacting securitised emergency measures in response to a global climate emergency, even though this would end up targeting the political ramifications, and not the socio-economic causes, of the underlying environmental crisis.

In a pluralist international society, we would expect such an ecological emergency to prompt decentralised emergency responses, with each state acting to protect its own national interest. But could it also trigger a collective response by major powers seeking to protect the survival of international society? In other words, could a pluralist international society move from a decentralised logic of *raison d'état* to one based on a shared sense of *raison de système*? In the following section, I explore the potential for the emergence of such a system of collective environmental hegemony, or what is referred to in ES terms as great power management.

Collective Environmental Hegemony: Towards an Environmental *Raison de Système?*

From Raison de Système to Great Power Management

The idea that international order requires some form of collective management by its members is well established in IR. Given the reality of persistent power inequality in international relations, one of the 'sheer facts' of international life (Bull, 1977: 205), this management role usually falls to the most powerful members of international society: the great powers in a multipolar system, or a single hegemon in a unipolar system. Materialist theories of IR make the standard assumption that the most powerful states are those that generally determine international outcomes; they therefore also carry out international management roles by default. Social theories accept the reality of asymmetrical power distribution but take this idea one step further and identify it as a *social,* as much as a material, fact. To be considered a great power requires both a certain distribution of power *and* social recognition by other states. While the sheer preponderance of power enables a state to carry out a management role, it is only social approval that can lend legitimacy to it. In the ES tradition, collective management by the great powers is therefore considered to be more than a consequence of material power structure, it is a social institution that forms part of the international normative structure. Bull (1977) identifies great power management (GPM) as one of the five classical primary institutions of international society (alongside diplomacy, international law, balance of power and war). Great powers accept special responsibilities for international order and provide certain public goods, and in exchange are given special rights that put them apart from other members of international society. The power inequality that underpins GPM is thus institutionalised in that these great power privileges and responsibilities are socially sanctioned (Clark, 2009: 207–20; Simpson, 2004; Suzuki, 2009: 50).

GPM in its classical form was focused mainly on security relations between states. Could it be mobilised to also take on a global environmental management role? Could an ecological *raison de système* emerge that would lead great powers down a path of taking on a special environmental leadership role as part of their great power responsibilities?

The traditional security focus comes out clearly from Bull's (1977: 207) list of functions that GPM fulfils: It includes the preservation of the balance of power, the avoidance or control of central crises, and the containment of central wars. Until the second half of the twentieth century, the scope of GPM was indeed limited to maintaining the peaceful coexistence of sovereign states. But the social purpose of GPM is not fixed and can be enlarged should the members of international society adopt a wider normative ambition. Indeed, signs of such a normative expansion can be found in the post-1945 era. The international security agenda has gradually been enlarged to include non-traditional or human security issues, such as economic, health and identity, as well as global environmental protection (Buzan, Wæver and De Wilde, 1998), and with the expanding scope of global security concerns have come demands for great powers to take on additional special responsibilities (Bukovansky et al., 2012: 47–8). Whether great powers are actually able to provide this growing list of global public goods remains an open question. Depending on the complexity of the problem structure, particularly in an age of increasing interaction density and interdependence, we may find that global cooperation needs to engage a wider range of relevant actors beyond the clique of great powers, which would bring GPM into closer contact with global governance (Cui and Buzan, 2016: 207–10).

Applying GPM thinking to the environmental field creates ambiguities and raises difficult questions. As is common in other areas of great power responsibility, the link between the possession of great power capabilities and acceptance of special responsibilities for the international order is not a straightforward one. Throughout history, great powers have repeatedly failed to take on special responsibilities in the pursuit of the global common good. They may have had the capacity for international leadership but lacked the will to perform such a role, or their national interest clashed fundamentally with what a responsible great power role demanded of them. To be sure, this central ambiguity in the great power concept is recognised in the ES literature. Bull (1980) himself memorably described the United States and the Soviet Union during the Cold War as behaving like 'great irresponsibles'. But even if a great power is willing to act as a responsible power, it may not find the approval of other states for taking on both special responsibilities and rights in the pursuit of international order. Other states may withhold their consent, because they disagree with the objectives that a great power is pursuing or because GPM is perceived to be ill-suited for producing the desired global public good. It is clear that GPM cannot be taken for granted as the primary institution that will always manage the affairs of international society.

Engaging GPM for environmental purposes thus raises important questions about what could trigger such a development in international relations, what role it would play in solving global environmental problems, and what legitimacy it could command among the wider membership of international society. In the following, I briefly review the historical record of how international society has dealt with asymmetrical power distributions and special rights and responsibilities of great powers, before exploring whether global securitisation of the environment might trigger some form of collective GPM in the search for a global environmental rescue.

Differentiation of Responsibilities in GEP

International power asymmetry has been a structural condition of global environmental politics from its origins in the nineteenth century, though it never came to be legitimised as the basis for organising international environmental policy. As discussed in Chapter 4, colonial empires provided an important context for the creation of ecological knowledge and experimentation with novel conservation practices (Grove, 1995). They also gave rise to a discourse of colonial powers' special environmental responsibility, with conservation in colonial territories increasingly seen as part of Europe's 'civilising mission' (Neumann, 2001). Environmentalist paternalism also characterised the efforts by some of the first international NGOs that campaigned for the protection of the flora and fauna in tropical regions. Their ideas of wildlife preservation often clashed with the interests of local populations whose livelihoods depended on access to their homelands (Domínguez and Luoma, 2020). Colonialism thus provided the first political and ideational framework within which certain ideas of global environmental responsibility emerged, among both states and non-state actors. Colonial environmental responsibilities of course lacked the free consent of colonised societies, and they never developed into a wider system of environmental GPM. If anything, the colonial legacy of green paternalism cast a dark shadow over twentieth-century environmentalism and efforts to establish environmental stewardship as a primary institution (see Chapter 6).

When environmental stewardship came to be recognised as a fundamental duty of GIS in the 1970s, the colonial legacy of environmentalism was very much on the mind of those Southern leaders that resisted the new global environmental agenda. As the difficult preparatory process for the 1972 Stockholm conference showed, developing countries viewed claims by Northern powers to provide international environmental leadership with deep scepticism. For some, Northern environmentalists' emphasis on pollution control and limits to growth was little more than a neocolonial ploy to prevent economic development in the Global South. The rifts at UNCHE left no doubt that if environmental stewardship was to become universally accepted it had to take into account the special needs of the developing world, and global environmental responsibilities would have to be defined and distributed along North–South lines (see Chapter 5).

While the Stockholm conference had established environmental stewardship as a general obligation, international society assigned special responsibilities to industrialised countries based on two criteria: their greater capacity to contribute to global environmental protection, which reflected their advanced economic and technological development; and their culpability in causing industrial pollution and resource depletion, even though leading industrialised countries were keen to avoid making explicit links between historical culpability and specific legal obligations to pay compensation. After Stockholm, a growing number of secondary institutions incorporated the principle of North–South differentiation, most notably the Montreal Protocol on ozone layer depletion (1987), which gave developing countries a delayed compliance schedule for phasing out ozone-depleting chemicals and offered them financial and technological aid. By the time of the 1992 UNCED, differentiation had evolved into a formal principle, CBDR. As had been noted in the UN General Assembly Resolution that convened the Rio Summit, 'the responsibility for containing, reducing and eliminating global environmental damage must be borne by the countries causing such damage, must be in relation to the damage caused and must be in accordance with their respective capabilities and responsibilities' (United Nations General Assembly, 1989). The question of Northern culpability remained a contested issue at the Rio Summit and beyond, and the leading industrialised countries have always resisted specific duties (e.g. financial transfers) that arise from their historical responsibility for environmental degradation. However, by adopting the CBDR norm in the Rio Declaration and the UNFCCC, GIS enshrined North–South differentiation as the key guiding principle in structuring global environmental responsibilities.

At first sight, CBDR appears to overlap somewhat with GPM in that it creates special environmental responsibilities for the wealthiest countries. However, these responsibilities apply not just to the great powers but to all industrialised countries, from the United States and Japan to Belgium and Denmark. Furthermore, secondary institutions in the environmental field do not establish any special rights or privileges for industrialised countries in recognition for the special responsibilities they have come to accept. Thus, whereas the UNSC gives formal veto power to its five permanent members as part of GPM for international security, and the IMF allocates voting rights that are weighted according to the size of countries' financial contributions, environmental regimes generally apply the 'one-member-one-vote' and consensus principle in decision-making.

Furthermore, environmental aid mechanisms (e.g. Multilateral Ozone Fund; GEF) have departed from the Bretton Woods model by creating a membership structure with equal representation of donor and recipient countries, giving each group a collective veto right in decision-making (Streck, 2001). The special responsibilities that the CBDR norm establishes in GEP are thus neither exclusive to the great powers, nor are they part of a quid pro quo arrangement in which special responsibilities are balanced with special rights (Bernstein, 2020).

This is not to suggest that some of the industrialised countries have not played an oversized role in the creation of international environmental rules. The United States and certain European countries have had a powerful influence over important environmental treaties, many of which originated from small-n negotiations that were dominated by leading powers (e.g. Montreal Protocol; see chapter 6). Over time, large-n environmental multilateralism has become the norm as participation in treaty negotiations broadened, but major powers continue to exercise considerable influence over outcomes in environmental rule-making. However, such forms of environmental leadership and influence do not equate with great power responsibility, and none of the major candidates for environmental leader status have ever claimed to be acting out of special environmental responsibilities that reflect their great power status. De facto power inequalities in GEP have thus never been legitimised by a de jure GPM system.

To be sure, the practice of minilateralism centred on a few powers has never disappeared from GEP. Even in the context of contemporary multilateral negotiations, the voices of the most powerful still carry a special weight in the crafting of delicate political compromises that underpin MEAs. However, minilateralism's main use in GEP has been to advance, rather than replace, environmental multilateralism. In the politically charged negotiations on climate change, for example, major powers have increasingly resorted to high-level discussions in minilateral forum (e.g. Major Economies Forum; G7/8; G20) to resolve some of their core conflicts of interest (Brenton, 2013; Kirton and Kokotsis, 2016). Calls for a more decisive shift towards climate minilateralism to break the deadlock in the UNFCCC process increased notably in the run-up to the 2009 Copenhagen conference and after (Victor, 2006; Naím, 2009; Antholis and Talbott, 2010). When the Copenhagen conference failed to agree a post-Kyoto treaty, a small group of major GHG emitters decided to circumvent traditional processes and negotiated a political compromise deal, the Copenhagen Accord, to avoid a collapse of the conference. In the end, this move by the informal 'climate club' of major emitters was challenged by several developing countries for its lack of legitimacy, which meant that the agreement was only noted at the final plenary session in Copenhagen (Dimitrov, 2010). Still, the main elements of the Copenhagen Accord were carried over into the new architecture of the Paris Agreement of 2015, which signalled a shift in the underlying logic of international climate policy to reflect more closely the preferences of the United States and leading emitters from the Global South (Falkner, 2016b). In this way, great power minilateralism played a decisive role in both rescuing the UNFCCC process and changing the core regulatory approach of the climate regime. Still, minilateral initiatives continue to be tied into the multilateral framework of the UNFCCC and are designed to reinforce it. Climate minilateralism continues to suffer from a severe legitimacy deficit, while none of the major powers has as yet signalled a desire to replace the UNFCCC process with an alternative climate club (Hjerpe and Nasiritousi, 2015; Falkner, 2016a). In climate change as much as in other environmental issue areas, the great powers remain reluctant to replace existing multilateral practices with a stronger focus on an exclusive GPM approach.

At the same time, the inherited model for North–South differentiation in GEP has come under strain due to the rise of emerging economies from the Global South. As discussed in Chapter 7, contestation around the CBDR norm has intensified since the early 2000s, especially within the climate regime. Due to their rapidly rising share of global GHG emissions, emerging powers (e.g. China, India, Brazil) have been asked to take on a greater climate mitigation burden in line with their growing ecological footprint. The shift towards the new logic of the Paris Agreement, which includes a more balanced climate mitigation commitment by *all* GHG emitters, has thus helped to erode the previous consensus on differentiation and brought in a more flexible approach to defining special responsibilities in climate politics. Unsurprisingly, given its dominant position within the group of emerging economies and self-identification as a rising great power, China has been confronted with growing external expectations that it should take on special global responsibilities for climate stability (Kopra, 2018). The shift in the distribution of power and ecological burdens in world politics has thus challenged the strict North–South differentiation of environmental responsibilities. While this has added a degree of urgency to the question of how great power status relates to special responsibilities for the global environment, it has yet to engage GPM as the basis for redefining the allocation of responsibility in GEP.

Environmental Securitisation: A Role for the UN Security Council?

As we have seen, a settled discourse of great power responsibility and management has yet to emerge from the existing international politics of the environment. In the context of a pluralist international society, environmental concerns have not taken on a wider systemic significance for maintaining international order. Could this change if certain environmental threats were to escalate and pose an existential threat to some or all states? Could a major ecological crisis create the conditions for a shared sense of environmental *raison de système* to arise? As discussed earlier, climate change is widely seen to be the one environmental threat that, if left unchecked, could induce a collective emergency response from great powers (Bernstein, 2020). In theory, successful securitisation of the environment at the international level could thus produce the conditions for a system of collective environmental hegemony.

Such a scenario is not entirely hypothetical, for attempts to securitize climate change at the highest level can be traced to the mid-2000s, when the UNSC was for the first time drawn into the global climate debate. The UNSC would be the natural choice within the UN system for establishing a system of environmental GPM. Its five permanent members already accept special responsibilities for international security and hold special rights in the form of a right to veto any substantive decisions by the UNSC. If its mandate to maintain international peace and security were to be extended to climate change, then the UNSC could use a range of measures, including the exceptional powers established in Chapter VII of the United Nations Charter, as part of a global climate emergency response. Following various securitisation moves at the national level, the United Kingdom initiated the first UNSC debate on climate change and energy in 2007. A second debate followed in 2011, which concluded with a presidential statement that expressed concern about climate change aggravating 'certain existing threats to international peace and security' (United Nations Security Council, 2011). Two further Council debates in 2018 and 2019, as well as other discussions in informal 'arria formula' meetings, have signalled the growing interest among UNSC members in exploring a climate role for the institution (Scott and Ku, 2018; Conca, 2019).

The number of countries that have supported UNSC discussions of climate change has risen dramatically since the United Kingdom first put climate security on the Council's agenda in 2007. In the last decade, various European countries (Germany, Spain, Italy, Sweden, the Netherlands, France, Ukraine) and developing countries (Pakistan, Malaysia, Senegal, Egypt, Maldives, Peru, Morocco, Dominican Republic) as well as Japan and the United States have initiated or co-chaired UNSC and UN General Assembly discussions on the climate security link. There is now broad recognition among UN members that climate change is a security concern, with both human security and traditional international security framings being used to justify the involvement of the UNSC. In 2018, the Council went as far as recognising climate change as a destabilising factor in Somalia, which was interpreted by observers as a 'historic' decision (Derler, 2018). International securitisation moves have thus shown some success in rallying a broad alliance of countries across the North–South divide behind the climate security agenda.

One question, however, remains unresolved in this environmental securitisation discourse: what kind of international action would be sanctioned once the UN's collective security system has been mobilised in support on an environmental *raison de système*. On this, opinions vary greatly among UN member states. A full-blown GPM approach would entail the UNSC accepting that climate change falls under Article 24(1) of the UN Charter, based on which the Council has 'primary responsibility for the maintenance of international peace of security'. This in turn would enable the UNSC to take action against the climate threat, with the options ranging from uncontroversial measures (e.g. support for climate mitigation and adaptation) to the more drastic but controversial chapter VII powers (establishing international criminal tribunals; use of sanctions; authorising military force) (Scott and Ku, 2018). As yet, most countries do not envisage full securitisation of climate change that would include coercive action by great powers based on a UNSC mandate. Some powers (notably Russia and China) are keen to keep climate change off the UNSC agenda altogether (Scott, 2018: 209; Conca, 2019: 9), while most other countries have sought to restrict the Council's climate agenda to measures that respect, in the words of a 2012 UN General Assembly resolution on human security, 'the sovereignty of States, territorial integrity and non-interference in matters that are essentially within the domestic jurisdiction of States' (United Nations General Assembly, 2012).

Just as in the case of national securitisation moves, global securitisation of climate change remains incomplete, despite the notable success in engaging the UNSC in climate-related debates. The security implications of global warming are generally accepted, and a broad North–South alliance of countries are pushing for a greater UN and UNSC role in addressing climate impacts. However, a shift towards global emergency responses based on a threat to international security remains only a theoretical possibility. Should climate change turn into a more immediate existential threat, the pluralist logic of coexistence could be mobilised to establish a collective emergency response led by the great powers, in pursuit of an ecological *raison de système*. However, the barriers to such a move remain stubbornly high. Irrespective of whether an environmental

GPM approach could ever hope to address the underlying causes of environmental crises, widespread concerns about its lack of legitimacy make a collective environmental hegemony an unlikely scenario for the foreseeable future.

Conclusions

Although environmental stewardship is, at its heart, a solidarist norm, it is entirely conceivable for it to emerge as a fundamental international norm within a pluralist international context. As I have argued in this chapter, the pluralist logic of coexistence can give rise to a certain degree of international environmental coordination. This is particularly the case when states face a shared environmental threat that can be dealt with in a manner that is compatible with the core pillars of a pluralist order: national sovereignty, non-intervention and diversity of values. Where states' pursuit of self-interest overlaps with the principle of coexistence, international society should be able to generate a base level of internationally coordinated environmental action. This will be a minimalist environmental order – Green Westphalia – that does not require convergence of values or a deeper level of international environmental institution-building and legalisation. Given the scale of the many ecological challenges, however, it is likely to fall short of the need for a more radical and sustained international response.

Could a future escalation of the environmental crisis trigger a stronger international response within the constraints of pluralist order? Could international society act in a collective manner against an existential environmental threat even without a universally shared vision for global environmental sustainability? I have argued in this chapter that pluralist international society holds two distinctive logics for escalating international environmental action: *raison d'état*, when states perceive environmental protection to be central to the pursuit of their own self-interest, including survival, and *raison de système*, when the leading powers decide to act together against a shared environmental threat, mainly in order to preserve the international order. For these logics to generate a more urgent international response, we would expect environmental problems to be perceived as an existential threat. Such a process of environmental securitisation is already underway in international politics, though it remains incomplete. Securitisation moves are strongest at the national level. Since the late twentieth century, nearly all major powers have identified climate change as a threat to national or human security. However, none of the existing securitisation discourses have resulted in actual emergency measures that target the root causes of climate change. Securitisation, apart from mobilising political support behind environmental objectives, has yet to deliver on its promise as an effective tool of environmental policy.

International-level securitisation has also gained momentum in recent years, with the UN Security Council as the main focus for delivering a securitised international environmental response. Again, we find that despite the growing popularity of the climate security link at international level, securitisation moves have failed to generate any substantial shift in the way GIS deals with the global climate threat. The link between ecological threats, environmental stewardship and international order remains weak (Bernstein, 2020), while existing security-based framings of global warming merely serve to increase momentum behind the UNFCCC's multilateral approach. As yet, the pluralist logic of coexistence has failed to generate an international environmental response that would go beyond the standard, shallow, forms of international policy coordination.

Unsurprisingly, therefore, the pluralist perspective holds only limited promise for a global project of building effective environmental governance, whether out of *raison d'état* or *raison de système*. Faced with a multitude of complex, long-term and diffuse environmental threats, the pluralist order has so far tended to act as a brake, rather than accelerator, for a global environmental rescue. Given the persistent strength of core pluralist institutions and an accelerating shift towards a post-Western form of deep pluralism (Acharya and Buzan, 2019: chapter 9), environmentalists would do well to accustom themselves with the inherent logic of environmental action within a pluralist international order. Still, should global ecological stresses intensify and turn into an existential threat to states or GIS, then it is not inconceivable that environmental securitisation might trigger a more effective pluralist response.

At the same time, the environmental failings of pluralist international society and the limited success of the state-centric solidarist project are redirecting our attention to potential sources for a global environmental response that are located outside the society of states. Could greater involvement of non-state actors in global environmental governance help overcome the limitations of state-centric approaches? In other words, could world society come to the rescue where international society has been found wanting? It is this question that I turn to in the next chapter.

10 World Society to the Rescue?

Environmental stewardship is a fundamental international norm that establishes environmental responsibilities for states and international society. Its origins lie in environmental discourses and campaigns in domestic and world society, and non-state actors – environmental campaigners, scientists and corporations – have played a critical role in its rise to prominence in international politics. Could it be that world society also holds the best hope of creating an effective system of global environmental governance? Do world society actors need to step in to shore up states' faltering environmental efforts? And what does greater involvement of non-state actors in GEP mean for global order and the relationship between international and world society?

There are good reasons for thinking that it is at the intersection of international and world society that we can find the sources of more effective global environmental action. After all, non-state actors have been critical norm entrepreneurs in the creation of the international environmental agenda, and they continue to exert important influence over international processes of environmental norm creation. Non-state actors are also increasingly able to generate mechanisms for environmental governance, with or without the involvement of state actors. Operating outside the state-centric confines of international society, such transnational governance initiatives have the potential to complement, strengthen and expand the existing problem-solving capacity of international society. Against the background of escalating environmental challenges, more and more scholars of GEP are therefore asking whether greater involvement of non-state actors can address the global environmental governance gap that states have been unable to fill. Could world society come to the rescue and help create a more viable global green order?

This chapter examines how international and world society relate to each other in GEP, and how world society-based environmentalism plays into state-centric efforts to create global environmental governance. It deals with what has rapidly become a central theoretical and empirical concern in ES research: world society's status as a distinct political realm and source of international norm creation, and the interaction between, and potential integration of, world society and international society. Although a central part of the ES's conceptual trinity, world society has received far less attention than the other two master concepts of international system and international society (Stivachtis and McKeil, 2018: 2). It is invariably described as the 'most problematic feature' (Little, 2000: 411) of the ES, a 'Cinderella' concept (Buzan, 2004: 11) that remains 'unclear and vague' (Buzan, 2018: 139). In the past, lack of theoretical reflection and empirical research held back the full development of the world society concept, though recent scholarship has begun to rectify the situation. Several ES scholars have developed a distinctive research programme on actors and norms that originate in world society (e.g. Buzan, 2004, 2018; Clark, 2007; Pella, 2013; Stroikos, 2018; Navari, 2018; Stivachtis and McKeil, 2018), and world society has now become a major site of theoretical innovation within the ES and beyond.

One of the central sources of confusion has been the ES's dual use of the world society concept, as a normative counterpoint to pluralist international society, and as an analytical concept for the empirical study of non-state actors. Many ES theorists typically combine both these dimensions. As with the solidarism–pluralism debate, this normative–analytical blending is certainly one of the strengths of the ES, but it has come at the cost of uncertainty over what world society stands for, and it may have also unnecessarily limited the scope of empirical research into world societal developments (Pella, 2013).

In the ES's normative debate, world society serves as a cipher for the 'moral cosmopolitanism' (Williams, 2014: 132) that is absent from pluralist international society. In this view, world society is largely synonymous with the solidarist position that expects humankind to unify around a common set of cosmopolitan values. Bull's *Anarchical Society* is a good example. Bull acknowledges world society's dual empirical and normative dimensions when stating that it represents 'not merely a degree of interaction linking all parts of the human community to one another, but a sense of common interest and common values, on the basis of which common rules and institutions may be built' (Bull, 1977: 279). Earlier on, however, Bull had already rejected the idea that world society exists as a significant political space outside the society of states. He is unequivocal in his judgement that world society matters mainly as a normative goal that is part of the solidarist imagination: It 'does not exist except as an idea or myth which may one day become powerful, but has not done so yet' (Bull, 1977: 85).

In contrast, treating world society as an empirical concept opens up an investigation into the observable growth in social interactions across boundaries, involving a wide range of actors: civil society organisations, corporations, scientific bodies, international organisations and subnational actors. In this perspective, the normative ambition that may be present in such interactions cannot be determined through a priori theorising but is subject to empirical exploration. This conceptualisation allows the ES to develop a more nuanced understanding of what specific normative positions world society actors actually promote and how these manifest themselves in given historical contexts. It also gets ES scholarship into closer contact with transnationalist research in IR that examines the activities of non-state actors, their interaction with the society of states and the impact they have on international outcomes. World society thus becomes 'transnational society' (Davies, 2017), which comprises both individuals and transnational groups (a combination of interhuman and transnational societies as defined by Buzan (2004: 133)).

Recent ES scholarship has managed to distinguish more clearly between normative and analytical perspectives while maintaining the benefits of keeping both in play. Clark (2007) and Ralph (2007) have sought to underpin normative claims about world society's cosmopolitan influence on international society with detailed case studies of international peace conferences and the ICC. Others (e.g. Weinert, 2018; Costa Buranelli, 2018; Stroikos, 2018; Linsenmaier, 2018; Stivachtis, 2018) have contributed to this effort and further elaborated the societal character of world society, that is the norms, institutions and identities that give seemingly random interactions among non-state actors the quality of belonging and contributing to a global societal structure. Having originally raised the need to re-think world society's societal foundation in *From International to World Society?* (2004), Buzan has recently returned to this question with a renewed effort to put the ES concept of world society on a sounder footing (Buzan, 2018).

Building on Buzan (2018) and Stivachtis and McKeil (2018), we can identify three potential uses of world society. The first, *normative world society*, represents the classical ES approach that treats Bull's 'great society of humankind' as an ethical ideal. It is based on the common human identity shared by all individuals, rather than any particular historical manifestation of collective agency by non-state actors. The second, *political world society*, captures the empirical dimensions of an existing, or emerging, political space outside state-centric international society. This space is populated by a plethora of non-state actors that operate independently of states and seek to influence outcomes in interstate as well as transnational relations. They draw on the ideational resources of normative world society, but it remains an empirical question as to which values and interests they promote. The third, *integrated world society*, is the most difficult to grasp empirically. It denotes the envisaged creation of a global social and political structure that merges interstate and transnational elements. It is a normative aspiration that remains empirically unfulfilled but serves as a referent in debates about the creation of more integrated forms of global governance.

In this chapter, I engage all three conceptual dimensions of world society as the basis for a closer examination of the role that non-state actors play in the greening of international society. In a first step, I discuss the normative aspirations of global environmentalism that animate non-state actors' engagement with international diplomacy: how it relates to Bull's 'great society of humankind', the solidarist ideal of a world of individuals held together by common environmental values and identities; and whether the existence of multiple environmental values instead points to a more pluralist structure of environmental world society. In a second step, I review the role that non-state actors play in shaping the outcomes in interstate environmental politics, but I also consider the reverse perspective of how engagement with the world of diplomacy shapes global environmental activism. In the final part, I consider the question of an emerging integrated world society, which grows out of the ever-deeper involvement of states and non-state actors in the creation of global environmental governance. This chapter thus brings together many of the different strands in the earlier parts of this book, to arrive at an overall interpretation of the growth of environmental world society, and how it is interwoven with international society.

Environmental World Society: Between Solidarist and Pluralist Values

The ES has been at the forefront of developing a normative account of world society within IR. In its classical form, as represented by Bull's writings, this account presents world society as an expression of a universal set of values – cosmopolitanism – which is based on people's shared identity as humans. It is for this reason that solidarism is often assumed to be hard-wired into world society (Williams, 2015: 34). Particularly when contrasted with the society of states, world society assumes the role of a solidarist counterpoint to international society's inherent pluralism. But whereas pluralists consider the cosmopolitan unity of humankind merely as an unachievable or undesirable dream, solidarists believe in the strengthening of world society as a way out of international society's pluralist predicament. Taking on an almost teleological function, world society thus becomes the endpoint of an international journey from the fractured world of sovereign states towards the 'social and political integration of humanity' (Linklater, 2011: 26).

The representation of world society as inherently and uniformly solidarist has been challenged more recently. Pella (2013) has argued that this classical ES perspective unnecessarily restricts the research agenda and ignores the diversity of normative influences that emanate from world society. In his own work on the emergence of the slave trade as an established practice in international politics, Pella shows how world society contains both progressive and regressive ideologies: The international trade in slaves was originally promoted by non-state actors and also came under attack by non-state actors, mainly religious and civil society groups that were campaigning for its abolition. We should therefore focus on the 'ongoing contestations between different ideologies in the non-state world, on the basis of which individuals form different types of social relationships to influence the international society of states' (Pella, 2013: 66). Buzan (2018: 128) similarly rejects the idea of world society's uniform normative character and introduces spatial differentiation and fragmentation. Human identities need not be universal in nature, they can be used to define any subset of humankind, from large-scale formations (religions, civilisations) to more restrictive transnational identities (e.g., Pan-Arabism, Pan-Russianism). Williams takes this one step further and argues that the inherent diversity of normative identities within world society is not necessarily a deficiency that is to be overcome but a source of strength. Arguing against the inherently solidarist nature of world society, he suggests that in fact 'a pluralist world society is potentially ethically desirable' (Williams, 2005: 19).

How then does global environmentalism map onto the normatively diverse landscape of world society? Does the environmental idea suggest a universal ethical outlook with humanity (or the planetary community of living beings) as its referent? Or do the different strands of environmental thinking that have emerged since the nineteenth century point to a more pluralist version of environmental world society? If 'collective identity' is the main primary institution of normative world society, as Buzan (2018) argues, then it should be possible to count environmentalism as one such collective identity on a potentially global scale. Collective identities exist outside the society of states, they are 'neither part of states nor dependent on states for their reproduction' (Buzan, 2018: 128). They can be global in scope, as in Bull's 'great society of humankind', but need not take on a universal character. As discussed earlier, religious and nationalist creeds as well as civilisations are transnational identities that represent only a subset of humanity. Environmentalism can easily be identified as one such collective identity. Whether it represents a universal or only sub-global community of individuals, however, hinges on the question of its appeal to their identity as members of humankind or some other normative referent. If the former, its underlying values are likely to be in the solidarist camp; if the latter, it opens up the possibility of multiple environmental identities across boundaries and a more pluralist context in which different environmentalisms have gone transnational.

Environmentalism as Solidarist Cosmopolitanism

There are good reasons for thinking of global environmentalism as a collective identity with universal scope. Environmentalists often appeal to fellow citizens on the basis that environmental protection is in humanity's interest, that maintaining a healthy and sustainable environment is a general, not special, interest. Two principal types of reasoning – necessity and global responsibility – stand behind this claim. For one, the universal interest in environmental protection is explained with reference to global ecological interdependence and the need for a globally coordinated response to environmental degradation. As discussed in chapter 3, the rise of ecological science in the twentieth century was accompanied by dramatic improvements in ecological measurement and modelling, which allowed scientists to detect the often hidden connections between local and global ecosystems (Warde, Robin and Sörlin, 2018: chapter 4). From transboundary air pollution to the build-up of toxic chemicals in marine food chains and the global threat of global warming, scientists have increasingly been able to demonstrate that global causes were behind many local environmental problems, and that therefore only a global response would be effective. The environmental crisis of the twentieth century thus created a powerful sense of 'common fate' that bound individuals from around the world together, as if they were travelling on 'Spaceship Earth'. In this way, global ecological interdependence helps to overcome the division of both international and world society into national identities by bringing the shared interest in survival to the fore.

Closely related to this argument is the second logic – global environmental responsibility – that reinforces the idea of environmentalism as a universal identity. This logic is driven by the rise of ethical arguments in the twentieth century that represent the planet and its natural environment as the 'common heritage of humankind', which creates a duty for individuals and states to protect the global environment in the interest of humanity. Underlying this framing of environmental responsibility is the notion of trusteeship or stewardship in ethical debates: Nature is no longer seen as an asset for consumption by current generations but is held in trust for current and future generations (and potentially other species) as its beneficiaries (Attfield, 2014: 188–92). Necessity and ethical duty thus combine to create the context in which environmentalism emerges as a political ideology and identity that is cast in universal terms, with 'the great society of humankind' at is main referent.

It is easy to see why this version of global environmentalism fits squarely into a solidarist conception of world society. Ecological representations of the planet as 'One Earth' or Gaia (Lovelock, 2009) strongly suggest that all humans need to come together to formulate a global political response. If indeed we live on a 'small, shared earth' (Guha, 2000: 141), then greater environmental awareness and concern can be expected to foster a new collective identity, a 'universal we', which extends our political loyalties beyond the nation-state. This solidarist vision was clearly present at the birth of the modern environmental movement after the Second World War and has since been promoted by scientists and campaigners. In response to the threat of nuclear weapons, concerned scientists called on their colleagues 'to move beyond their parochial loyalties' in order to avert nuclear Armageddon (Allitt, 2014: 18). Rising industrial pollution and resource scarcity in the post-war era reinforced the global imaginary of nature as 'an interrelated whole of global concern' (Hironaka, 2014: 27), and by the 1960s 'Spaceship Earth' (Ward, 1966) had become widely used as a metaphor for humanity's vulnerability but also the need to reimagine the world as a political community (Höhler, 2015). By highlighting the 'planetary scope of human affairs' (Bartelson, 2009: 173), environmentalists were aligning themselves with other solidarists who challenged the nationalisation of the concept of community and sought to rekindle a sense of belonging to a world community. Many of the first modern environmental NGOs (e.g. Greenpeace) were explicitly cosmopolitan in their political outlook, in contrast to the often nationally and locally rooted conservation groups of the nineteenth century. This is also how much of the IR literature came to define the normative position of environmental organisations in international affairs. Environmental NGOs are often portrayed as representing 'constituencies that are not bound by territory but by common values, knowledge, and/or interests related to a specific issue' (Betsill and Correll, 2008: 2); they are 'bound together by shared values, a common discourse, and dense exchanges of information and services' (Keck and Sikkink, 1998: 2), and they 'appear to be motivated by a cosmopolitanism which transcends the nation-state and even presents challenges to international organisations (Kellow, 2000: 1).

Not only can global environmentalism be presented in solidarist terms, it is also possible to find empirical evidence of environmental 'cosmopolitanisation' in world society. As defined by Beck and Sznaider (2006), cosmopolitanisation happens when social interactions that cross the boundaries between political spaces – from the local and national to the regional and global – lead to the growth of more cosmopolitan orientations, identities and institutions, whether intentionally or unintentionally. Despite having their roots in often local struggles over natural resources and environmental protection, many environmental NGOs have over time become embedded in transnational activist networks. These networks promote horizontal linkages and flows of ideas and information between individual NGOs, partly to promote learning and capacity building among the members and partly to connect them in support of global campaigns. Over the course of the twentieth century, the environmental movement has thus become increasingly globalised in its outlook and activism, particularly when it is seeking to bring about a broader cultural change towards environmental sustainability (Wapner, 2002). At least four channels can be identified through which environmentalism has spread throughout world society (Wang and Hosoki, 2016): flows of monetary resources provided by INGOs and philanthropic foundations (Bartley, 2007; Delfin Jr and Tang, 2008), vertical organisational linkages that promote homogenised cultural models of environmentalism (Schofer and Hironaka, 2005; Tarrow, 2005a), the diffusion of expert knowledge (Hironaka, 2014) and horizontal organisational linkages that lead to the diffusion of activist practices and discourses (Giugni, 1998). Through these and other transnational networks of ideational and cultural exchange, environmental ideas and practices have spread throughout world society. They lend at least some empirical support to the normative representation of global environmentalism as a coherent cosmopolitan project with global reach, even if the global integration of environmental activism remains an ongoing project with many challenges and constraints.

A Plurality of Environmentalisms

While the idea of global environmentalism as a cosmopolitan movement is widespread, it captures only one facet of environmental world society. Alternative representations point to a plurality of normative beliefs and organisational forms in the global environmental movement. In fact, closer analysis suggests that eco-globalism and its solidarist vision make up only one strand of normative world society, which coexists with alternative, eco-localist, visions of a more pluralist form of environmental politics. I argue in this section that the normative structure of environmental world society is indeed complex and diverse, with solidarist ideas coexisting with other, more particularist and localised versions of environmentalism.

As discussed in Chapter 3, environmentalism is a broad church that accommodates a wide range of traditions of environmental thinking. Environmentalists disagree about many things, not least the meaning of 'nature', what they are trying to protect and how this can be achieved. These differences reflect a variety of factors, from societal values and culture to national identity, political ideology and geography. The idea of wilderness protection, for example, is more deeply ingrained in North American thinking than in the European environmental tradition, where cultivated landscapes and national monuments dominated the nineteenth-century environmental imagination. In the post-1945 era, Northern environmental movements focused on industrial pollution and the limits to growth, while Southern environmentalists were preoccupied with local struggles over access to environmental resources and services. Conservative environmentalists see ecological sensibility rooted in humans' relationship with their local environment, reflecting a sense of place, tradition and belonging. In contrast, for radical environmentalists environmental protection is part of a larger social and political struggle, over justice and power, and the future of capitalism. Environmentalists may thus share a general sense of caring and concern for nature, but the environmental problematique, indeed the very concept of 'environment' itself, is contested.

The North–South divide has produced one of the most enduring political fault-lines in not only international environmental politics but also environmental world society. Ever since their first forays into transnational campaigning in colonial contexts (see chapters 3 and 4), environmental NGOs have faced accusations of environmental paternalism, a bias for Northern environmental issues and even 'conservation imperialism' (Guha, 1997: 19). Unable to 'escape the web of national and regional interests' (Kellow, 2000: 3), transnational civil society today is said to 'reproduce many of the power dynamics of a neocolonial world system' (Bandy and Smith, 2005: 239). Such conflicts have surfaced repeatedly in the history of the global environmental movement. During and after decolonisation, wildlife conservation groups sought to defend existing nature parks but faced criticisms for ignoring the rights of local communities and alternative approaches to biodiversity protection (Guha, 1997). Similar charges have been laid against international NGOs ever since, for example when Mac Chapin (2004: 17) accused Conservation International, WWF and The Nature Conservancy of 'a disturbing neglect of the indigenous peoples whose land they are in business to protect'. North–South divisions have also erupted within international NGOs that are more explicitly global in orientation. In the early 2000s, Friends of the Earth International was gripped by internal tensions over accusations of a Northern bias in its international campaigns. This simmering conflict came to a head in 2002, when Ecuadorian NGO *Acción Ecológica* resigned from FoEI's federal structure, accusing the international NGO of prioritising Northern campaign issues, such as corporate accountability and climate change (Doherty and Doyle, 2013: 25). Climate change is no longer seen as a predominantly Northern issues, and environmental NGOs from the Global South have since engaged more fully with international climate politics. However, the emergence of the climate justice movement in the run up to the 2009 Copenhagen climate summit has provided a new focal point for developing a distinctive Southern agenda within environmental world society, focused on the North's ecological debt and a more radical restructuring of global capitalism (Chatterton, Featherstone and Routledge, 2013; Warlenius, 2018).

The persistent North–South divide in environmental world society reflects not only competing environmental interests but also a more profound divergence in environmental identities. Generally speaking, Northern environmentalists are more likely to endorse eco-globalist visions, framing nature as the common heritage of humankind and demanding environmental protection in the interest of future generations. In contrast, Southern environmentalists tend to emphasise eco-localist ideals, in particular the need to protect ecological sovereignty for local communities and indigenous peoples. Their political demands may include global economic redistribution as part of a global justice agenda, but their conception of environmentalism is more likely to be informed by place-based environmental identities at the local or national level. Northern and Southern environmentalists are all part of a loosely structured global environmental movement, but more in the sense of a 'movement of movements' (Mertes, 2004) that each maintain their particular cultures and values. The solidarist cosmopolitan identity that is often associated with the idea of an environmental world society is thus merely one facet of a more pluralist reality of global environmentalist association.

Apart from the normative differences within the global environmental movement, we also need to take into account the continued contestation around the very idea of environmentalism in world society. Any account of transnational activism that ignores ongoing resistance to environmentalists' values provides a reductionist and ultimately misleading account of world society. In the past, the study of transnationalism may have tended to ignore the full normative spectrum among non-state actors, particularly in normatively guided studies that focus on 'good' global norms (Checkel, 1998: 339), but the growing literature on global norm contestation (Wiener, 2004; 2018) and norm anti-preneurs (Bloomfield and Scott, 2017) has served to correct this one-sided representation of the normative spectrum in world society. According to Marlies Glasius (2010), the reality of global civil society comes closer to a 'post-modern version' of an 'arena or collection of actors in (uneven) contestation from a plurality of normative perspectives, not engaged in any one single master project'. Indeed, across a wide range of global issues, from the establishment of the ICC to the ban on landmines and small arms trade, progressive transnational campaigns face opposition from organised groups on the right of the political spectrum that oppose global norm creation (Bob, 2013). The same is true in the field of environmental sustainability campaigns. The growth of the global environmental movement from the 1970s onwards has provoked opposition from other organised interests, in the global business community and civil society. Multinationals and international business associations have organised transnational campaigns against new international environmental regulations, whether in the field of ozone layer depletion, biosafety or whaling (Levy, 1997; Reifschneider, 2002; Blok, 2008). The decades-long international climate negotiations have been the focus of particularly intense contestation between opposing groups of non-state actors, pitting various environmentalist campaign groups against a well-funded and internationally coordinated lobbying effort by fossil fuel industries (Falkner, 2008; Dunlap and McCright, 2011). Environmental values have spread throughout world society and may have been in the ascendancy since the late twentieth century, but they are far from universally accepted.

Undoubtedly, there are close links between global environmentalism and the solidarist notion of 'the great society of humankind'. If normative world society is taken as a merely abstract idea, a philosophical construct that serves as a 'referent object against which to judge the society of states' (Buzan, 2018: 127), then environmentalism can count as a major source of moral inspiration for ecological one-worldism and planetary stewardship. However, the normative landscape of global environmentalism is more complicated than this narrowly solidarist ES lens would suggest. Instead, we need to recognise that different versions of collective environmental identity operate in world society, not to mention ongoing anti-environmental discourses and identities. Only some of these fall squarely into a global cosmopolitan category, while others espouse alternative sub-global versions of environmental community. The normative debate within environmental world society includes positions that can be found along the spectrum from eco-globalism to eco-localism, from arguments that consider humanity and planetary health as the main referent to those that seek to strengthen local communities' political autonomy and ecological sovereignty. The normative dimension of environmental world society thus expresses itself in pluralist form, it is fragmented rather than united, multiple rather than singular (cf. Buzan, 2018: 128). Solidarist ecological visions are an important source of collective identity formation across national boundaries, but they are only one of several varieties of global environmental identity.

Environmental Activism in Political World Society

When we shift the focus from normative to *political* world society, we move from the collective identities that define non-state actors' purpose to the empirical realm in which they interact with each other and with the society of states. Classical ES theory has been accused of not providing clear concepts and evidence for the empirical reality of world society (McKeil, 2018). Although Manning (1962: 177) spoke of the 'nascent society of all mankind [sic]' as an emerging empirical phenomenon, most ES writers instinctively followed Wight's depiction of world society as a normative aspiration or, as was the case with Bull, relegated it to a marginal existence on the fringes of international society. More recently, Clark (2007) and Ralph (2007) have delved more deeply into the empirical manifestations of world society as a separate realm of political action as well as the interactions and interdependencies between international and world society.

Normative and political world society are closely connected in that the non-state actors that make up the latter draw on the ideational resources provided by the former (Buzan, 2018: 129–30). Thus, the idea of human rights, which derives from a cosmopolitan normative outlook and has as its subjects 'not members of this or that society, but of the community of humankind' (Vincent, 1986: 9), informs the campaigns of NGOs such as Amnesty International and Human Rights Watch. Similarly, it is the idea of collective human responsibility for the planet's health that provides the ideological foundation for transnational environmental campaigns. While collective identity is the primary institution that constitutes the actors of, and gives structure to, normative world society, *advocacy* is political world society's main primary institution (Buzan, 2018: 135). That non-state actors should express public opinion and influence inter-state relations first came to be legitimised in the nineteenth century (Clark, 2007), and the level of regular interaction between states and non-state actors increased dramatically in the twentieth century, particularly with the expansion of international organisations and global governance after 1945. Thanks to the intensifying interaction between both types of actors, international society cannot be thought of as existing in isolation from world society. As non-state actors 'have worked to shape the domestic and international normative contexts in which states constitute their identities, define their interests, and conduct their relations' (Reus-Smit, 2002: 504), international and world society have become highly interdependent, even mutually constitutive. In this section, I explore the role that environmental NGOs play in interstate relations, but also the degree to which engagement with international environmental diplomacy has shaped their advocacy role and political identity.

Environmental Advocacy in International Society

NGO engagement in world politics is frequently portrayed as a relatively novel phenomenon, the result of the latest wave of globalisation, but as Davies (2013) shows that the origins of transnational civil society can be traced back as far as the late eighteenth century. As discussed in Chapter 4, environmentalists began to organise the first transnational environmental campaigns from the late nineteenth century onwards. International NGOs such as the World League for Protection of Animals (founded in Germany in 1898) and SPWFE (founded in Britain in 1903) are prominent examples of the nascent environmental world society at the turn of the century. During the early decades of the twentieth century, these international NGOs made sporadic interventions in international politics, relying mainly on their members' privileged access to social and political elites. But it is fair to conclude that it was not until late in the century that NGOs acquired the necessary capabilities to stage sustained transnational campaigns and engage more extensively with intergovernmental processes (Keck and Sikkink, 1998: 10–11). Chief among these increases in organisational capacity was the communication technology revolution of the 1990s, which helped reduce communication and networking costs and also produced a dramatic growth in the global orientation of social activism (Bennett, 2005). The difficulties of maintaining international campaigns had held back even the most globally oriented groups of the modern environmental movement. For example, ten years into its existence Friends of the Earth was still little more than a loosely organised federation of national NGOs. International communication was too sporadic and expensive to allow for greater global coordination of its campaigns. 'The fax was still a dream at this time and the telephone was used very sparingly' is how an Italian member of Friends of the Earth remembered the state of communication in the late 1970s (quoted in Doherty and Doyle, 2013: 62).

The early history of GEP (see chapters 4 and 5) shows that, despite the constraints on transnational campaigning, non-state actors played a critical role in introducing environmental ideas and values into international society. In the early twentieth century, when environmental problems were not yet routinely dealt with in international diplomacy, environmentalists and scientists helped stimulate discussion of pollution and conservation issues at the international level. They created awareness of the global environment at international scientific congresses and lobbied state bureaucracies and international organisations, though often with limited impact. Once most states had accepted a formal environmental responsibility by the 1970s, environmental activists and scientists provided critical expertise in international negotiations and were thus able to have some influence over international environmental rule-making. The dramatic expansion of environmental multilateralism after Stockholm created new international arenas for non-state actors to operate in, and international environmental negotiations and UN summits became focal points for the creation and deepening of transnational activist networks. Given the high costs of travel and communication, however, it was not until the 1990s that NGOs were to play a more active role in UN forums.

The increase in the number of observers at UN environmental conferences offers a good indicator of the dramatic expansion of environmental advocacy by non-state actors. The 1972 Stockholm conference, which is widely credited for having boosted NGO engagement in international environmental politics (Willetts, 1996), was attended by just 258 NGO observers (United Nations, 1972), though this was many more than had attended the 1968 Biosphere Conference. The real 'watershed moment' (Burgiel and Wood, 2012: 127) came with UNCED in 1992. With more than 1,400 NGOs having gained accreditation to UNCED and over 25,000 individuals participating in the parallel NGO Forum (Reilly, 1993; Morphet, 1996; Willetts, 1996), the Rio Earth Summit became the then biggest gathering of world leaders and non-state actors of its kind. UNCED Secretary-General Maurice Strong, conscious of the need to 'anchor action to promote sustainability firmly in society' (Kjellén, 2008: 46), reached out to civil society groups to engage them in the Rio process and ensured they had extensive access to the UNCED meetings. After 1992, there was no way back from the high level of non-state actor engagement that UNCED had achieved, and many more environmental NGOs were to acquire consultative status at the UN through the newly created CSD. Their involvement in international environmental negotiations became the new normal, and indeed negotiating processes that had started prior to UNCED subsequently changed their procedures so as to allow for more NGO participation (Betsill, 2008b: 196). When the UN convened a new world sustainability summit ten years after Rio, more than 3,200 NGOs were accredited to the international conference in Johannesburg (Betsill and Correll, 2008: 1–2). Even higher numbers of NGO observers started to flock to the annual COP meetings of the UNFCCC. At COP-6 in 2000, when the rules of the Kyoto Protocol were to be finalised, over 3,500 NGO participants had registered for the event. Nearly a decade later, at COP-15 in Copenhagen, the number of NGO observers shot up to over 20,000 (Fisher, 2010: 13).

To understand the role that advocacy by world society actors has played in international environmental politics, we need to focus on both sides of the equation: the goals and strategies of non-state actors and how international society reacts to their interventions. Environmental NGOs and business actors pursue multiple goals and strategies when they engage with environmental diplomacy. First and foremost is the desire to influence the outcomes of international processes and shape the behaviour of states. Initially, NGOs often put novel environmental threats on the international agenda and create global awareness. Once negotiations on an environmental treaty are underway, environmental campaigners push for ambitious targets and stringent rules. Finally, after a new international treaty has entered into force, NGOs will monitor states' compliance with the agreement and put the spotlight on those that fail to live up to treaty commitments (Betsill and Correll, 2008; Burgiel and Wood, 2012; Park, 2013). In the past, business actors may have resisted the rise of international environmental regulation, but with the strengthening of global environmentalism and the expansion of MEAs a more diverse range of business interests has come to the fore, with some opposing, and others supporting, specific forms of environmental rule-making. Indeed, the emergence of business conflict in international environmental politics has given rise to new transnational alliances between progressive state actors, environmental NGOs and pro-regulatory business sectors (Falkner, 2008; Vormedal, 2011; Meckling, 2011). International environmental diplomacy is thus conducted in an increasingly diverse and complex field of actors that all seek to shape the outcomes of international negotiations.

In their efforts to influence state practices, environmental NGOs rely on a range of intervention strategies: they create publicity for global environmental issues; they apply moral pressure in intergovernmental processes; they provide states with scientific, legal or political expertise during international negotiations; and they use public 'naming and shaming' strategies against those states that fail to live up to international commitments. Business actors similarly rely on their extensive financial and organisational resources, economic and technological expertise, and close connections to business-friendly state ministries in their effort to ensure that new international regulations reflect business interests. To some extent, this is a straightforward story of interest representation and lobbying in international processes. But world society actors' impact on international outcomes goes beyond this transactional dimension. Through strategies of persuasion and socialisation, non-state actors also 'condition the normative environment in which states act, clarifying and promoting international norms and exposing the dissonance between standards and state conduct' (Reus-Smit, 2002: 500). In other words, by engaging in the political struggles around international environmental regulations and their implementation, non-state actors reinforce environmental stewardship as a fundamental global norm.

Non-state actor influence in GEP ebbs and flows but there are many well-documented instances when environmental NGOs and businesses have had a lasting impact on intergovernmental processes. In some cases, environmentalists played a critical role in setting the agenda for international action and initiated work on international legal instruments. The CBD of 1992, for example, has its roots in various biodiversity strategies developed by IUCN and WWF in collaboration with UNEP and other international organisations from the late 1970s onwards. The first proposal for an international biodiversity treaty was made at the Third World Conference on National Parks, held in Bali in 1982, and IUCN's Environmental Law Centre and Commission on Environmental Law began drafting specific articles for what was to become the CBD (Holdgate, 1999: 170, 213). In other cases, NGO campaigns had a powerful influence over the framing of environmental issues and the positions that major players took in international processes. In the negotiations leading to the 2000 Cartagena Protocol on Biosafety, the transnational environmental campaign against genetically modified (GM) foods played an important role in defining the issue as one of global environmental concern that required a precautionary approach. NGOs were particularly influential in Europe, where their protests succeeded in closing off the European market to GM food in the late 1990s, which in turn helped to harden the European Union's negotiation stance from 1998 onwards (Falkner, 2007). And even in the slow-paced negotiations on climate change, which have proved to be a source of repeated frustration for environmentalists for over two decades, NGOs were able to have some discernible influence over how the agenda was framed (e.g. emissions trading and sinks in Kyoto Protocol; concept of net-zero emissions in Paris Agreement) (Betsill, 2008a; Darby, 2019).

In contrast to NGOs, business actors have been less successful in international agenda-setting but have played a more influential role in shaping state preferences and negotiation outcomes. Their significant economic weight and key role in technological innovation have given global corporations a privileged position in GEP, even if this does not mean that they always get their way (Levy and Newell, 2005; Clapp and Meckling, 2013). During the early history of GEP, business actors exercised a mostly negative role, seeking to prevent or water down the creation of international environmental regulation. At least since the 1992 Rio Earth Summit and certainly since the 2000s, business engagement with environmental diplomacy has become more nuanced and less antagonistic, with a growing number of business voices advocating certain forms of environmental regulation. This emerging business conflict in GEP has opened up opportunities for new alliances between pro-regulatory forces from across the spectrum in world society (Falkner, 2008). Thus, the fossil fuel industry was a dominant anti-regulatory voice in the early climate negotiations of the 1990s, but renewable energy firms, manufacturing companies and the services sector are now broadly aligned in their support for the Paris Agreement (Gies, 2017). Companies possess a range of power capabilities to influence negotiations, but these 'do not equal actual political influence' (Clapp and Meckling, 2013: 295). The pharmaceutical sector, for example, succeeded in having its products excluded from the Cartagena Protocol on Biosafety, but the agricultural biotechnology industry largely failed in its lobbying effort against the international agreement (Falkner, 2008: chapter 5).

Apart from individual success stories, the overall verdict on non-state actors influence, whether by civil society or business groups, is more mixed. In their review of five major global environmental issues, Betsill and Correll (2008: 183) point to only one case, the Convention to Combat Desertification, where NGOs had a consistently high influence on both the negotiation process and its outcome. In other cases, their influence was more modest, often limited to certain aspects of the negotiations, and varied over time. Indeed, although the GEP literature generally points to greater enmeshment between international and world society when it comes to international environmental rule-making, it has struggled to find a definitive answer to the question of just how much influence non-state actors have acquired in state-centric processes. What we do know is that non-state actors' capabilities and leverage vary across issue areas, as does their coherence and consistency in pushing for certain outcomes. Moreover, despite having gained in legitimacy as observers and participants in state-centric processes, non-state actors still depend on international society providing them with a permissive environment to operate in. States can restrict access to international negotiations, and they routinely do so (Clark, Friedman and Hochstetler, 1998; Fisher, 2010), particularly when politically salient issues are at stake or sensitive bargains have to be struck in the 'detail phase' (Zartman and Berman, 1982: chapter 5). Non-state actors may try to circumvent such restrictions and use other lobbying channels to exert indirect pressure on states (Betsill, 2008b: 191–2), but the state-centric nature of UN processes has largely remained intact.

However, when focusing solely on non-state actors' observable influence over specific outcomes in interstate bargaining, we are at risk of missing the bigger picture of how the rise of transnational politics has altered the relationship between international and world society. Their impact goes well beyond relational concepts of influence-seeking, via lobbying and information-sharing. Non-state actors, and environmental campaigners in particular, are often at their most effective not when they target states but when they shape the ideational context in which states and other actors operate. Indeed, as even those GEP authors that focus on international negotiations highlight, environmental campaigners are often most influential when setting the agenda for international rule-making, especially if we conceive agenda-setting as an 'ongoing process rather than a distinct stage of policy making that ends once negotiations begin' (Betsill, 2008b: 193). But this agenda and norm-setting process extends beyond state-centric realms. By creating environmental awareness, defining environmental problems and promoting value change in domestic and world society, environmentalists have not only altered how people think about the relationship between society and nature, they have also reshaped understandings of international legitimacy and rightful membership in international society. They are engaged in what Wapner (1996) calls 'world civic politics'. Business actors similarly shape the political and economic context within which GEP outcomes are determined. Their structural and ideational power often allows them to ensure that proposed international environmental policies are business and market-friendly and tie in with their global competitive strategies (Levy and Newell, 2005; Falkner, 2008). From an ES and constructivist perspective, the impact of non-state actors thus shows itself both in terms of specific negotiation outcomes and in the way global issues are defined and normative structures are changed. In this sense, Clark's (2007: 7) focus on critical international conferences at which international society came to redefine its principles of legitimacy is too narrowly conceived, even if such episodes of high politics reveal most clearly world society's 'traceable history'.

Mutual Shaping and Accommodation

In investigating the links between world and international society, we should also reverse the perspective and focus on how the latter has shaped the former. For as Clark (2007: 7, 10) rightly points out, 'the two have shared a common and overlapping history', which also involves some form of 'accommodation between international and world society'. GEP offers an interesting example of how both have grown increasingly interdependent as environmental stewardship has hardened into a global norm. More specifically, the rise of environmental diplomacy and the expansion of international environmental rule-making has provided a major boost for environmental world society, in terms of its visibility, organisational strength and political legitimacy. For one, engagement in international negotiations has been a vital factor in the creation of larger and more durable transnational advocacy coalitions and networks, and it has also strengthened the global orientation of the environmental movement. Ever since the Stockholm conference, UN summits, MEA negotiations and annual COPs have provided a platform for non-state actors to form transnational alliances and develop common positions and campaigns. In some cases, the need to maintain a permanent presence in international negotiations prompted the creation of new international coalitions of national and regional NGOs. One of the longest standing and most prominent umbrella organisations is the Climate Action Network (CAN), which first became involved in the international climate talks in 1990 and has since grown into a coalition of over 500 national and international NGOs (Duwe, 2001; Rietig, 2016). The Global Forest Coalition, which was founded in 2000 by nineteen NGOs and indigenous peoples' organisations, has intervened in forest-related debates at the UNFCCC, CBD and UN Forum on Forests. Indeed, many NGOs attend international environmental negotiations to not only lobby states but also develop deeper networks with other NGOs (Clark, Friedman and Hochstetler, 1998). In climate politics, the side events at the UNFCCC's annual COP meetings have become a major forum for connecting environmental groups from around the world. They serve as a vehicle for not only lobbying state representatives but information sharing, capacity-building and networking within environmental world society (Hjerpe and Linnér, 2010; Schroeder and Lovell, 2011).

Participation in state-centric diplomacy has also served as a legitimation tool for world society actors. Through interaction with states in formal conference settings, they have gained recognition from international society (Clark, 2007: 24) as well as credibility within transnational and domestic contexts. To some extent, this is a reflection of the fact that states have increasingly come to rely on the environmental expertise and problem-solving capacity of non-state actors (Raustiala, 1997; Rietig, 2014). The Rio Earth Summit proved to be a key event in the formal recognition of world society's role. Both the Rio Declaration and Agenda 21 acknowledge the critical role that citizens and civil society organisations play in implementing the sustainable development agenda. In the context of secondary institutions, scientists and environmentalists are routinely involved in providing relevant expertise during the negotiation and implementation phase, and NGOs may also monitor states' compliance with regime rules (Andresen and Gulbrandsen, 2005; Van Asselt, 2016). In some regimes, such as the Montreal Protocol on ozone layer depletion, industry groups and technical experts have been invited to join technical assessment panels that advise on regulatory restrictions on ozone-depleting chemicals (Falkner, 2008: 89). Arguably more so than in other areas of global governance (Stec, 2010: 373; Fisher, 2004), non-state actors are thus widely recognised to strengthen the problem-solving capacity of international environmental regimes.

Environmental campaigners' success in influencing outcomes in environmental diplomacy has come at price, however. In order to be recognised by states as legitimate actors and participants in intergovernmental forums, environmentalists had to adapt and professionalize their campaign strategies. The growing NGO engagement with international society has shifted the balance of power among activist groups in favour of those that take a more pragmatic, reformist, approach. Providing policy-relevant knowledge is more likely to give campaigners access to negotiators than noisy protest and confrontation. Activists that argue for a more radical green agenda have therefore viewed involvement in intergovernmental processes with greater suspicion, pointing to the exclusionary logic of elite discourses and the political taming of the environmental movement (Newell, 2020: 147–8). This concern has been expressed in other areas of NGO–state interaction too (Dany, 2014). Critics point to the constraining effect that engagement in diplomatic forums has on the range of political views non-state actors can express. Rather than empowering NGOs, it forces them into an incrementalist reform agenda that can only be delivered by the slow-paced mechanisms of state-centric international governance. For some activists, this poses a 'serious and growing danger of being co-opted to serve watered-down intergovernmental agendas rather than advancing their own visions and objectives' (McKeown, 2009: 11).

Thus, while moderate environmental groups have thrived and grown in numbers as states have welcomed them into multilateral rule-making, the radical wing of the environmental movement has for the most part been sidelined in international processes. In a sense, many of the mainstream environmental NGOs have followed an eco-globalist strategy of supporting global environmental governance, while radical groups have been more closely aligned with an eco-localist strategy, focused on building bottom-up support for a decentralised green political–economic order (Newell, 2020: 140–1). This divergence in political outlook and strategy has led to repeated tensions and conflicts in environmental world society, both within international NGOs and among NGOs that engage with international society. As discussed earlier, Friends of the Earth underwent a difficult process of internal strategic repositioning as it confronted long-standing North–South divisions that broke out into the open in the early 2000s. Other NGOs have also struggled with the ideological divides that exist in world society. At times, such divisions spilled over into international environmental processes, most notably in the post-Kyoto climate negotiations. During the 1990s and until around the time of the Kyoto Protocol's entry into force in 2005, reformist NGOs were by far the dominant voice at the annual climate COPs. This began to change in the second half of the 2000s as international society prepared to replace the Kyoto Protocol with a new treaty and some environmental activists became increasingly disillusioned with the slow pace and low ambition in the negotiations. The conflict came out in the open during the 2009 Copenhagen conference. Whereas CAN's more pragmatic approach dominated NGO engagement with the negotiations and focused on nudging states in the direction of a new climate treaty (Hadden, 2014: 11), a new climate justice movement took to the streets outside the conference centre and mobilised 'against the climate regime and global capitalism more broadly' (Fisher, 2010: 15). It can be argued that this divide reflects merely different approaches to global campaigning – insider versus outside tactics – that are to some extent mutually supportive and suggest a division of labour within the environmental movement (Tarrow, 2005b). But the rift between traditional climate NGOs and climate justice groups is about more than merely questions of tactics. It brings to the fore the very question of whether world society's close involvement in international diplomacy legitimises and reinforces state-centric approaches in GEP, and whether 'advocates working out of high-rise offices in New York, Brussels or Geneva lend legitimacy to these institutions and deprive the people they claim to represent of an authentic voice' (Tarrow, 2005b: 55).

To conclude, advocacy by non-state actors, which can be seen as a primary institution of world society (Buzan, 2018), has become a firmly established practice in multilateral processes, arguably more so in the environmental field than in most other international issue areas. Having gained a foothold in international diplomacy and legitimacy as partners in global governance, world society actors have been able to exert a notable influence over international environmental rule-making. Most states have largely welcomed the increasingly open nature of environmental diplomacy, particularly as they have struggled to find effective answers to the many complex problems on the international agenda. To some extent, therefore, international and world society have entered into a kind of symbiosis in GEP, depending on each other to perform core governance functions. This has not afforded world society actors co-equal status, however. The fact that vast numbers of non-state actors now routinely gather as observers at high-profile environmental conferences has not altered the fundamental reality of a state-centric international policy-making process. States have retained exclusive control over intergovernmental processes, particularly in areas characterised by rising political salience and intense distributive conflicts.

One important side effect of the growing acceptance of non-state actor involvement in environmental diplomacy has been the professionalisation of NGO campaigning, which has gone hand in hand with the marginalisation of world society's more radical voices. Engagement with international society has given a boost to the pragmatic and reform-minded wing of the environmental movement, providing it with an international platform, transnational networking opportunities and even material support. At the same time, the intense focus of major NGOs on international environmental politics has reinforced the centrality of the state-centric international order to the search for a global environmental response. Despite failing to get a grip on major environmental problems, states and international organisations continue to be at the centre of many non-state actors' environmental campaigns. Just as NGOs have benefitted from a legitimation effect and recognition as responsible international stakeholders, so too have states and intergovernmental institutions seen their central role in global environmental governance reaffirmed. The state system may not have proven itself to be capable of solving many of the worst environmental problems the world faces, but ironically it is world society actors that have inadvertently helped to alleviate the nation-state's and international society's legitimacy crisis in the face of a widening ecological crisis. Despite the dramatic expansion of transnational environmental activism and governance, it remains the case that 'environmental issues will still of necessity be managed within the constraints of a political system in which sovereign states play a major part' (Hurrell and Kingsbury, 1992b: 9).

Towards an Integrated World Society?

The first two dimensions of world society discussed earlier are concerned with its normative identity and political role in dealing with interstate society. Both assume a clear separation of the interstate and interhuman/transnational domains. They explore how political world society, using the ideological resources provided by normative world society, intersects with the society of states. The third dimension, integrated world society, is about the confluence of these domains and the (potential) emergence of an integrated social structure at the global level. As Buzan (2018: 130) notes, this is 'an ideal-type for a prospective future', not a descriptor for an existing reality. It would involve a move away from the current, strictly hierarchical, social ordering based on the principles of sovereignty and territory and a privileging of states, towards a future social organisation that differentiates state and non-state actors along functional lines. Clearly, we are still far from reaching such a state of affairs, even if some forms of global governance point in this direction (Acharya, 2016; Weiss, 2016). Still, the concept of integrated world society can be used as an analytical yardstick, not quite for measuring how far the world has already travelled down this path, but at least for identifying the first steps in this direction. It is in this sense that I explore in this last section the ways in which international and world society have become more closely aligned in the pursuit of global environmental sustainability.

The ES tradition contains some preliminary ideas of the kind of transformation that such a move towards an integrated world society would involve. For the most part, they are closely associated with the solidarist ambition to overcome international society's hierarchical structure that privileges states over non-state actors. Bull alluded to this perspective in his notion of a 'New Mediaevalism', which would replace a sovereignty-based international society with a more fluid system of 'overlapping authorities and criss-crossing loyalties that hold all peoples together in a universal society, while at the same time avoiding the concentration of power inherent in a world government' (1977: 255). While Bull's pluralism led him to conclude that such a political structure is inherently unstable, with 'more ubiquitous and continuous violence and insecurity' than the modern states system, solidarists broadly welcome it. Vincent (1986) sees it as a desirable move to ensure that the rights of individuals are fully established internationally. Buzan (2004: 202–3) takes this one step further and envisages an integrated world society to be based on a wholesale social transformation, though without expecting it to materialise 'any time soon' (Buzan, 2018: 131). This transformation replaces hierarchical differentiation with a functional one between different types of actors – states, NGOs, global corporations – that would all become subject to international law. The path towards such an integrated world society leads through a series of profound changes that find a parallel in constructivist notions of an emerging 'world political system' (Keck and Sikkink, 1998: 212) or global society: 'An increasingly dense fabric of international law, norms and rules that promote forms of association and solidarity, the growing role of an increasingly dense network of state and non-state actors that are involved in the production and revision of multi-layered governance structures, and the movement towards forms of dialogue that are designed to help identify shared values of "humankind"' (Barnett and Sikkink, 2009: 749). Social theories of IR are thus well attuned to the merging of international and world society, even if it remains a largely unfulfilled ambition.

Could global environmentalism be a major driving force behind this social integration of the state-centric and non-state realms? As discussed earlier, interaction levels between state and non-state actors have increased dramatically in GEP. World society actors have played a key role as norm entrepreneurs and lobbyists seeking to influence outcomes in multilateral processes, while states have increasingly recognised scientists, environmentalists and business actors as legitimate stakeholders and even partners in the crafting of global environmental solutions. This would suggest that the environmental field is a prime candidate for ever-closer integration between international and world society. What does the history of global environmentalism tell us about the potential for, and indeed reality of, such a development?

There are good reasons for thinking that we can expect a comparatively high level of interplay between states and non-state actors in GEP. Many environmental problems are transboundary in nature, requiring international cooperation by states, but because they tend to have roots in complex social and economic processes they are difficult to address through environmental diplomacy alone. To tackle global environmental challenges, from climate change to marine pollution and biodiversity loss, profound changes have to be made to global industrial production, energy systems, transportation networks and food production, and all these require the involvement of a multitude of actors and authorities outside the states system. Simply put, the global ecological crisis far exceeds the problem-solving capacity of a purely state-centric management approach. International and world society are locked into a form of functional interdependence in environmental politics that necessitates a higher degree of cooperation between states and non-state actors than in many other areas (e.g. national security). States need to engage non-state actors because they bring different capabilities to the table: Environmental organisations play a key role in mobilising citizens behind environmental causes and promoting norm change in society, while corporations have the technological know-how and market power to reduce industrial pollution and promote economy-wide innovation. At the same time, non-state actors depend on states to put in place the required legal and regulatory frameworks that can guide and sustain green transformations. Addressing the complexity of global environmental challenges thus requires different actors to work together at different scales of action, building multiactor and multilevel governance (Andonova, Betsill and Bulkeley, 2009: 52).

International and world society have taken some steps towards this model of a functionally interdependent system of global environmental governance. As discussed earlier, non-state actors have been welcomed by states as lobbyists in international environmental negotiations and as advisers in the implementation of environmental treaties. Increasingly, states and international organisations have also invited non-state actors to perform governance functions themselves. This shift has been in the making for some time. The first transnational climate initiatives date back to the early 1990s, when the Rio Earth Summit established a framework for international cooperation on climate change (Roger, Hale and Andonova, 2017: 6). At around the same time, non-state actors also began to establish transnational initiatives to protect tropical rainforests, for which no international regime has ever come into existence. Founded in 1993, the Forest Stewardship Council (FSC) became one of the first transnational bodies to bring together NGOs, businesses and indigenous community groups to develop private environmental norms and rules (Pattberg, 2005). By developing sustainable forestry standards and enforcing them through a certification scheme, the FSC developed a model for private environmental governance that has been applied in other areas of environmental sustainability, from fisheries to agricultural commodities (Gulbrandsen, 2008: 62; Ponte and Cheyns, 2013; Schleifer, 2017). The number and scope of transnational environmental governance initiatives has since expanded dramatically, with hundreds of transnational initiatives now clustered around the intergovernmental climate regime (Chan et al., 2018; Streck, 2020) and over 4,000 PPPs linked to the UN's SDGs (Marx, 2019: 1).

This complexity of 'global governance beyond the state' (Hurrell, 2007) can also be seen from the links that have emerged between public and private environmental rules at the global level. Not only have non-state actors become more active in providing transnational governance functions, they have also created new private norms and rules that are then adopted by states. Environmental management systems, for example, were first developed as private standards (e.g. ISO 14,000) but were subsequently adopted by a number of powerful countries, including by the EU (Delmas, 2002). In other areas, too, states have come to rely on private standards created by non-state actors as part of an enlarged and complex governance field. In the field of biofuels governance, for example, EU policy-makers have established framework regulation that uses private biomass certification schemes as qualifying standards for the EU (Schleifer, 2013). This blending of public and private standards, which can be observed across a wide range of issue areas, has led to 'new hybrid governance forms where public and private come together in complex configurations that include civil society, business and a plethora of non-traditional actors' (Ponte and Daugbjerg, 2015). Some of these transnational activities may be a form of 'delegated private authority' (Green, 2013: 33), where states or international organisations explicitly transfer governance authority to non-state actors. In other cases, the authority that firms and NGOs develop will be 'entrepreneurial' (ibid.), in that self-regulation is provided autonomously but then adopted by other actors.

A significant literature has recently grown up around the notion of 'orchestration' of transnational governance by international society. As Abbott et al. (2015a) demonstrate in their comparative study of international policy arenas from public health, finance and trade to conflict minerals, climate finance and responsible investment, both states and international organisations now routinely seek to engage non-state actors as partners in a larger network of public and private governance. Orchestration happens when public actors 'enlist intermediary actors on a voluntary basis, by providing them with ideational and material support, to address target actors in pursuit of [their] governance goals' (Abbott et al., 2015b: 3). As such, orchestration cannot rely on the authority of states over non-state actors in a hierarchical form of functional differentiation, it is an indirect and soft mode of governance (ibid., 4) that relies on the horizontal expansion of the global governance field. Such orchestration efforts are particularly widespread in GEP, where non-state actors possess critical capabilities that support public policy objectives (Abbott, 2012; Schleifer, 2013; Hale and Roger, 2014). As I have argued earlier (Falkner, 2003), the growing interweaving of public and private authority in global environmental governance thus makes it questionable whether private environmental governance can be conceived of as an entirely 'private' activity. International society increasingly relies on self-governance in world society to expand, complement and strengthen formal international rules. Even where non-state actors take the initiative, transnational governance usually emerges within a wider institutional context that is actively shaped by states, and private norms and rules are frequently adopted, codified or reinforced by later state action. In this way, international and world society have become entwined in a much closer two-way relationship of mutual influencing and shaping.

One of the implications of this interweaving of public and private environmental governance is the growing normative integration of international and world society in GEP. Environmental stewardship, a fundamental norm that originated in world society (see chapter 4), emerged as a primary institution in international society in the 1970s but is equally present in transnational/political and interhuman/normative world society. Just like nationalism, which has moved in a 'two-way street' between world and international society (Buzan, 2018: 136), environmentalism too has shaped identities and ideational structures in both global domains. As I have argued throughout this book, the normative influence of environmentalism has moved in both ways, from world to international society and vice versa. Non-state actors have been important norm entrepreneurs in the greening of international society, creating environmental awareness, promoting norm change and putting pressure on states to accept global environmental responsibilities. At the same time, however, states have themselves played a major role in embedding environmental values in international institutional structures, which have in turn become a focal point for transnational environmental advocacy and have shaped and nurtured new forms of transnational environmental governance. This interpretation goes beyond Clark's (2007) and Ralph's (2007) interpretation of world society as a 'parasite' that is seeking to shape interstate norms but is ultimately dependent on the institutional framework provided by the society of states (Pella, 2013: 68). Instead, GEP has witnessed the emergence of an increasingly close and interdependent relationship between states and non-state actors. This relationship is still far from the ideal of an interdependent world society, but it suggests a level of integration that calls into question standard ES and other interpretations of world society as merely a residual category of marginal consequence for international relations.

Conclusions

World society plays an important, indeed central, role in the greening of international society. The environmental stewardship norm has no natural source within international society but was introduced into interstate relations out of environmental discourses and campaigns that originate from domestic and transnational societal contexts. World society actors have continued to influence and shape the growth and evolution of environmental stewardship in international relations. Indeed, non-state actor advocacy has become a ubiquitous feature of environmental multilateralism, all the way from the setting of the international agenda to treaty negotiation, implementation and compliance monitoring. States have in fact provided a permissive environment to facilitate enhanced involvement by non-state actors, but they have never relinquished control over international processes of rule-making. In this sense, the states system has not become obsolete as pluralists once feared (Bull, 1977: chapter 12). The prospects for global environmental management are still heavily affected by international conflicts and the difficulties of creating lasting interstate cooperation. Non-state actors play an active role in this, but reducing world society's significance to the impact it has on international regime-building would miss its wider and more consequential role in shaping the ideational and normative context in which global rule-making takes place.

In ES theory, the world society concept also takes on a normative dimension, which goes beyond conventional transnationalist research in IR. Viewed through this lens, we find that environmentalism offers the basis for a coherent collective identity that extends beyond the nation-state and is rooted in cosmopolitan values. In this solidarist interpretation, environmental world society represents the global community of humans that shares an interest in the prudent management of its environmental heritage, for both current and future generations. Yet, a closer reading of the environmental discourses that make up the environmental tradition suggest a different interpretation, one that emphasises multiple versions of environmentalism and a plurality of normative beliefs. In this view, world society is profoundly pluralist in nature, with eco-globalist identities coexisting with eco-localist ones. This reduces solidarist environmentalism to just one among many versions of global environmentalism, some of which are rooted in particularist and localised identities. Furthermore, the very idea of environmentalism, that the natural environment deserves a special place within humanity's moral landscape and ought to be treated with a sense of respect and care, remains contested. Environmentalism may have spread throughout world society and become a global ideational force, but anti-environmental beliefs continue to thrive, even if they have gradually lost ground as environmental beliefs have spread. Normative representations of global environmentalism as a form of cosmopolitan solidarism thus capture only part of the ideational spectrum that can be found in world society.

As far as the third use of world society in ES theory is concerned – the notion of an integrated world society that has brought together world and international society in a functionally differentiated world political system – it is clear that this remains an unfulfilled promise of global societal development. GEP has witnessed the emergence of an increasingly intertwined system of public and private global governance, with states and non-state actors recognising each other's contribution to global environmental management and seeking to weave together the fragmented elements of decentralised environmental governance that have come into existence at the global level. While it is fair to conclude that interplay between states and non-state actors is now at low to moderate levels in GEP, well above the minimal level that characterises a Green Westphalian order, we are still far from the maximal level that characterises an integrated world society 'based on acknowledged and legitimate functional differentiation amongst different types of units' (Buzan, 2018: 136).

Part IV

Conclusions

11 Conclusions

International Relations in the Anthropocene

The Rise of Environmentalism in International Relations

The rise of global environmentalism from a fringe social movement to a primary institution of global international society represents a remarkable normative change in world politics. Environmentalism evolved from a few scattered initiatives in the nineteenth century into a global movement in the twentieth century that introduced a green dimension to established understandings of international legitimacy and the moral purpose of the state. By the 1970s, it had succeeded in establishing a general duty of the state to protect the planet's health. This is all the more remarkable as environmentalism did not arise out of a systemic need for international society to take on an environmental role, for example in order to safeguard international order and stability. Instead, the rise of environmental stewardship reflects the creation of a new social purpose within GIS, a collective response to the environmentally destructive potential of modern industrialism and capitalism. As such, it is a prime example of norm transfer from world society to the society of states. Environmentalism is thus part of a group of international norms – from the abolition of slavery to the idea of racial equality, social justice and human rights – that can be traced back to the activities of moral entrepreneurs and transnational networks that operate outside international society (Clark, 2007). In all these instances of exogenous normative development, including in the environmental case, international society has shown itself to be porous, open to new social purposes that pull it away from its narrowly defined pluralist roots and move it towards an enlarged and ambitious role of world ordering.

I have argued that environmental stewardship has been accepted by the society of states as an integral part of its normative structure. States have come to adopt environmentalism as a consensual norm, and this book has tracked its social consolidation in international society through changes at the level of secondary institutions and constitutive changes in member states. The first UN environmental conference in 1972 can be seen as the 'constitutional moment' in the greening of GIS, though it is part of a longer history of attempts by world society actors to establish environmental concerns on the international agenda. In fact, the first such attempts date back to the early twentieth century. Established in 1913, just before the outbreak of the First World War, the Consultative Commission for the International Protection of Nature became the first international environmental body but was never able to take up its work. Environmental campaigners sought to resuscitate the body after the war and lobbied the League of Nations to take over its role, though without any success. Similar efforts to create an explicit environmental mandate for the United Nations failed as leading powers continued to reject the notion of a general responsibility for the global environment. Only the environmental revolution of the 1960s and 1970s, which transformed environmentalism from an elite concern into a broader mass movement, was able to mobilise enough political support for a change of heart among core members of GIS.

The 1972 Stockholm conference serves as a marker for the emergence of environmental stewardship as a primary institution of international society. The Stockholm Declaration included several principles that spoke to the solidarist notion of humans' shared interest in a healthy planet and nature as the 'common heritage of humanity'. For the first time, states had collectively recognised a responsibility to protect the global environment, though calls for a radical break with the Westphalian basis for international environmental action went unanswered. The Declaration balanced solidarist ambition with an unambiguous commitment to pluralist principles, in particular national sovereignty and developmentalism. The cosmopolitan notion of humanity living on 'Spaceship Earth' may have provided an important impetus for the Stockholm conference; however, in order to be accepted as a fundamental norm in international relations, environmental stewardship had to be framed in conformity with the pluralist constraints of the existing society of states.

Despite the successful Stockholm conference, it could not be taken for granted that environmental stewardship would become recognised as a universal norm for *global* international society. During the preparatory process for the UN conference, deep divisions between the Global North and the Global South characterised the debate on how to define states' global environmental responsibility. The conference managed to bridge this divide to some extent, not least by accommodating some of the developing world's concerns around the right to economic development, but it was not until UNCED in 1992 that environmental stewardship shed its Western origins and became fully accepted around the world. The Rio Declaration reinforced countries' sovereign right to use their natural resources and determine their domestic environmental priorities, and it also established unambiguously the principle of North–South differentiation in environmental obligations and financing. By reconciling the normative demands of both developed and developing countries, the global environmental compromise became an integral part of the emergence of a *global* international society in the aftermath of decolonisation.

By the end of the twentieth century, the social consolidation of the environmental norm had become evident from the rapid increase in the number and scope of environmental secondary institutions. It also showed itself in the way states developed the required capacity to participate in environmental diplomacy and implement international agreements. Environmental norms also began to seep into the normative and institutional structures of other international policy areas. Many international organisations adopted environmental mandates and integrated environmental principles into their operations. Moreover, world society actors (NGOs, business) reinforced environmental stewardship, by supporting the mainstreaming of the environment in international governance and by creating transnational environmental initiatives that complemented and expanded intergovernmental regimes. In this way, global environmentalism became increasingly entrenched in state practices, intergovernmental bodies and transnational governance institutions.

The international commitment to environmental stewardship has certainly hardened since the 1970s, but this is not to suggest that normative contestation of the environmental norm also disappeared. Far from it, key environmental and equity principles (e.g. common heritage; precaution; polluter pays; North–South differentiation) have come up against persistent, and in some case increasing, resistance. Some normative fault lines still run between the Global North and Global South while others have pitted different industrialised countries against each other (e.g. EU vs the United States). The rise of emerging economies and their rapidly growing environmental footprint has called into question the established principle of North–South differentiation, while the rise of populist leaders in major powers has undermined their commitment to important environmental treaties. However, this ongoing contestation has not weakened GIS's commitment to environmental stewardship as such. If anything, fundamentalist opposition to environmentalism has declined in international and world society as certain environmental threats have become more evident (e.g. global warming), and where it exists it is largely a debate within, not between, states.

The story of the rise of global environmentalism seems to suggest a progressive, almost linear, normative development in GIS. Such an interpretation contrasts, however, with the limited success in bringing global environmental problems under control. In fact, key ecological indicators, from the continued loss of biodiversity to rising greenhouse gas emissions, point to an escalating environmental crisis that GIS has failed to come to terms with. It is important, therefore, to note that the international commitment to environmental stewardship is stronger in its procedural dimension, but weaker in terms of the substantive commitment to goal achievement. Virtually all states have accepted the need to participate in global environmental policy-making. Environmental multilateralism, understood as a procedural norm, is indeed firmly established. Compliance with international commitments, however, remains largely at the discretion of individual states. States can still breach international environmental obligations without facing serious sanctions or other forms of punitive action. Failure to live up to the substantive expectations of environmental stewardship may provoke international opprobrium but does not yet call into question a state's rightful membership in GIS.

Interplay with Other Primary Institutions of GIS

That environmental stewardship has successfully established itself as a new primary institution of GIS is all the more remarkable as other new international norms have struggled to gain global traction. In contrast to human rights and democracy, for example, global environmentalism has transcended its early roots in Western society and has come to be redefined as a global responsibility for all societies and states. But to what extent has environmental stewardship affected the international normative structure, and especially the moral purpose of the state and GIS? To better understand environmentalism's transformative potential, it is important to consider how it interacts with other fundamental international norms.

How environmental stewardship relates to the existing primary institutions of GIS depends on the kind of environmental ideas it represents. As discussed in chapter 3, environmentalism is a broad church, with views ranging from a radical and ecocentric tradition to a more reformist and anthropocentric approach. The former identifies the protection of nature as an intrinsic value and treats the non-human environment as a moral referent in its own right. Radical greens demand a more profound restructuring of the international order. They challenge the division of the world into territorially defined sovereign political entities, reject claims by states and international organisations to represent the planetary interest and instead call for 'the decentralisation of power and radical subsidiarity' (Newell, 2020: 140). The latter, pragmatic, view is that environmental protection is about securing the human right in a liveable natural environment. In its anthropocentric solidarist perspective, the pluralist strictures of the Westphalian order are to be transcended by globalising environmental policy-making and institutionalising international cooperation. While radical views continue to resonate in environmental debates (Scarce, 2016), it is mainly the pragmatic and reformist wing that drives environmental stewardship within GIS.

Environmental stewardship offers a good fit with certain primary institutions that have reinforced the growth of environmental norms and practices in GIS. This is the case with *diplomacy*, which provides many of the mechanisms for global environmental policy-making and has itself benefitted from the rise of the environmental agenda. As participation in multilateral environmental negotiations has become a matter of routine, states have had to enlarge their foreign policy machinery to include environmental and technical expertise. The creation of over 1,300 environmental treaties has led to a dramatic expansion of diplomatic representation in annual COPs, subsidiary bodies and compliance committees. To cope with the increasing demands of environmental diplomacy, most countries had to open their diplomatic services to other governmental and non-governmental experts, which now routinely serve on national delegations in MEAs and populate international environmental bureaucracies. Environmental stewardship is also a smooth fit with *international law* and has opened up a new functional area for extensive legal development in GIS. International law provides mechanisms and procedures for creating international environmental rules, monitoring their implementation and promoting compliance with treaty obligations. It offers regulative instruments for reducing environmental degradation and has also begun to strengthen individuals' rights with regard to environmental information and protection. Some legal developments push beyond the anthropocentric tradition of conventional international law, but these have yet to fundamentally alter the nature of international environmental law.

As regards the primary institution of *human rights*, which has struggled to gain recognition as a norm of *global* international society, environmental stewardship has provided some support for a human rights-based interpretation of states' global environmental responsibility. This is particularly so with the pragmatic, anthropocentric, tradition of environmentalism that regards a healthy environment as one of the fundamental rights of individuals and communities. Environmental rights have found their way into international declarations and national constitutions, and special treaties exist to promote procedural rights to environmental information (e.g. Aarhus Convention). However, despite these advances, the very notion of a judicable human right to a clean environment remains controversial and heavily contested in GIS.

With regard to some other primary institutions, the rise of environmental stewardship has been of limited or no consequence. This is the case with *nationalism* and the *balance of power,* and to some extent also *great power management.* Although nationalism stands in the way of realising cosmopolitan visions for a green global order, it has largely been unaffected by the rise of environmentalism. As discussed in chapter 3, there are good reasons to suggest that nationalism might work in favour of certain forms of environmentalism, particularly when it promotes ideas of heritage, a sense of place-based environmental concern and obligations towards future generations. Furthermore, the reality of a pluralist world society suggests that national differences in the way societies conceive of nature and environmental duties are likely to persist. At the same time, where the urgency of addressing global environmental problems reinforces the solidarist logic of deeper international cooperation, global environmentalism can still pose a direct challenge to nationalism's normative strength in GIS. As yet, however, no such challenge has materialised in GEP. The same can be said for the balance of power, which is not directly impacted by the international environmental agenda. An escalating ecological crisis could potentially affect the distribution of material capabilities among the great powers, and it might call into question the legitimacy of great power balancing as a means to securing international order. Should the environmental crisis escalate and pose an existential threat not just to planetary health but also to GIS, then it is entirely conceivable that the need to manage ecological threats could emerge as a task for great power management. However, environmental issues have as yet not taken on systemic relevance for the maintenance of international order and stability, and great power management has not been mobilised for environmental governance purposes. In any case, the balance of power has weakened as a core pluralist institution, and great power management faces an uncertain future in a GIS increasingly characterised by deepening pluralism (Buzan, 2014: 143–7).

Environmental stewardship has posed a stronger challenge to two primary institutions that are not part of the core pluralist institutions of the Westphalian order but rose to prominence in the twentieth century as part of an enlarged core purpose of the state: the *market* and *development.* In the case of the market norm, environmentalism challenges the idea of the market as an efficient mechanism for managing the depletion of natural resources and the degradation of ecosystems. Some environmentalists have called into question the inherent growth imperative in market-based economic systems, and many advocate stricter regulations and state intervention in the global economy, to ensure that international trade and investment flows do not undermine environmental protection measures. However, while the rise of environmental stewardship has led to new environmental standards and rules that regulate international economic transactions, it has not posed a fundamental challenge to the market as the dominant organising principle for the global economy. In fact, both the Stockholm and Rio Declarations reinforce the need to make global environmental governance market-friendly, and MEAs and other international environmental regulations have increasingly come to favour market-based instruments. The compromise of liberal environmentalism (Bernstein, 2001) that emerged at the time of the 1992 Rio Earth Summit has remained largely intact.

The same can be said for *developmentalism*, the norm that the economic development of postcolonial societies is a national and global imperative that deserves support from developed economies. At the time when environmental stewardship came to be accepted in international society, many developing countries viewed environmentalism as a threat to their developmental aspirations. Once environmental stewardship had been reinterpreted as sustainable development, however, the North–South conflict started to subside, and repeated references in international treaties to the sovereign rights of states to pursue their own economic development agenda have helped to weaken the challenge that GEP poses to developmentalism. Environmentalism has certainly added a sustainability dimension to developmental practices of states and international organisations, but this has come in exchange for recognition of the developmental and growth aspirations of poorer societies.

Environmentalism has always posed a challenge to the core pluralist norms of *sovereignty* and *territoriality*, and especially so today when the deterioration of major global ecological trends (global warming; biodiversity loss) reinforces the perception that a world divided into politically sovereign states is unfit for meeting these challenges. Both primary institutions have therefore softened somewhat as states have accepted ever more legal obligations arising from environmental treaties and have built a web of international organisations and regulatory networks that have made global policy-making on the environment the norm. In this sense, the rise of GEP has made it difficult for most states to defend strong interpretations of both sovereignty and territoriality. Yet, most major powers and especially the developing world have been careful to restrict the extent to which international environmental policies erode national sovereignty. Efforts to internationalise the environmental management of natural resources or ecosystems that can be found within national boundaries (e.g. tropical forests) have been rebuffed time and again. Likewise, the implementation of MEAs has not been backed up by powerful compliance mechanisms and enforcement tools. Environmental stewardship has gradually chipped away at the Westphalian foundations of GIS, but sovereigntist concerns have returned in GEP as in other areas of international politics, making it difficult to accelerate the transformation of sovereignty along ecological lines.

As this review of the interaction between environmental stewardship and other primary institutions shows, environmentalism has been inserted into the normative structure of GIS but without fundamentally rewriting the moral purpose of the state. Environmental protection is now recognised alongside other global social purposes but has not taken precedence over other core responsibilities. Global discourses around economic growth and development have taken on a green hue but without radically calling into question the underlying growth imperative, and states have recognised global ecological responsibilities without these imposing serious limitations on national sovereignty. The state's moral purpose has expanded to include the role of environmental guardian, which suggests a shift 'from a construction of the state as owner/overload of its territory and toward that of caretaker/trustee of territory, with custodial and management obligations owed not only to other states but also to citizens and the global community' (Eckersley, 2004a: 135). Yet, this initial transformation is still far from reaching a settled notion of ecologically responsible statehood. Environmental stewardship has arrived as a primary institution, but a deeper and more comprehensive greening of GIS remains a work in progress.

Broadening the focus beyond state-centric international society, it has also become clear that the rise of environmental stewardship as a fundamental norm has brought about a closer relationship between international and world society in environmental matters. The gradual shift in understandings of international legitimacy could not have happened without the normative changes in domestic and global society and the persistent political advocacy by non-state actors, primarily environmental NGOs but increasingly also businesses. Environmental stewardship is now firmly established in state-centric international relations as well as in the political structures of world society. Indeed, both realms of global environmental action are closely entwined and reinforce each other through processes of mutual norm transfer and cross-fertilisation. World society has grown into an important site of global environmental governance in its own right, though it would be misleading to conceptualise this form of environmental action in opposition to, or isolation from, state-centric processes. To a large extent, non-state environmental governance operates under the shadow of hierarchy, with states and international organisations providing a critical context within which world society actors create and exercise new forms of authority. In this sense, we have witnessed small steps towards greater integration between the state-centric and transnational political spheres, though the vision of an integrated world society remains an unfulfilled ambition.

The discussion in this book has also highlighted the need to avoid treating solidarist and pluralist dynamics in GEP in isolation from each other. As recent ES scholarship has emphasised, solidarism and pluralism are not mutually exclusive global normative structures but co-exist and overlap to a significant degree. This is the case for the various primary institutions, whether pluralist or solidarist in nature, that define a given international society's normative structure. It also applies to world society, which contains a large variety of ideological stances on environmental questions. The non-state political sphere shows clear signs of pluralist value diversity, in that environmentalists are confronted by organised anti-environmentalism, and even within the environmentalist tradition significant differences between eco-globalist and eco-localist perspectives persist until today. Environmentalism has always been a movement of movements, not a homogenous ideology. Just as in other areas of norms research in IR, the study of GEP needs to pay greater attention to normative fluidity in environmental politics and the 'complex normative terrain that actors actually navigate' (Reus-Smit, 2018: 134).

Rethinking Normative Change in Global International Society

In this book I have deployed and developed the ES approach to studying long-term normative change in international society. In contrast to much of the GEP literature in IR, my analysis has sought to track environmental norm change in the long run, over a century and a half, from the late nineteenth to the early twenty-first century. My argument is that adopting an ES perspective helps correct the historical short-sightedness of GEP research and directs our attention to deeper levels of international change that manifest themselves within the normative structure of GIS, and not simply in the accretion of international environmental institutions and governance mechanisms. The ES's core conceptual toolkit – from primary and secondary institutions to the triad of international system/international society/world society and the pluralism–solidarism distinction – offers a useful framework for tracking this fundamental norm change and assessing how the rise of global environmentalism affects the foundational norms of GIS. Contrary to the widespread perception of the ES as a state-centric approach, I have also shown how it can be employed to investigate the growing interplay between state and non-state actors and assess the degree to which international and world society have integrated. Taking an ES approach can thus help connect GEP research with wider debates in IR, about change in international normative structures and the processes that initiate, sustain and also resist such changes.

My intention behind this study was also to correct the long-standing failure of the ES to engage with environmental politics as a major empirical field of international policy-making. While human rights and humanitarian intervention has been ES scholars' preferred empirical focus, I have shown how closer engagement with GEP provides a rich case study of solidarist norm change. Indeed, the rise of global environmentalism and its import from world society into international society offers valuable insights into the nature of normative change in international society. First, it underlines and further illustrates the malleability of the international normative order, an argument that ES and constructivist scholars have put forward in other contexts. More specifically, international society is not static but is open to the inclusion of new social purposes. Conventional theorising of normative change, in ES theory as much as in other IR approaches, has focused on the arrival, change or decline of international norms that originate within international society, reflecting their role in maintaining international order and stability. The case study of environmentalism points to the close interaction between state and non-state actors, a perspective that chimes with arguments by Clark (2007) and Buzan (2018) that locate critical shifts in international legitimacy at the interface of international and world society. It also speaks to Allan's (2018: 43) argument that identifies the roots of international change 'in activist networks and epistemic communities' as much as in state-centric contexts.

Second, the case of environmentalism provides strong evidence of the ability of GIS to evolve not only in an autopoietic manner, out of its own normative resources and functional needs, but also in response to broader societal purposes that are increasingly defined and negotiated in a global context. To some extent, this extended moral purpose of the state and international society reflects a gradual transition from pluralist to solidarist principles. But whereas many solidarists in the ES tradition have focused on global justice and human rights, I have shown that safeguarding planetary health needs to be added to the list of major solidarist causes. Just as the state's moral purpose has expanded in response to the growing complexity of industrial and post-industrial societies, so has international society come to adopt an enlarged set of social purposes that go beyond the narrow question of regulating interstate relations. GIS may not be well equipped to deliver an effective environmental response – a point repeatedly stressed by environmentalists – but this has not stopped it from being drawn ever further into the difficult task of creating global environmental norms and governance structures. IR needs to further open its analytical perspective to changes in domestic and transnational societal values that percolate up and insert themselves into the international normative structure. Because of the endlessly inventive nature of human societies, the only way we can establish the direction of normative evolution in GIS is through empirical research rather than deductive reasoning.

The historical analysis of the rise of global environmentalism also suggests a more diffuse process of international norm change than is commonly assumed in IR. State-centric approaches tend to focus on 'order-building moments' (Allan, 2018: 6) when great powers renegotiate the institutional and normative structure of international society, usually after major international crises or wars. Even when it comes to incidents of norm transfer from world to international society, most scholars zoom in on equivalent encounters between state and non-state actors at which international order is rebuilt (Clark, 2007). Advocacy by non-state actors in international forums and negotiations is certainly the most visible form of attempted norm entrepreneurship, and this is one reason why it attracts a great deal of attention in transnationalist studies. As discussed earlier, GIS has come to formalise non-state actors' participation within intergovernmental processes, and especially so in the environmental field, which has enabled NGOs to score some notable successes in international policy development. However, the focus on specific outcomes in international bargaining risks distracting attention from the more important but diffuse role of norm innovation and framing that world society is engaged in. Clark (2007), who adopts a narrow focus on international peace conferences following major international conflicts, concludes that 'world-society influence generally takes the form of the normative framing of the issues, rather than of policy determination'. Indeed, the 1972 Stockholm conference and the 1992 Rio Summit count as 'constitutional moments' in the greening of international society but have not been the main occasions for norm transfer from world to international society. Instead, we need to see the greening of GIS through the analytical lens of the *longue durée*, a gradual accumulation of social and ideological changes in response to the great transformation of the nineteenth century (Buzan and Lawson, 2015). Environmental stewardship did not arise in response to a major international crisis, a rapture in the international normative structure brought about by great power war leading to an order-building moment. Instead, it was is the result of gradual shifts in societal discourses, which started in leading industrialised countries and eventually spread worldwide. State power and leadership were instrumental in the creation of the international environmental agenda in the 1970s, but the success of the environmental revolution in GIS is down to the changes in, and convergence of, societal values (belief) and interests (calculation), not international coercion.

The ES provides a useful conceptual toolkit for analysing the *longue durée* of international norm change. The distinction between primary and secondary institutions, in particular, allows us to separate out different levels of change and assess the degree to which new norms become embedded in GIS. It is in the interplay between primary and secondary institutions that we can identify processes of norm emergence, evolution, strengthening, contestation and decay. To be sure, primary institutions are not the most fundamental units of the international social structure. Reus-Smit (1999) argues that it is the moral purpose of the state that determines why and how international societies are structured, including primary institutions. Allan (2018: 33) takes this one step further and identifies cosmological ideas as the 'discursive resources that constitute discourses of state purpose'. Whether we need to follow these ideational structures 'all the way down' depends on the kind of international change we seek to explain. Within the environmental field, it is certainly possible to point to deeper ideational shifts that underpin the rise of environmental stewardship as a primary institution. As discussed in Chapters 3, 4 and 5, the emergence of environmentalism is predicated on at least two novel discursive formations: the development of ecological knowledge that allowed humans to understand the fragility and global interconnectedness of ecological systems; and the rise of ethical ideas that demand a duty of care towards the environment, whether by individuals, communities, states or international society. In this sense, the rise of environmental stewardship that I have tracked in this book begs the question of which prior discursive shifts facilitated its emergence in the first place. These more fundamental shifts would have to be identified in other realms: within science, ethics and indeed cosmological ideas about the relationship between humans and their natural environment. My study followed a different direction and purpose, namely to track the emergence of the environmental norm in its historical context and to explore its institutionalisation in international society. This has required me to focus on primary institutions and how they are reproduced through secondary institutions and state practices, and the adoption of new values and identities by relevant actors in international and world society. The

ES focus on primary institutions is thus only one, albeit important, way to open up an investigation of fundamental international change.

The case study of global environmentalism also provides insights into the indeterminacy of normative meanings and open-endedness of norm change. Environmentalism originates in a diverse set of ideas that coalesced around a shared sense of environmental care and responsibility. These ideas gradually merged into a global movement that drove policy change at the national and international level, with environmental stewardship emerging as a 'clearly defined value or principle applicable across international society' – one of the key criteria for determining the arrival of new primary institutions (Falkner and Buzan, 2019: 136). At the same time, the environmental movement itself was a broad church and represented a diverse field of societal values and policy ideas. Environmentalism evolved over time and also took on different meanings in diverse geographical and cultural contexts, something that is common to all norms, values and practices (Reus-Smit, 2018: 206). Unsurprisingly, therefore, environmental stewardship remained subject to competing interpretations even after it had been accepted by a majority of states at the Stockholm conference. New international norms don't arrive fully formed and with a fixed meaning, they remain malleable and open to re-negotiation. As new norms become part of the international normative structure, they interact with existing primary institutions, leading to normative accommodation or dissonance. Political struggles over their precise meaning and practical application carry on even as new norms consolidate their social status. Meaning indeterminacy and norm contestation are thus part and parcel of any process of international normative change. As a consequence, new norms are marked by a 'tension between continuity and change' (Niemann and Schillinger, 2017: 30). This may complicate the task of identifying the arrival of new primary institutions, but does not make it a futile endeavour. Global norms are adopted and reinterpreted in localised contexts (Acharya, 2004), which increases meaning indeterminacy but can help to strengthen their global recognition. Likewise, ongoing international contestation opens up the possibility of norm decay and decline, but depending on the nature of contestation it may also end up confirming its basic social validity (Deitelhoff and Zimmermann, 2020). As the case of global environmentalism has shown, indeterminacy and contestation do not stand in the way of declaring novel fundamental norms a social fact in international society.

The history of the rise of environmental stewardship also lends support to the growing recognition in ES and constructivist IR of the need for a more spatially differentiated account of international norm change.

Recent work has identified the inherent Eurocentrism in the expansion story of the ES and sought to replace it with a more geographically and culturally nuanced globalisation story (Dunne and Reus-Smit, 2017). Buzan and Schouenborg (2018) propose a comprehensive framework for theorising the coexistence of different normative structures at regional, sub-global and global levels. In this perspective, geography is a central principle of differentiation for social structures. New norms in international relations cannot be assumed to acquire global recognition by default, their geographical reach is an empirical question. Some norms will develop resonance only within a geographically contiguous (regional) or noncontiguous (sub-global) context, while others may eventually outgrow such spatial limitations to achieve global recognition. The environmental norm's trajectory in international relations shows a clear pattern of globalisation that helped transcend its geographic roots. Originating in a regionally specific context (mainly North America and Europe), environmental stewardship emerged as an international norm in the 1970s. At this time, its global reach was still limited by fundamental opposition among communist countries and strong opposition from many developing countries. Following a process of normative accommodation and re-negotiation, the environmental norm eventually reached global recognition by the 1990s. However, despite this seemingly linear progression towards universal applicability within GIS, the strength with which environmental stewardship is recognised and institutionalised today still varies across the world. European international society has gone furthest in establishing environmental stewardship as part of its constitutional order, with the EU's institutional architecture providing a strong basis for implementing global environmental commitments. Other regional entities (NAFTA, ASEAN) also provide an additional institutional home in support of environmental sustainability, though their impact on state practices and regional environmental cooperation is less pronounced than in Europe. In this way, environmental stewardship is both universally recognised in GIS and geographically differentiated, in terms of the 'thinness' and 'thickness' of regional versions of environmental international society.

Looking to the Future: Global International Society in the Anthropocene

This book has taken the long view of the history of global environmental politics. It has analysed over a century and a half of global environmentalism and its impact on the international normative structure. The ES provides a conceptual toolkit to make sense of states' and international society's deepening commitment to environmental protection and to place this trend in the context of wider normative developments in international relations. Can this perspective also be employed to help us come to terms with the likely futures of GEP? What does the history of international and world society's engagement with environmental issues tell us about how we might deal with worsening or newly emerging ecological threats? It is this more speculative question that I turn to in this final section.

The starting point in this reflection on future trends has to be a recognition that humanity and international society have so far failed to tackle some of the worst forms of environmental degradation. Although GEP has managed to bring some environmental problems under control (e.g. ozone layer depletion, acid rain), the 'great acceleration' of natural resource depletion, greenhouse gas emissions and population growth since the mid-twentieth century carries on largely unimpeded (McNeill and Engelke, 2016). Scientists have coined the term 'Anthropocene' to describe the new reality of a planet profoundly shaped by human activity. In contrast to the Holocene epoch, the Anthropocene represents a new geological era in which humanity has come to exert a dominant influence on global ecology. Traces of large-scale human interference with global ecosystems can be found in the global climate system, the stratospheric ozone layer, marine environments, forest coverage, nitrogen cycles and biodiversity. Many of these transformative trends are destabilising ecosystems and threaten to impair the welfare of future generations. Facing up to the challenges of the Anthropocene could force humanity to rethink and reshape the foundations of its national and international political–economic orders.

Human-induced climate change provides one of the best-understood scenarios for thinking through the political consequences of a global ecological crisis that is not just accelerating but spinning out of control. We know that human activities have so far led to a global warming trend of around 1°C since the beginning of the Industrial Revolution, and GIS has agreed to keep the overall warming trend to under 2°C by the end of the century. Current GHG emission trends suggest that the world is likely to overshoot this target and may face a 3°C or even 4°C warming scenario. Such an outcome would speed up already existing climate change impacts, such as the melting of glaciers and Arctic ice caps, the acidification of oceans and coral bleaching, shifts in precipitation and more extreme weather events. It is also likely to bring the world closer to ecological tipping points, that is 'large-scale discontinuities' in the climate system that would lead to potentially catastrophic and irreversible forms of global environmental change. Global sea levels, for example, could rise well beyond the currently expected increase of between 0.5 m and 1 m by the end of the century; ecosystems, such as the Amazonian rainforests and coral reefs, could be destroyed more rapidly, leading to large-scale losses of vital biospheres; and the world could soon reach a new 'hothouse' climate state that would reduce agricultural yields and undermine transport, energy and urban infrastructures (Lenton et al., 2019). Undoubtedly, the combination of dramatic sea level rises, extreme weather events, agricultural output loss and accelerated destruction of biodiversity would pose an existential threat to most societies on the planet. The odds of reaching such a state of runaway global warming have shortened dramatically in recent years. Global GHG emissions have been on a rising trajectory until the outbreak of the COVID-19 crisis, and despite a temporary fall in emissions in 2020 it is proving difficult to radically alter the emissions trajectory that would allow the world to stay below the internationally agreed 2°C target.

How would GIS be impacted by the onset of an existential ecological crisis of this magnitude, and how would it respond to its political and economic ramifications? In one scenario, escalating climate change would force humanity to recognise the shared nature of the ecological threat and the need for a collective response. The sharp deterioration of environmental conditions would underline the shortcomings of the existing international approach to environmental management – the weakness of international organisations, the lack of enforcement of international treaties and the inadequacy of a system of environmental governance built on pluralist principles of sovereignty and non-intervention. Applying a 'Spaceship Earth' logic, with humans as passengers travelling together on a finite planet and no exit option, international society would come to the realisation that only a solidarist response can secure humanity's future on a warming planet. In this perspective, the existential threat to humanity would accelerate the transformation of GIS, away from its pluralist roots and towards global environmental sustainability and justice. In this way, the accelerating ecological crisis would hasten the full realisation of environmental stewardship's solidarist potential, a shift towards the full greening of the state and international society. The future green global order would be built on the confluence of strong belief and calculation, and would require only a modest degree of coercion.

This outcome cannot be taken for granted, however. A rapidly worsening ecological situation would also put the international architecture for environmental governance under greater strain. Catastrophic global warming would have differential effects around the world, creating distributional conflicts among nations as competition for scarce natural resources intensifies and environmentally induced migration increases. Humanity could find itself not in a 'spaceship', with a captain ably steering its passengers to safety, but in a 'lifeboat' that floats precariously on the ocean, without a leader in charge. In Garrett Hardin's infamous analogy, those in the lifeboat (representing rich countries) fight for their survival against those still in the water (representing poor countries) (Höhler, 2015: 100). The desire to survive would not become a shared, collective, imperative for humanity, it would translate into a zero-sum logic that drives people and communities apart. Applied to the international context, such a scenario reinforces rather than weakens the pluralist principles of GIS. As the global climate crisis spirals out of control, collective management via international cooperation to reduce emissions is abandoned as nations individually seek to adapt to rising temperatures and safeguard their survival. The ecological *raison d'état* that is triggered by the existential threat would thus strengthen the pluralist foundations of GIS, making it harder than ever to build global environmental governance.

Could a pluralist GIS prevent the disintegration of international order amidst rising environmental nationalism and conflict? The pluralist scenario holds the promise for a collective response against the global ecological threat, which requires leading powers to accept a shared responsibility for global system management – *raison de système*. Such great power management has not played a prominent role in GEP in the past, despite the persistence of minilateral approaches within environmental multilateralism and repeated calls for great powers to shoulder greater environmental responsibilities. However, attempts to securitise the environment have increased in recent years, including at the UN Security Council, and should an ecological catastrophe become a systemic threat to international society it is at least conceivable that some great powers will feel compelled to take environmental action in the global interest. This could be in the form of unilateral action by individual powers (e.g. unilateral deployment of geo-engineering) or a collective green hegemony to manage the worst excesses of environmental damage. Either way, full securitisation of the global environment would be required to lend legitimacy to such an emergency response. A mixture of calculation and coercion, rather than belief, would be the main basis for such a pluralist environmental order.

Without further evidence of how political and economic systems will react to catastrophic climate change, such scenarios remain in the realm of speculation. We can identify different logics of international action along the solidarist–pluralist spectrum, as well as different motivations behind international order creation, involving mixtures of belief, calculation and coercion. We can chart the different pathways and outcomes for GEP, from a progressive strengthening of international environmental institutions and law to a regression into environmental nationalism and conflict, with collective or unilateral environmental hegemony somewhere in between these poles. And we can identify different forms of interaction between international and world society, including a future strengthening of the legitimacy and authority of non-state actors as providers of global environmental functions. For all this, the ES perspective offers valuable insights into long-term trends of normative development in international relations. While it cannot predict likely outcomes, it offers, at a minimum, a conceptual language and analytical framework that allows us to make sense of humanity's future on a planet threatened by ecological collapse but divided by national boundaries.

References

Abbott, Kenneth W. (2012). Engaging the Public and the Private in Global Sustainability Governance. *International Affairs*, 88(3), 543–64.

Abbott, Kenneth W., Genschel, Philipp, Snidal, Duncan, & Zangl, Bernhard (2015a). *International Organizations as Orchestrators*. Cambridge: Cambridge University Press.

(2015b). Orchestration: Global Governance through Intermediaries. In Kenneth W. Abbott, Philipp Genschel, Duncan Snidal, & Bernhard Zangl (Eds.), *International Organizations as Orchestrators*. Cambridge: Cambridge University Press, pp. 3–36.

Acharya, Amitav (2004). How Ideas Spread: Whose Norms Matter? Norm Localization and Institutional Change in Asian Regionalism. *International Organization*, 58(2), 239–75.

(2009). *Whose Ideas Matter? Agency and Power in Asian Regionalism*. Ithaca, NY: Cornell University Press.

(2014). *The End of American World Order*. Cambridge: Polity Press.

Acharya, Amitav (Ed.) (2016). *Why Govern? Rethinking Demand and Progress in Global Governance*. Cambridge: Cambridge University Press.

Acharya, Amitav, & Buzan, Barry (2019). *The Making of Global International Relations: Origins and Evolution of IR at Its Centenary*. Cambridge: Cambridge University Press.

Acte De Fondation d'une Commission Consultative Pour La Protection Internationale De La Nature (1913). Available at: https://iea.uoregon.edu/treaty-text/3708-0, accessed 29 January 2021.

Adger, W. Neil (2010). Climate Change, Human Well-Being and Insecurity. *New Political Economy*, 15(2), 275–92.

Ahrens, Bettina (2017). The Solidarisation of International Society: The EU in the Global Climate Change Regime. GLOBUS Research Paper 5. Available at: https://papers.ssrn.com/sol3/papers.cfm?abstract_id=3059873, accessed 29 January 2021.

Allan, Bentley B. (2018). *Scientific Cosmology and International Orders*. Cambridge: Cambridge University Press.

Allitt, Patrick (2014). *A Climate of Crisis: America in the Age of Environmentalism*. New York: Penguin.

Ameli, Nadia, Drummond, Paul, Bisaro, Alexander, Grubb, Michael, & Chenet, Hugues (2020). Climate Finance and Disclosure for Institutional Investors: Why Transparency Is Not Enough. *Climatic Change*, 160(4), 565–89.

Andonova, Liliana B. (2004). *Transnational Politics of the Environment: The European Union and Environmental Policy in Central and Eastern Europe*. Cambridge, MA: MIT Press.

(2017). *Governance Entrepreneurs: International Organizations and the Rise of Global Public–Private Partnerships*. Cambridge: Cambridge University Press.

Andonova, Liliana B., Betsill, Michele M., & Bulkeley, Harriet (2009). Transnational Climate Governance. *Global Environmental Politics*, 9(3), 52–73.

Andonova, Liliana B., Hale, Thomas N., & Roger, Charles B. (2017). National Policy and Transnational Governance of Climate Change: Substitutes or Complements? *International Studies Quarterly*, 61(2), 253–68.

Andonova, Liliana B., & Mitchell, Ronald B. (2010). The Rescaling of Global Environmental Politics. *Annual Review of Environment and Resources*, 35, 255–82.

Andresen, Steinar (2013). International Regime Effectiveness. In Robert Falkner (Ed.), *The Handbook of Global Climate and Environment Policy*. Cheltenham: John Wiley & Sons, pp. 304–19.

Andresen, Steinar, & Gulbrandsen, Lars H. (2005). The Role of Green NGOs in Promoting Climate Compliance. In Olav Schram Stokke, Jon Hovi, & Geir Ulfstein (Eds.), *Implementing the Climate Regime: International Compliance*. London: Earthscan, pp. 169–86.

Anker, Peder (2002). *Imperial Ecology: Environmental Order in the British Empire, 1895–1945*. Cambridge, MA: Harvard University Press.

Anon (1935). International Office for the Protection of Nature. *Nature*, 135 (3408), 301.

Antholis, William, & Talbott, Strobe (2010). *Fast Forward: Ethics and Politics in the Age of Global Warming*. Washington, DC: Brookings Institution Press.

Anton, Donald K. (2012). "Treaty Congestion" in Contemporary International Environmental Law. In Erika J. Techera (Ed.), *Routledge Handbook of International Environmental Law*. London: Routledge, pp. 681–96.

Arculus, Ronald (1970). Environment: ECE and the United Nations, Folio 82. FCO 55/384. Kew National Archives.

Armiero, Marco, & Sedrez, Lise Fernanda (2014). Introduction. In Marco Armiero & Lise Fernanda Sedrez (Eds.), *A History of Environmentalism: Local Struggles, Global Histories*. London: Bloomsbury, pp. 1–19.

Aronova, Elena, Baker, Karen S., & Oreskes, Naomi (2010). Big Science and Big Data in Biology: From the International Geophysical Year through the International Biological Program to the Long Term Ecological Research (LTER) Network, 1957–Present. *Historical Studies in the Natural Sciences*, 40(2), 183–224.

Attfield, Robin (2014). *Environmental Ethics*. Cambridge: Polity Press.

Avalle, Oscar A. (1994). The Decision-Making Process from a Developing Country Perspective. Paper presented at the Negotiating International Regimes: Lessons Learned from the United Nations Conference on Environment and Development (UNCED).

Bäckstrand, Karin, & Kronsell, Annica (Eds.) (2015). *Rethinking the Green State: Environmental Governance towards Climate and Sustainability Transitions*. London: Routledge.

Bahro, Rudolf (1989). *Logik der Rettung: Wer kann die Apokalypse aufhalten? Ein Versuch über die Grundlagen ökologischer Politik*. Stuttgart: Edition Weilbrecht.

Bailey, Jennifer L. (2008). Arrested Development: The Fight to End Commercial Whaling as a Case of Failed Norm Change. *European Journal of International Relations*, 14(2), 289–318.

Bain, William (2014). The Pluralist–Solidarist Debate in the English School. In Cornelia Navari & Daniel M. Green (Eds.), *Guide to the English School in International Studies*. Chichester: John Wiley & Sons, pp. 159–69.

Bandy, Joe, & Smith, Jackie (2005). Factors Affecting Conflict and Cooperation in Transnational Movement Networks. In Joe Bandy & Jackie Smith (Eds.), *Coalitions across Borders: Transnational Protest and the Neoliberal Order*. Lanham, MD: Rowman & Littlefield, pp. 231–52.

Bang, Guri, Froyna, Camilla Bretteville, Hovia, Jon, & Menza, Fredric C. (2007). The United States and International Climate Cooperation: International 'Pull' versus Domestic 'Push'. *Energy Policy*, 35, 1282–91.

Bansard, Jennifer S., Pattberg, Philipp H., & Widerberg, Oscar (2017). Cities to the Rescue? Assessing the Performance of Transnational Municipal Networks in Global Climate Governance. *International Environmental Agreements: Politics, Law and Economics*, 17(2), 229–46.

Barkdull, John, & Harris, Paul G. (2002). Environmental Change and Foreign Policy: A Survey of Theory. *Global Environmental Politics*, 2(2), 63–91.

Barkin, J. Samuel, & Cronin, Bruce (1994). The State and the Nation: Changing Norms and Rules of Sovereignty in International Relations. *International Organization*, 48(1), 107–30.

Barnett, Jon (2003). Security and Climate Change. *Global Environmental Change*, 13(1), 7–17.

Barnett, Michael N., & Sikkink, Kathryn (2009). From International Relations to Global Society. In Robert E. Goodin (Ed.), *The Oxford Handbook of Political Science*. Oxford: Oxford University Press. Available at: www.oxfordhandbooks.com/view/10.1093/oxfordhb/9780199604456.001.0001/oxfordhb-9780199604456-e-035, accessed 29 January 2021.

Barry, John, & Eckersley, Robyn (Eds.) (2005). *The State and the Global Ecological Crisis*. Cambridge, MA: MIT Press.

Bartelson, Jens (2009). *Visions of World Community*. Cambridge: Cambridge University Press.

Bartley, Tim (2007). How Foundations Shape Social Movements: The Construction of an Organizational Field and the Rise of Forest Certification. *Social Problems*, 54(3), 229–55.

Barton, Gregory A. (2002). *Empire Forestry and the Origins of Environmentalism*. Cambridge: Cambridge University Press.

Bauer, Steffen (2013). Strengthening the United Nations. In Robert Falkner (Ed.), *The Handbook of Global Climate and Environment Policy*. Cheltenham: John Wiley & Sons Ltd, pp. 320–38.

Bayne, Nicholas, & Woolcock, Stephen (Eds.) (2017). *The New Economic Diplomacy: Decision-making and Negotiation in International Economic Relations*. London: Routledge.

Beattie, James (2011). *Empire and Environmental Anxiety: Health, Science, Art and Conservation in South Asia and Australasia, 1800–1920*. Basingstoke: Palgrave Macmillan.

Beck, Ulrich, & Sznaider, Natan (2006). Unpacking Cosmopolitanism for the Social Sciences: A Research Agenda. *The British Journal of Sociology*, 57(1), 1–23.

Beckerman, Wilfred (1995). *Small Is Stupid: Blowing the Whistle on the Greens*. London: Duckworth.

Beers, Diane L. (2006). *For the Prevention of Cruelty: The History and Legacy of Animal Rights Activism in the United States*. Athens, OH: Ohio University Press.

Beisheim, Marianne (2012). Partnerships for Sustainable Development: Why and How Rio+20 Must Improve the Framework for Multi-stakeholder Partnerships. *SWP Research Paper*. Berlin: Stiftung Wissenschaft und Politik.

Bellamy, Alex J. (Ed.) (2005). *International Society and Its Critics*. Oxford: Oxford University Press.

Benedick, Richard E. (1991). *Ozone Diplomacy: New Directions in Safeguarding the Planet*. Cambridge, MA: Harvard University Press.

Bennett, W. Lance (2005). Social Movements Beyond Borders: Organization, Communication, and Political Capacity in Two Eras of Transnational Activism. In Donatella Della Porta & Sidney Tarrow (Eds.), *Transnational Protest and Global Activism*. Lanham, MD: Rowman & Littlefield, pp. 203–26.

Bergandi, Donato, & Blandin, Patrick (2012). From the Protection of Nature to Sustainable Development: The Genesis of an Ethical and Political Oxymoron. *Revue d'histoire des sciences*, 65(1), 103–42.

Bernauer, Thomas, Kalbhenn, Anna, Koubi, Vally, & Spilker, Gabriele (2010). A Comparison of International and Domestic Sources of Global Governance Dynamics. *British Journal of Political Science*, 40(03), 509–38.

Bernauer, Thomas, & Moser, Peter (1996). Reducing Pollution of the River Rhine: The Influence of International Cooperation. *The Journal of Environment & Development*, 5(4), 389–415.

Bernstein, Steven (2001). *The Compromise of Liberal Environmentalism*. New York: Columbia University Press.

(2002). Liberal Environmentalism and Global Environmental Governance. *Global Environmental Politics*, 2(3), 1–16.

(2013). Global Environmental Norms. In Robert Falkner (Ed.), *The Handbook of Global Climate and Environment Policy*. Cheltenham: John Wiley & Sons Ltd, pp. 127–45.

(2020). The Absence of Great Power Responsibility in Global Environmental Politics. *European Journal of International Relations*, 26(1), 8–32.

Berry, Robert James (2006). Stewardship: A Default Position? In Robert James Berry (Ed.), *Environmental Stewardship: Critical Perspectives – Past and Present*. London: T&T Clark, pp. 1–13.

Bess, Michael D. (1995). Ecology and Artifice: Shifting Perceptions of Nature and High Technology in Postwar France. *Technology and Culture*, 36(4), 830–862.

Betsill, Michele M. (2008a). Environmental NGOs and the Kyoto Protocol Negotiations: 1995 to 1997. In Michele M. Betsill & Elisabeth Corell (Eds.), *NGO Diplomacy: The Influence of Nongovernmental Organizations in International Environmental Negotiations*. Cambridge, MA: MIT Press, pp. 43–66.

(2008b). Reflections on the Analytical Framework and NGO Diplomacy. In Michele M. Betsill & Elisabeth Corell (Eds.), *NGO Diplomacy: The Influence of Nongovernmental Organizations in International Environmental Negotiations*. Cambridge, MA: MIT Press, pp. 177–206.

Betsill, Michele M., & Bulkeley, Harriet (2006). Cities and the Multilevel Governance of Global Climate Change. *Global Governance*, 12(2), 141–59.

Betsill, Michele M., & Correll, Elisabeth (Eds.) (2008). *NGO Diplomacy: The Influence of Nongovernmental Organizations in International Environmental Negotiations*. Cambridge, MA: MIT Press.

Biedenkopf, Katja (2017). Gubernatorial Entrepreneurship and United States Federal-State Interaction: The Case of Subnational Regional Greenhouse Gas Emissions Trading. *Environment and Planning C: Politics and Space*, 35 (8), 1378–400.

Biermann, Frank (2014). *Earth System Governance: World Politics in the Anthropocene*. Cambridge, MA: MIT Press.

Biermann, Frank, & Bauer, Steffen (Eds.) (2005). *A World Environment Organization: Solution or Threat for Effective International Environmental Governance?* Farnham: Ashgate.

Biermann, Frank, Davies, Olwen, & van der Grijp, Nicolien (2009). Environmental Policy Integration and the Architecture of Global Environmental Governance. *International Environmental Agreements: Politics, Law and Economics*, 9(4), 351–69.

Biermann, Frank, Kanie, Norichika, & Kim, Rakhyun E. (2017). Global Governance by Goal-Setting: The Novel Approach of the UN Sustainable Development Goals. *Current Opinion in Environmental Sustainability*, 26, 26–31.

Biermann, Frank, & Siebenhüner, Bernd (2009). The Influence of International Bureaucracies in World Politics: Findings from the MANUS Research Program. In Frank Biermann & Bernd Siebenhüner (Eds.), *Managers of Global Change: The Influence of International Environmental Bureaucracies*. Cambridge, MA: MIT Press, pp. 319–49.

Birnie, Patricia, & Boyle, Alan (2002). *International Law and the Environment*. Oxford: Oxford University Press.

Björkbom, Lars (1988). Resolution of Environmental Problems: The Use of Diplomacy. In John E. Carroll (Ed.), *International Environmental Diplomacy: The Management and Resolution of Transfrontier Environmental Problems*. Cambridge: Cambridge University Press, pp. 123–40.

Blay, Samual K. N. (1992). New Trends in the Protection of the Antarctic Environment: The 1991 Madrid Protocol. *American Journal of International Law*, 86(2), 377–99.

Block, Walter (1998). Environmentalism and Economic Freedom: The Case for Private Property Rights. *Journal of Business Ethics*, 17(16), 1887–99.

Blok, Anders (2008). Contesting Global Norms: Politics of Identity in Japanese Pro-Whaling Countermobilization. *Global Environmental Politics*, 8(2), 39–66.

Bloomfield, Alan, & Scott, Shirley V. (Eds.) (2017). *Norm Antipreneurs and the Politics of Resistance to Global Normative Change*. London: Routledge.

Boardman, Robert (1981). *International Organization and the Conservation of Nature*. London: Macmillan.

Bob, Clifford (2013). The Global Right Wing and Theories of Transnational Advocacy. *The International Spectator*, 48(4), 71–85.

Bodansky, Daniel (1993). The U.N. Framework Convention on Climate Change: A Commentary. *Yale Journal of International Law*, 18, 451–558.

(2007). Legitimacy. In Daniel Bodansky, Jutta Brunnée, & Ellen Hey (Eds.), *The Oxford Handbook of International Environmental Law*. Oxford: Oxford University Press, pp. 704–23.

(2010). *The Art and Craft of International Environmental Law*. Cambridge, MA: Harvard University Press.

Bodansky, Daniel, Brunnée, Jutta, & Hey, Ellen (2007). International Environmental Law: Mapping the Field. In Daniel Bodansky, Jutta Brunnée, & Ellen Hey (Eds.), *The Oxford Handbook of International Environmental Law*. Oxford: Oxford University Press, pp. 1–25.

Bomberg, Elizabeth (2017). Environmental Politics in the Trump Era: An Early Assessment. *Environmental Politics*, 26(5), 956–63.

Borowy, Iris (2019). Before UNEP: Who Was in Charge of the Global Environment? The Struggle for Institutional Responsibility 1968–72. *Journal of Global History*, 14(1), 87–106.

Boulding, Kenneth E. (1966). The Economics of the Coming Spaceship Earth. In H. Jarrett (Ed.), *Environmental Quality in a Growing Economy*. Baltimore, MD: Resources for the Future/Johns Hopkins University Press, pp. 3–14.

Bowman, Megan, & Minas, Stephen (2019). Resilience Through Interlinkage: The Green Climate Fund and Climate Finance Governance. *Climate Policy*, 19(3), 342–53.

Boyd, David R. (2011). *The Environmental Rights Revolution: A Global Study of Constitutions, Human Rights, and the Environment*. Vancouver: UBC Press.

Boyle, Alan (2012). Human Rights and the Environment: Where Next? *European Journal of International Law*, 23(3), 613–42.

Brack, Duncan, & Hyvarinen, Joy (2002). *Global Environmental Institutions: Arguments for Reform*. London: Royal Institute of International Affairs.

Bradford, Anu (2020). *The Brussels Effect: How the European Union Rules the World*. New York: Oxford University Press.

Brain, Stephen (2016). The Appeal of Appearing Green: Soviet-American Ideological Competition and Cold War Environmental Diplomacy. *Cold War History*, 16(4), 443–62.

Brenton, Anthony (2010). Interviewed at His Home in Cambridge by Malcolm BcBain, 6 May. Available at: www.chu.cam.ac.uk/media/uploads/files/Brenton.pdf

(2013). 'Great Powers' in Climate Politics. *Climate Policy*, 13(5), 541–46.

Brenton, Tony (1994). *The Greening of Machiavelli. The Evolution of International Environmental Politics*. London: Earthscan/RIIA.

Brown, Lester R. (1977). *Redefining National Security*. Worldwatch Paper, 14. Washington, DC: Worldwatch Institute.

Brunnée, Jutta (2002). COPing with Consent: Law-Making under Multilateral Environmental Agreements. *Leiden Journal of International Law*, 15(1), 1–52.

(2004). The United States and International Environmental Law: Living with an Elephant. *European Journal of International Law*, 15(4), 617–49.

(2006). Enforcement Mechanisms in International Law and International Environmental Law. In Ulrich Beyerlin, Peter-Tobias Stoll, & Rüdiger Wolfrum (Eds.), *Ensuring Compliance with Multilateral Environmental Agreements: A Dialogue between Practitioners and Academia*. Leiden: Martinus Nijhoff, pp. 1–24.

(2007). Common Areas, Common Heritage, and Common Concern. In Daniel Bodansky, Jutta Brunnée, & Ellen Hey (Eds.), *The Oxford Handbook of International Environmental Law*. Oxford: Oxford University Press, pp. 550–73.

Brunnée, Jutta, & Toope, Stephen J. (2010). *Legitimacy and Legality in International Law: An Interactional Account*. Cambridge: Cambridge University Press.

Bukovansky, Mlada, Clark, Ian, Eckersley, Robyn, Price, Richard, Reus-Smit, Christian, & Wheeler, Nicholas J. (2012). *Special Responsibilities: Global Problems and American Power*. Cambridge: Cambridge University Press.

Bulkeley, Harriet, Andonova, Liliana, Betsill, Michele M., Compagnon, Daniel, Hale, Thomas, Hoffmann, Matthew J., Newell, Peter, Paterson, Matthew, Roger, Charles, & VanDeveer, Stacy D. (2014). *Transnational Climate Change Governance*. Cambridge: Cambridge University Press.

Bull, Hedley (1966). The Grotian Conception of International Society. In Herbert Butterfield & Martin Wight (Eds.), *Diplomatic Investigations: Essays in the Theory of International Politics*. London: Allen & Unwin, pp. 51–73.

(1977). *The Anarchical Society. A Study of Order in World Politics*. London: Macmillan.

(1980). The Great Irresponsibles? The United States, the Soviet Union, and World Order. *International Journal*, 35(3), 437–47.

(1984). The Revolt Against the West. In Hedley Bull & Adam Watson (Eds.), *The Expansion of International Society*. Oxford: Clarendon Press, pp. 217–28.

Bull, Hedley, & Watson, Adam (Eds.) (1984). *The Expansion of International Society*. Oxford: Clarendon Press.

Bunce, Michael F. (1994). *The Countryside Ideal: Anglo-American Images of Landscape*. London: Routledge.

Burgiel, Stanley W., & Wood, Peter (2012). Witness, Architect, Detractor. The Evolving Role of NGOs in International Environmental Negotiations. In Pamela S. Chasek & Lynn M. Wagner (Eds.), *The Roads from Rio. Lessons Learned from Twenty Years of Multilateral Environmental Negotiations*. London: Routledge, pp. 127–48.

Burnett, D. Graham (2012). *The Sounding of the Whale: Science and Cetaceans in the Twentieth Century*. Chicago: University of Chicago Press.

Busch, Per-Olof, & Jörgens, Helge (2005). The International Sources of Policy Convergence: Explaining the Spread of Environmental Policy Innovations. *Journal of European Public Policy*, 12(5), 1–25.

(2012). Europeanization Through Diffusion? Renewable Energy Policies and Alternative Sources for European Convergence. In Francesc Morata & Israel Solorio Sandoval (Eds.), *European Energy Policy: An Environmental Approach*. Cheltenham: Edward Elgar, pp. 66–82.

Busch, Per-Olof, Jörgens, Helge, & Tews, Kerstin (2005). The Global Diffusion of Regulatory Instruments: The Making of a New International Environmental Regime. *The ANNALS of the American Academy of Political and Social Science*, 598(1), 146–67.

Butterfield, Herbert, & Wight, Martin (Eds.) (1966). *Diplomatic Investigations*. London: Allen & Unwin.

Buzan, Barry (1999). The English School as a Research Program: An Overview, and a Proposal for Reconvening. Paper presented at the BISA Annual Conference 1999, Manchester.

(2004). *From International to World Society? English School Theory and the Social Structure of Globalisation*. Cambridge: Cambridge University Press.

(2014). *An Introduction to the English School of International Relations*. Cambridge: Polity Press.

(2018). Revisiting World Society. *International Politics* 55(1): 125–40.

Buzan, Barry, & Gonzalez-Pelaez, Ana (Eds.) (2009). *International Society and the Middle East*. Basingstoke: Palgrave.

Buzan, Barry, & Lawson, George (2015). *The Global Transformation: History, Modernity and the Making of International Relations*. Cambridge: Cambridge University Press.

Buzan, Barry, & Little, Richard (2000). *International Systems in World History: Remaking the Study of International Relations*. Oxford: Oxford University Press.

Buzan, Barry, & Schouenborg, Laust (2018). *Global International Society: A New Framework for Analysis*. Cambridge: Cambridge University Press.

Buzan, Barry, & Wæver, Ole (2009). Macrosecuritisation and Security Constellations: Reconsidering Scale in Securitisation Theory. *Review of International Studies*, 35(2), 253–76.

Buzan, Barry, Wæver, Ole, & De Wilde, Jaap (1998). *Security: A New Framework for Analysis*. Boulder, CO: Lynne Rienner Publishers.

Buzan, Barry, & Zhang, Yongjin (Eds.) (2014). *Contesting International Society in East Asia*. Cambridge: Cambridge University Press.

Calderwood, Kevin J. (2019). Discourse in the Balance: American Presidential Discourse About Climate Change. *Communication Studies*, 70(2), 235–52.

Caldwell, Lynton Keith (1996). *International Environmental Policy: From the Twentieth to the Twenty-First Century*. 3rd ed. Durham: Duke University Press.

Callenbach, Ernest (1975). *Ecotopia: The Notebooks and Reports of William Weston*. Berkeley, CA: Banyan Tree Books.

Camilleri, Joseph A., & Falk, Jim (1992). *The End of Sovereignty: The Politics of a Shrinking and Fragmenting World*. London: Edward Elgar.

Campiglio, Emanuele, Dafermos, Yannis, Monnin, Pierre, Ryan-Collins, Josh, Schotten, Guido, & Tanaka, Misa (2018). Climate Change Challenges for Central Banks and Financial Regulators. *Nature Climate Change*, 8(6), 462–8.

Carvalho, Fernanda Viana de (2012). The Brazilian Position on Forests and Climate Change from 1997 to 2012: From Veto to Proposition. *Revista brasileira de política internacional*, 55(SPE), 144–69.

Ceballos, Gerardo, Ehrlich, Paul R., & Dirzo, Rodolfo (2017). Biological Annihilation via the Ongoing Sixth Mass Extinction Signaled by Vertebrate Population Losses and Declines. *Proceedings of the National Academy of Sciences*, 114(30), E6089–96.

Chambers, Bradnee W. (2008). *InterLinkages and the Effectiveness of Multilateral Environmental Agreements*. Tokyo: United Nations University Press.

Chan, Nicholas (2018). 'Large Ocean States': Sovereignty, Small Islands, and Marine Protected Areas in Global Oceans Governance. *Global Governance: A Review of Multilateralism and International Organizations*, 24(4), 537–55.

Chan, Sander, Falkner, Robert, Goldberg, Matthew, & van Asselt, Harro (2018). Effective and Geographically Balanced? An Output-Based Assessment of Non-State Climate Actions. *Climate Policy*, 18(1), 24–35.

Chapin, Mac (2004). A Challenge to Conservationists. *World Watch Magazine* (11–12), 17–31.

Charnovitz, Steve (2002). A World Environmental Organization. *Columbia Journal of Environmental Law*, 27, 323–62.

(2007). The WTO's Environmental Progress. *Journal of International Economic Law*, 10(3), 685–706.

Chasek, Pamela S. (2001). NGOs and State Capacity in International Environmental Negotiations: The Experience of the Earth Negotiations Bulletin. *Review of European, Comparative & International Environmental Law*, 10, 168.

Chasek, Pamela S., Downie, David L., & Brown, Janet Welsh (2017). *Global Environmental Politics*. 7th ed. London: Routledge.

Chatterton, Paul, Featherstone, David, & Routledge, Paul (2013). Articulating Climate Justice in Copenhagen: Antagonism, the Commons, and Solidarity. *Antipode*, 45(3), 602–20.

Checkel, Jeffrey T. (1998). The Constructivist Turn in International Relations Theory. *World Politics*, 50(2), 324–48.

(2001). Why Comply? Social Learning and European Identity Change. *International Organization*, 55(3), 553–88.

CITES (no year). What Is CITES? Available at: www.cites.org/eng/disc/what.php

Clapp, Jennifer (2001). *Toxic Exports: The Transfer of Hazardous Wastes from Rich to Poor Countries*. Ithaca: Cornell University Press.

Clapp, Jennifer, & Meckling, Jonas (2013). Business as a Global Actor. In Robert Falkner (Ed.), *The Handbook of Global Climate and Environment Policy*. Cheltenham: John Wiley & Sons Ltd, pp. 286–303.

Clark, Ann Marie, Friedman, Elisabeth J., & Hochstetler, Kathryn (1998). The Sovereign Limits of Global Civil Society: A Comparison of NGO Participation in UN World Conferences on the Environment, Human Rights, and Women. *World Politics*, 51(1), 1–35.

Clark, Gregory (2007). *A Farewell to Alms: A Brief Economic History of the World*. Princeton: Princeton University Press.

Clark, Ian (2005). *Legitimacy in International Society*. Oxford: Oxford University Press.

(2007). *International Legitimacy and World Society*. Oxford: Oxford University Press.

(2009). Towards an English School Theory of Hegemony. *European Journal of International Relations*, 15(2), 203–28.
(2011). *Hegemony in International Society*. Oxford: Oxford University Press.
Clayre, Alasdair (Ed.) (1977). *Nature and Industrialization*. Oxford: Oxford University Press.
Cole, Daniel H. (1993). Marxism and the Failure of Environmental Protection in Eastern Europe and the U.S.S.R. *Legal Studies Forum*, 17(1), 35–72.
Commoner, Barry (1971). *The Closing Circle: Nature, Man, and Technology*. New York: Alfred Knopf.
Conca, Ken (2015). *An Unfinished Foundation: The United Nations and Global Environmental Governance*. New York: Oxford University Press.
(2019). Is There a Role for the UN Security Council on Climate Change? *Environment: Science and Policy for Sustainable Development*, 61(1), 4–15.
Cook, Kate (2002). Liability: 'No Liability, No Protocol'. In C. Bail, R. Falkner, & H. Marquard (Eds.), *The Cartagena Protocol on Biosafety: Reconciling Trade in Biotechnology with Environment and Development?* London: RIIA/ Earthscan, pp. 371–84.
Cosgrove, Denis (1994). Contested Global Visions: One-World, Whole-Earth, and the Apollo Space Photographs. *Annals of the Association of American Geographers*, 84(2), 270–94.
Costa Buranelli, Filippo (2018). World Society as a Shared Ethnos and the Limits of World Society in Central Asia. *International Politics*, 55(1), 57–72.
Craig, Cambell (2011). The Resurgent Idea of World Government. *Ethics & International Affairs*, 22(2), 133–42.
Crosby, Alfred W. (2003). *The Columbian Exchange: Biological and Cultural Consequences of 1492*. Westport, CT: Praeger Publishers.
(2004). *Ecological Imperialism: The Biological Expansion of Europe, 900–1900*. Cambridge: Cambridge University Press.
Crossley, Noële (2016). *Evaluating the Responsibility to Protect: Mass Atrocity Prevention as a Consolidating Norm in International Society*. London: Routledge.
Cui, Shunji, & Buzan, Barry (2016). Great Power Management in International Society. *The Chinese Journal of International Politics*, 9(2), 181–210.
Cutler, A. Claire (1991). The 'Grotian Tradition' in International Relations. *Review of International Studies*, 17, 41–65.
Dafermos, Yannis, Nikolaidi, Maria, & Galanis, Giorgos (2018). Climate Change, Financial Stability and Monetary Policy. *Ecological Economics*, 152, 219–34.
D'amato, Anthony, & Chopra, Sudhir K. (1991). Whales: Their Emerging Right to Life. *American Journal of International Law*, 85(1), 21–62.
Damro, Chad (2015). Market Power Europe: Exploring a Dynamic Conceptual Framework. *Journal of European Public Policy*, 22(9), 1336–54.
Dant, Sara (2016). *Losing Eden: An Environmental History of the American West*. Chichester: John Wiley & Sons.
Dany, Charlotte (2014). Janus-Faced NGO Participation in Global Governance: Structural Constrains for NGO Influence. *Global Governance*, 20(3), 419–36.

Darby, Megan (2019). Net Zero: The Story of the Target That Will Shape Our Future. Climate Home News, 16 September. Retrieved from www.climatechangenews.com/2019/09/16/net-zero-story-target-will-shape-future/

Dauvergne, Peter (2016). *Environmentalism of the Rich*. Cambridge, MA: MIT Press.

(2018). Why Is the Global Governance of Plastic Failing the Oceans? *Global Environmental Change*, 51, 22–31.

Davies, Thomas (2013). *NGOs: A New History of Transnational Civil Society*. London: Hurst.

(2017). Institutions of World Society: Parallels with the International Society of States. Paper presented at the International Studies Association Annual Convention, Baltimore, MD.

Davis, Diana (2000). Environmentalism as Social Control? An Exploration of the Transformation of Pastoral Nomadic Societies in French Colonial North Africa. *The Arab World Geographer*, 3(3), 182–98.

Davis, Janet M. (2013). Bird Day: Promoting the Gospel of Kindness in the Philippines during the American Occupation. In Erika Marie Bsumek, David Kinkela, & Mark Atwood Lawrence (Eds.), *Nation-States and the Global Environment: New Approaches to International Environmental History*. Oxford: Oxford University Press, pp. 181–206.

Davis, Mike (2002). *Late Victorian Holocausts*. London: Verso.

De Almeida, João M. (2006). Hedley Bull, 'Embedded Cosmopolitanism', and the Pluralist-Solidarist Debate. In Richard Little & John Williams (Eds.), *The Anarchical Society in a Globalized World*. Basingstoke: Palgrave Macmillan, pp. 51–73.

De Steiguer, J. Edward (2006). *The Origins of Modern Environmental Thought*. Tucson: The University of Arizona Press.

Deitelhoff, Nicole, & Zimmermann, Lisbeth (2020). Things We Lost in the Fire: How Different Types of Contestation Affect the Robustness of International Norms. *International Studies Review*, 22(1), 51–76.

Delfin Jr, Francisco G., & Tang, Shui-Yan (2008). Foundation Impact on Environmental Nongovernmental Organizations: The Grantees' Perspective. *Nonprofit and Voluntary Sector Quarterly*, 37(4), 603–25.

Delmas, Magali A. (2002). The Diffusion of Environmental Management Standards in Europe and in the United States: An Institutional Perspective. *Policy Sciences*, 35(1), 91–119.

Depledge, Joanna, & Chasek, Pamela S. (2012). Raising the Tempo: The Escalating Pace and Intensity of Environmental Negotiations. In Pamela S. Chasek & Lynn M. Wagner (Eds.), *The Roads from Rio: Lessons Learned from Twenty Years of Multilateral Environmental Negotiations*. New York: RFF Press, pp. 19–38.

Derkx, Boudewijn, & Glasbergen, Pieter (2014). Elaborating Global Private Meta-Governance: An Inventory in the Realm of Voluntary Sustainability Standards. *Global Environmental Change*, 27, 41–50.

Derler, Zak (2018). UN Security Council Makes "Historic" Warning on Climate Threat to Somalia. *Climate Home News*, 28 March 2018.

De-Shalit, Avner (2006). Nationalism. In Andrew Dobson & Robyn Eckersley (Eds.), *Political Theory and the Ecological Challenge*. Cambridge: Cambridge University Press, pp. 75–90.

DeSombre, Elizabeth R. (2000). *Domestic Sources of International Environmental Policy: Industry, Environmentalists, and U.S. Power*. Cambridge, MA: MIT Press.

Deudney, Daniel (1990). The Case Against Linking Environmental Degradation and National Security. *Millennium*, 19(3), 461–76.

Diez, Thomas (2017). Diplomacy, Papacy, and the Transformation of International Society. *The Review of Faith & International Affairs*, 15(4), 31–8.

Diez, Thomas, Von Lucke, Franziskus, & Wellmann, Zehra (2016). *The Securitisation of Climate Change: Actors, Processes and Consequences*. London: Routledge.

Diez, Thomas, & Whitman, Richard G. (2002). Analysing European Integration: Reflecting on the English School – Scenarios for an Encounter. *Journal of Common Market Studies*, 40(1), 43–67.

Dimitrov, Radoslav S. (2005). Hostage to Norms: States, Institutions and Global Forest Politics. *Global Environmental Politics*, 5(4), 1–24.

(2010). Inside Copenhagen: The State of Climate Governance. *Global Environmental Politics*, 10(2), 18–24.

Doel, Ronald E. (2003). Constituting the Postwar Earth Sciences: The Military's Influence on the Environmental Sciences in the USA after 1945. *Social Studies of Science*, 33(5), 635–66.

Doherty, Brian, & Doyle, Timothy (2013). *Environmentalism, Resistance and Solidarity: The Politics of Friends of the Earth International*. Basingstoke: Palgrave.

Domínguez, Lara, & Luoma, Colin (2020). Decolonising Conservation Policy: How Colonial Land and Conservation Ideologies Persist and Perpetuate Indigenous Injustices at the Expense of the Environment. *Land*, 9(3), 65.

Dorsey, Kurk (1995). Scientists, Citizens, and Statesmen: US-Canadian Wildlife Protection Treaties in the Progressive Era. *Diplomatic History*, 19(3), 407–30.

(1998). *The Dawn of Conservation Diplomacy: U.S.-Canadian Wildlife Protection Treaties in the Progressive Era*. Seattle: University of Washington Press.

(2013). National Sovereignty, the International Whaling Commission, and the Save the Whales Movement. In Erika Marie Bsumek, David Kinkela, & Mark Atwood Lawrence (Eds.), *Nation-States and the Global Environment: New Approaches to International Environmental History*. Oxford: Oxford University Press, pp. 43–61.

Dotto, Lydia, & Schiff, Harold (1978). *The Ozone War*. New York: Doubleday.

Dryzek, John S., Downes, David, Hunold, Christian, Schlosberg, David, & Hernes, Hans-Kristian (2003). *Green States and Social Movements: Environmentalism in the United States, United Kingdom, Germany, and Norway*. Oxford: Oxford University Press.

Duch, Raymond M., & Taylor, Michaell A. (1993). Postmaterialism and the Economic Condition. *American Journal of Political Science*, 37(3), 747–79.

Duffy, Rosaleen (2013). Global Environmental Governance and North–South Dynamics: The Case of the CITES. *Environment and Planning C: Government and Policy*, 31(2), 222–39.

Duit, Andreas, Feindt, Peter H, & Meadowcroft, James (2016). Greening Leviathan: The Rise of the Environmental State? *Environmental Politics*, 25 (1), 1–23.

Dunlap, Riley E., & McCright, Aaron M. (2011). Organized Climate Change Denial. In John S. Dryzek, Richard B. Norgaard, & David Schlosberg (Eds.), *The Oxford Handbook of Climate Change and Society*. Oxford: Oxford University Press, pp. 144–60.

Dunne, Tim (1998). *Inventing International Society: A History of the English School.* Basingstoke: Macmillan.

(2010). The English School. In R. E. Goodin (Ed.), *The Oxford Handbook of Political Science*. Oxford: Oxford University Press.

Dunne, Tim, & Reus-Smit, Christian (Eds.) (2017). *The Globalization of International Society*. Oxford: Oxford University Press.

Dunne, Tim, & Wheeler, Nicholas J. (Eds.) (1999). *Human Rights in Global Politics*. Cambridge: Cambridge University Press.

Dupont, Alan (2008). The Strategic Implications of Climate Change. *Survival*, 50(3), 29–54.

Duwe, Matthias (2001). The Climate Action Network: A Glance Behind the Curtains of a Transnational NGO Network. *Review of European Community & International Environmental Law*, 10(2), 177–89.

Dwivedi, O.P. (1997). *India's Environmental Policies, Programmes and Stewardship*. Basingstoke: Macmillan.

Eckersley, Robyn (1992). *Environmentalism and Political Theory: Toward an Ecocentric Approach*. London: UCL Press.

(2004a). *The Green State: Rethinking Democracy and Sovereignty*. Cambridge, MA: MIT Press.

(2004b). The Big Chill: The WTO and Multilateral Environmental Agreements. *Global Environmental Politics*, 4(2), 24–50.

(2006). Communitarianism. In Andrew Dobson & Robyn Eckersley (Eds.), *Political Theory and the Ecological Challenge*. Cambridge: Cambridge University Press, pp. 91–108.

(2007). Ambushed: The Kyoto Protocol, the Bush Administrations Climate Policy and the Erosion of Legitimacy. *International Politics*, 44(2–3), 306–24.

Eckley, Noelle, & Selin, Henrik (2004). All Talk, Little Action: Precaution and European Chemicals Regulation. *Journal of European Public Policy*, 11(1), 78–105.

Economy, Elizabeth C. (2004). *The River Runs Black: The Environmental Challenge to China's Future*. Ithaca: Cornell University Press.

Economy, Elizabeth C., & Schreurs, Miranda A. (1997). Domestic and International Linkages in Environmental Politics. In Miranda A. Schreurs & Elizabeth Economy (Eds.), *The Internationalization of Environmental Protection*. Cambridge: Cambridge University Press, pp. 1–18.

Ehrlich, Paul (1968). *The Population Bomb: Population Control or Race to Oblivion.* New York: Ballantine.

Elbe, Stefan (2006). Should HIV/AIDS Be Securitized? The Ethical Dilemmas of Linking HIV/AIDS and Security. *International Studies Quarterly*, 50(1), 119–44.

Elliott, Lorraine (2006). Cosmopolitan Environmental Harm Conventions. *Global Society*, 20(3), 345–63.

Emmers, Ralf (2003). ASEAN and the Securitization of Transnational Crime in Southeast Asia. *The Pacific Review*, 16(3), 419–38.

Engels, Jens Ivo (2006). *Naturpolitik in der Bundesrepublik: Ideenwelt und politische Verhaltensstile in Naturschutz und Umweltbewegung 1950–1980*. Paderborn: Verlag Ferdinand Schöningh.

Epstein, Charlotte (2006). The Making of Global Environmental Norms: Endangered Species Protection. *Global Environmental Politics*, 6(2), 32–54.

(2008). *The Power of Words in International Relations: Birth of an Anti-Whaling Discourse*. Cambridge, MA: MIT Press.

Epstein, Charlotte, & Barclay, Kate (2013). Shaming to 'Green': Australia–Japan Relations and Whales and Tuna Compared. *International Relations of the Asia-Pacific*, 13(1), 95–123.

Esty, Daniel C., & Ivanova, Maria H. (2002). *Global Environmental Governance: Options and Opportunities*. New Haven: Yale School of Forestry & Environmental Studies.

Evans, David (1997). *A History of Nature Conservation in Britain*. London: Routledge.

Falk, Richard A. (1971). *This Endangered Planet*. New York: Random House.

(1973). Environmental Warfare and Ecocide – Facts, Appraisal, and Proposals. *Bulletin of Peace Proposals*, 4(1), 80–96.

Falkner, Robert (2000). Regulating Biotech Trade: The Cartagena Protocol on Biosafety. *International Affairs*, 76(2), 299–313.

(2003). Private Environmental Governance and International Relations: Exploring the Links. *Global Environmental Politics*, 3(2), 72–87.

(2005). American Hegemony and the Global Environment. *International Studies Review*, 7(4), 585–99.

(2007). The Political Economy of 'Normative Power' Europe: EU Environmental Leadership in International Biotechnology Regulation. *Journal of European Public Policy*, 14(4), 507–26.

(2008). *Business Power and Conflict in International Environmental Politics*. Basingstoke: Palgrave Macmillan.

(2012). Global Environmentalism and the Greening of International Society. *International Affairs*, 88(3), 503–22.

(2013). The Nation-State, International Society, and the Global Environment. In Robert Falkner (Ed.), *The Handbook of Global Climate and Environment Policy*. Cheltenham: John Wiley & Sons Ltd., pp. 251–67.

(2016a). A Minilateral Solution for Global Climate Change? On Bargaining Efficiency, Club Benefits, and International Legitimacy. *Perspectives on Politics*, 14(01), 87–101.

(2016b). The Paris Agreement and the New Logic of International Climate Politics. *International Affairs*, 92(5), 1107–125.

(2017a). International Climate Politics Between Pluralism and Solidarism: An English School Perspective. In H. Stevenson & O. Corry (Eds.), *Traditions and Trends in Global Environmental Politics International Relations and the Earth*. London: Routledge, pp. 26–44.

(2017b). The Anarchical Society and Climate Change. In H. Suganami, M. Carr, & A. Humphreys (Eds.), *The Anarchical Society at 40: Contemporary Challenges and Prospects*. Oxford: Oxford University Press, pp. 198–215.

(2019). The Unavoidability of Justice – and Order – in International Climate Politics: From Kyoto to Paris and Beyond. *The British Journal of Politics and International Relations*, 21(2), 270–8.

Falkner, Robert, & Buzan, Barry (2019). The Emergence of Environmental Stewardship as a Primary Institution in Global International Society. *European Journal of International Relations*, 25(1), 131–55.

Falkner, Robert, & Jaspers, Nico (2012). Regulating Nanotechnologies: Risk, Uncertainty and the Global Governance Gap. *Global Environmental Politics*, 12(1), 30–55.

Falkner, Robert, Stephan, Hannes, & Vogler, John (2010). International Climate Policy after Copenhagen: Towards a 'Building Blocks' Approach. *Global Policy*, 1(3), 252–62.

Finnemore, Martha (2001). Exporting the English School? *Review of International Studies*, 27(3), 509–13.

Finnemore, Martha, & Sikkink, Kathryn (1998). International Norm Dynamics and Political Change. *International Organization*, 52(4), 887–917.

Fisher, Dana R. (2004). Civil Society Protest and Participation: Civic Engagement Within the Multilateral Governance Regime. In Norichika Kanie & Peter M. Haas (Eds.), *Emerging Forces in Environmental Governance*. Tokyo: United Nations University Press, pp. 176–99.

(2010). COP-15 in Copenhagen: How the Merging of Movements Left Civil Society Out in the Cold. *Global Environmental Politics*, 10(2), 11–17.

Flippen, J. Brooks (2008). Richard Nixon, Russell Train, and the Birth of Modern American Environmental Diplomacy. *Diplomatic History*, 32(4), 613–38.

Floyd, Rita (2010). *Security and the Environment: Securitisation Theory and US Environmental Security Policy*. Cambridge: Cambridge University Press.

(2016). Extraordinary or Ordinary Emergency Measures: What, and Who, Defines the 'Success' of Securitization? *Cambridge Review of International Affairs*, 29(2), 677–94.

Food and Agriculture Organization (2018). *The State of the World's Forests 2018*. Rome: Food and Agriculture Organization.

Ford, Caroline (2004). Nature, Culture and Conservation in France and Her Colonies, 1840–1940. *Past & Present*, (183), 173–98.

Founex Report (1971). The Founex Report on Development and Environment. Available at: www.unedforum.org/fileadmin/files/Earth%20Summit%202012new/Publications%20and%20Reports/founex_report_on_development_and_environment_1972.pdf

Frank, David John, Longhofer, Wesley, & Schofer, Evan (2007). World Society, NGOs and Environmental Policy Reform in Asia. *International Journal of Comparative Sociology*, 48(4), 275–95.

Freeden, Michael (1996). *Ideologies and Political Theory: A Conceptual Approach*. Oxford: Clarendon Press.

Frost, Warwick, & Hall, C. Michael (2012). American Invention to International Concept: The Spread and Evolution of National Parks. In Warwick Frost &

C. Michael Hall (Eds.), *Tourism and National Parks: International Perspectives on Development, Histories and Change*. London: Routledge, pp. 30–59

Garcia-Johnson, Ronie (2000). *Exporting Environmentalism: U.S. Multinational Chemical Corporations in Brazil and Mexico*. Cambridge, MA: MIT Press.

Garfield, Seth (2013). The Brazilian Amazon and the Transnational Environment, 1940–1990. In Erika Marie Bsumek, David Kinkela, & Mark Atwood Lawrence (Eds.), *Nation-States and the Global Environment: New Approaches to International Environmental History*. Oxford: Oxford University Press, pp. 228–51.

Gauger, Anja, Rabatel-Fernel, Mai Pouye, Kulbicki, Louise, Short, Damien, & Higgins, Polly (2012). *Ecocide Is the Missing 5th Crime Against Peace*. London, Human Rights, Consortium, School of Advanced Study, University of London.

Gereke, Marika, & Brühl, Tanja (2019). Unpacking the Unequal Representation of Northern and Southern NGOs in International Climate Change Politics. *Third World Quarterly*, 40(5), 870–89.

German Advisory Council on Global Change (2008). *Climate Change as a Security Risk*. London: Earthscan.

Gies, Erica (2017). Businesses Lead Where US Falters. *Nature Climate Change*, 7 (8), 543–46.

Gillespie, Alexander (2005). *Whaling Diplomacy: Defining Issues in International Environmental Law*. Cheltenham: Edward Elgar.

Giugni, Marco (1998). The Other Side of the Coin: Explaining Crossnational Similarities Between Social Movements. *Mobilization: An International Quarterly*, 3(1), 89–105.

Glasius, Marlies (2010). Dissecting Global Civil Society: Values, Actors, Organisational Forms. Blogpost. Retrieved from www.opendemocracy.net/en/5050/dissecting-global-civil-society-values-actors-organisational-forms/

Gleick, Peter H. (1991). Environment and Security: Clear Connections. *Bulletin of the Atomic Scientists*, 47(3), 17–21.

Goeteyn, Nils, & Maes, Frank (2011). Compliance Mechanisms in Multilateral Environmental Agreements: An Effective Way to Improve Compliance? *Chinese Journal of International Law*, 10(4), 791–826.

Golley, Frank Benjamin (1993). *A History of the Ecosystem Concept in Ecology: More Than the Sum of the Parts*. New Haven: Yale University Press.

Goodman, James (2009). From Global Justice to Climate Justice? Justice Ecologism in an Era of Global Warming. *New Political Science*, 31(4), 499–514.

Goossen, Benjamin W. (2020). A Benchmark for the Environment: Big Science and 'Artificial' Geophysics in the Global 1950s. *Journal of Global History*, 15 (1), 149–168.

Gordon, Gwendolyn J. (2018). Environmental Personhood. *Columbia Journal of Environmental Law*, 43(1), 49–91.

Graaf, Nan Dirk De, & Evans, Geoffrey (1996). Why Are the Young More Postmaterialist? A Cross-National Analysis of Individual and Contextual Influences on Postmaterial Values. *Comparative Political Studies*, 28(4), 608–35.

Graff, Laurence (2002). The Precautionary Principle. In Christoph Bail, Robert Falkner, & Helen Marquard (Eds.), *The Cartagena Protocol on Biosafety: Reconciling Trade in Biotechnology with Environment and Development?* London: Earthscan, pp. 410–22.

Green, Jessica F. (2013). *Rethinking Private Authority: Agents and Entrepreneurs in Global Environmental Governance*. Princeton, NJ: Princeton University Press.

Grim, John, & Tucker, Mary Evelyn (2014). *Ecology and Religion*. Washington, DC: Island Press.

Grove, Richard H. (1995). *Green Imperialism: Colonial Expansion, Tropical Island Edens and the Origins of Environmentalism, 1600–1860*. Cambridge: Cambridge University Press.

Grove, Richard, & Damodaran, Vinita (2006). Imperialism, Intellectual Networks, and Environmental Change: Origins and Evolution of Global Environmental History, 1676–2000: Part I. *Economic and Political Weekly*, 41(41), 4345–54.

Guha, Ramachandra (1997). The Authoritarian Biologist and the Arrogance of Anti-Humanism. *The Ecologist*, 27(1), 14–20.

(2000). *Environmentalism: A Global History*. New Delhi: Oxford University Press.

Guha, Ramachandra, & Martinez-Alier, Juan (Eds.) (1997). *Varieties of Environmentalism: Essays North and South*. London: Earthscan.

Gulbrandsen, Lars H. (2008). Organizing Accountability in Transnational Standards Organizations: The Forest Stewardship Council as a Good Governance Model. In Magnus Boström & Christina Garsten (Eds.), *Organizing Transnational Accountability* (Vol. 61–79). Cheltenham: Edward Elgar, pp. 61–79.

(2010). *Transnational Environmental Governance: The Emergence and Effects of the Certification of Forests and Fisheries*. Cheltenham: Edward Elgar.

Gupta, Aarti (2013). Biotechnology and Biosafety. In Robert Falkner (Ed.), *The Handbook of Global Climate and Environment Policy*. Cheltenham: John Wiley & Sons Ltd, pp. 89–106.

Gutner, Tamar (2005). World Bank Environmental Reform: Revisiting Lessons from Agency Theory. *International Organization*, 59(3), 773–83.

Haas, Peter M. (1995). Epistemic Communities and the Dynamics of International Environmental Co-operation. In Volker Rittberger (Ed.), *Regime Theory and International Relations*. Oxford: Clarendon Press, pp. 168–201.

(2002). UN Conferences and Constructivist Governance of the Environment. *Global Governance*, 8(1), 73–91.

Haas, Peter M., Keohane, Robert O., & Levy, Marc A. (Eds.) (1993). *Institutions for the Earth. Sources of Effective International Environmental Protection*. Cambridge, MA: The MIT Press.

Hadden, Jennifer (2014). Explaining Variation in Transnational Climate Change Activism: The Role of Inter-Movement Spillover. *Global Environmental Politics*, 14(2), 7–25.

Hale, Thomas (2020). Transnational Actors and Transnational Governance in Global Environmental Politics. *Annual Review of Political Science*, 23(1).

Hale, Thomas, Held, David, & Young, Kevin (2013). *Gridlock: Why Global Cooperation is Failing When We Need It Most.* Cambridge: Polity Press.

Hale, Thomas, & Roger, Charles (2014). Orchestration and Transnational Climate Governance. *The Review of International Organizations*, 9(1), 59–82.

Harris, Katie (2012). Climate Change in UK Security Policy: Implications for Development Assistance? *ODI Working Paper*, 342. London, ODI.

Harrop, Stuart (2013). Biodiversity and Conservation. In Robert Falkner (Ed.), *The Handbook of Global Climate and Environment Policy.* Cheltenham: John Wiley & Sons Ltd, pp. 37–52.

Hayes, Jarrod, & Knox-Hayes, Janelle (2014). Security in Climate Change Discourse: Analyzing the Divergence Between US and EU Approaches to Policy. *Global Environmental Politics*, 14(2), 82–101.

Hays, Samuel P. (1959). *Conservation and the Gospel of Efficiency: The Progressive Conservation Movement, 1890–1920.* Cambridge, MA: Harvard University Press.

Hayward, Tim (2005). *Constitutional Environmental Rights.* Oxford: Oxford University Press.

He, Guizhen, Lu, Yonglong, Mol, Arthur P.J., & Beckers, Theo (2012). Changes and Challenges: China's Environmental Management in Transition. *Environmental Development*, 3, 25–38.

Heggelund, Gørild, & Backer, Ellen Bruzelius (2007). China and UN Environmental Policy: Institutional Growth, Learning and Implementation. *International Environmental Agreements: Politics, Law and Economics*, 7(4), 415–38.

Hey, Ellen (2003). *Teaching International Law: State-Consent as Consent to a Process of Normative Development and Ensuing Problems.* The Hague: Kluwer Law International.

Hicks, Robert L., Parks, Bradley C., Roberts, J. Timmons, & Tierney, Michael J. (2008). *Greening Aid? Understanding the Environmental Impact of Development Assistance.* Oxford: Oxford University Press.

Higgins, Polly, Short, Damien, & South, Nigel (2013). Protecting the Planet: a Proposal for a Law of Ecocide. *Crime, Law and Social Change*, 59(3), 251–66.

Hilton, Isabel, & Kerr, Oliver (2017). The Paris Agreement: China's 'New Normal' Role in International Climate Negotiations. *Climate Policy*, 17(1), 48–58.

Hironaka, Ann (2014). *Greening the Globe: World Society and Environmental Change.* New York: Cambridge University Press.

Hjerpe, Mattias, & Linnér, Björn-Ola (2010). Functions of COP Side-Events in Climate-Change Governance. *Climate Policy*, 10(2), 167–80.

Hjerpe, Mattias, & Nasiritousi, Naghmeh (2015). Views on Alternative Forums for Effectively Tackling Climate Change. *Nature Climate Change*, 5(9), 864–7.

Hochstetler, Kathryn, & Keck, Margaret E. (2007). *Greening Brazil: Environmental Activism in State and Society.* Durham: Duke University Press.

Hochstetler, Kathryn, & Milkoreit, Manjana (2015). Responsibilities in Transition: Emerging Powers in the Climate Change Negotiations. *Global Governance*, 21(2), 205–26.

Hoffmann, Matthew J. (2010). Norms and Social Constructivism in International Relations. In Robert A. Denemark (Ed.), *The International Studies Encyclopedia*. Oxford: Wiley-Blackwell, pp. 5410–26.

(2011). *Climate Governance at the Crossroads: Experimenting with a Global Response after Kyoto*. New York: Oxford University Press.

Höhler, Sabine (2015). *Spaceship Earth in the Environmental Age, 1960–1990*. London: Routledge.

Holdgate, Martin (1999). *The Green Web: A Union for World Conservation*. London: Earthscan.

Hollis, Martin, & Smith, Steve (1991). *Explaining and Understanding International Relations*. Oxford: Clarendon Press.

Holsti, Kal J. (2004). *Taming the Sovereigns: Institutional Change in International Politics*. Cambridge: Cambridge University Press.

Holzinger, Katharina, Knill, Christoph, & Sommerer, Thomas (2008). Environmental Policy Convergence: The Impact of International Harmonization, Transnational Communication, and Regulatory Competition. *International Organization*, 62(04), 553–87.

Hopewell, Kristen (2019). How Rising Powers Create Governance Gaps: The Case of Export Credit and the Environment. *Global Environmental Politics*, 19(1), 34–52.

Hopgood, Stephen (1998). *American Foreign Environmental Policy and the Power of the State*. Oxford: Oxford University Press.

Hünemörder, Kai F. (2004). *Die Frühgeschichte der globalen Umweltkrise und die Formierung der deutschen Umweltpolitik (1950–1973)*. Stuttgart: Franz Steiner Verlag.

Hünemörder, Kai (2010). Environmental Crisis and Soft Politics: Détente and the Global Environment, 1968–1975. In John R. McNeill & Corinna R. Unger (Eds.), *Environmental Histories of the Cold War*. Cambridge: Cambridge University Press, pp. 257–76.

Humphreys, David (2013). Deforestation. In Robert Falkner (Ed.), *The Handbook of Global Climate and Environment Policy*. Cheltenham: John Wiley & Sons Ltd, pp. 72–88.

Humphreys, John, & Clark, Robert W.E. (2020). A Critical History of Marine Protected Areas. In John Humphreys & Robert W.E. Clark (Eds.), *Marine Protected Areas: Science, Policy and Management*. Amsterdam: Elsevier, pp. 1–12.

Humphreys, Stephen (Ed.) (2010). *Human Rights and Climate Change*. Cambridge: Cambridge University Press.

Hunold, Christian, & Dryzek, John S. (2002). Green Political Theory and the State: Context Is Everything. *Global Environmental Politics*, 2(3), 17–39.

Hunter, David B. (2007). Civil Society Networks and the Development of Environmental Standards at International Financial Institutions. *Chicago Journal of International Law*, 8, 437–77.

Hurd, Ian (2008). Constructivism. In Christian Reus-Smit & Duncan Snidal (Eds.), *The Oxford Handbook of International Relations*. Oxford: Oxford University Press, pp. 298–316.

Hurrell, Andrew (1994). A Crisis of Ecological Viability? Global Environmental Change and the Nation State. *Political Studies*, 42 (Special Issue), 146–65.

(1995). International Political Theory and the Global Environment. In Ken Booth & Steve Smith (Eds.), *International Relations Theory Today*. Cambridge: Polity Press, pp. 129–53.

(2001). Keeping History, Law and Political Philosophy Firmly within the English School. *Review of International Studies*, 27(3), 489–94.

(2007). *On Global Order: Power, Values, and the Constitution of International Society*. Oxford: Oxford University Press.

(2014). Order and Justice. In Cornelia Navari & Daniel M. Green (Eds.), *Guide to the English School in International Studies*. Chichester: John Wiley & Sons, pp. 143–58.

Hurrell, Andrew, & Kingsbury, Benedict (Eds.) (1992a). *The International Politics of the Environment: Actors, Interests, and Institutions*. Oxford: Clarendon Press.

Hurrell, Andrew, & Kingsbury, Benedict (1992b). The International Politics of the Environment: An Introduction. In Andrew Hurrell & Benedict Kingsbury (Eds.), *The International Politics of the Environment*. Oxford: Clarendon Press, pp. 1–47.

Hurrell, Andrew, & Sengupta, Sandeep (2012). Emerging Powers, North-South Relations and Global Climate Politics. *International Affairs*, 88(3), 463–84.

Huxley, Julian (1946) *UNESCO: Its Purpose and Its Philosophy*. Preparatory Commission of the United Nations Educational, Scientific and Cultural Organisation.

Huysmans, Jef (2000). The European Union and the Securitization of Migration. *JCMS: Journal of Common Market Studies*, 38(5), 751–77.

Inglehart, Ronald (1977). *The Silent Revolution: Changing Values and Political Styles Among Western Publics*. Princeton: Princeton University Press.

International Conference on the Conservation of Wetlands and Waterfowl (1972). International Conference on the Conservation of Wetlands and Waterfowl: Convention on Wetlands of International Importance. *International Legal Materials*, 11(5), 963–76.

International Convention for the High Seas Fisheries of the North Pacific Ocean (1952), 9 May. Available at: https://sedac.ciesin.columbia.edu/entri/texts/fisheries.north.pacific.1952.html

International Union for the Protection of Nature (1955). *Proceedings and Papers of the Fourth General Assembly. Held at Copenhagen (Denmark), 25 August to 3 September 1954*. Statutory Meetings. Brussels.

Isaac, Grant E., & Kerr, William A. (2003). Genetically Modified Organisms and Trade Rules: Identifying Important Challenges for the WTO. *World Economy*, 26(1), 29–42.

Ivanova, Maria (2010). UNEP in Global Environmental Governance: Design, Leadership, Location. *Global Environmental Politics*, 10(1), 30–59.

(2012). Institutional Design and UNEP Reform: Historical Insights on Form, Function and Financing. *International Affairs*, 88(3), 565–84.

(2013). The Contested Legacy of Rio+20. *Global Environmental Politics*, 13(4), 1–11.

Ivanova, Maria, & Esty, Daniel C. (2008). Reclaiming US Leadership in Global Environmental Governance. *SAIS Review of International Affairs*, 28(2), 57–75.

Jackson, Robert (1992). Pluralism in International Political Theory. *Review of International Studies*, 18(3), 271–81.
Jackson, Robert H. (1996). Can International Society Be Green? In Rick Fawn & Jeremy Larkins (Eds.), *International Society after the Cold War: Anarchy and Order Revisited*. Houndmills: Macmillan, pp. 172–92.
(2000). *The Global Covenant: Human Conduct in a World of States*. Oxford: Oxford University Press.
(2009). International Relations as a Craft Discipline. In Cornelia Navari (Ed.), *Theorising International Society*. Basingstoke: Palgrave, pp. 21–38.
James, Alan (1993). System or Society. *Review of International Studies*, 19(3), 269–88.
Jaspers, Nico, & Falkner, Robert (2013). International Trade, the Environment, and Climate Change. In Robert Falkner (Ed.), *The Handbook of Global Climate and Environment Policy*. Cheltenham: John Wiley & Sons Ltd, pp. 412–27.
Javaudin, Enora (2017). Environmental Problem-Solvers? Scientists and the Stockholm Conference. In Wolfram Kaiser & Jan-Henrik Meyer (Eds.), *International Organizations and Environmental Protection: Conservation and Globalization in the Twentieth Century*. New York: Berghahn Books, pp. 74–102.
Jinnah, Sikina (2011). Climate Change Bandwagoning: The Impacts of Strategic Linkages on Regime Design, Maintenance, and Death. *Global Environmental Politics*, 11(3), 1–9.
(2014). *Post-Treaty Politics: Secretariat Influence in Global Environmental Governance*. Cambridge, MA: MIT Press.
Jinnah, Sikina, & Morgera, Elisa (2013). Environmental Provisions in American and EU Free Trade Agreements: A Preliminary Comparison and Research Agenda. *Review of European, Comparative & International Environmental Law*, 22(3), 324–39.
Johann, Elisabeth (2006). Historical Development of Nature-Based Forestry in Central Europe. In Jurij Diaci (Ed.), *Nature-Based Forestry in Central Europe: Alternatives to Industrial Forestry and Strict Preservation*. Ljubljana: Department of Forestry and Renewable Forest Resources, pp. 1–17.
Jones, Roy E. (1981). The English School of International Relations: A Case for Closure. *Review of International Studies*, 7(1), 1–13.
Jörgens, Helge (1996). Die Institutionalisierung von Umweltpolitik im internationalen Vergleich. In Martin Jänicke (Ed.), *Umweltpolitik der Industrieländer: Entwicklung – Bilanz – Erfolgsbedingungen*. Berlin: Edition Sigma, pp. 59–111.
Josephson, Paul, Dronin, Nicolai, Mnatsakanian, Ruben, Cherp, Aleh, Efremenko, Dmitry, & Larin, Vladislav (Eds.) (2013). *An Environmental History of Russia*. Cambridge: Cambridge University Press.
Kalb, Deborah, Peters, Gerhard, & Woolley, John T. (Eds.) (2007). *State of the Union: Presidential Rhetoric from Woodrow Wilson to George W. Bush*. Washington, DC: CQ Press.
Kean, Hilda (1998). *Animal Rights: Social and Political Change Since 1800*. London: Reaction.

Keck, Margaret E., & Sikkink, Kathryn (1998). *Activists Beyond Borders: Advocacy Networks in International Politics*. Ithaca, NY: Cornell University Press.

Keenan, Jesse M. (2019). A Climate Intelligence Arms Race in Financial Markets. *Science*, 365(6459), 1240–3.

Keene, Edward (2002). *Beyond the Anarchical Society: Grotius, Colonialism and Order in World Politics*. Cambridge: Cambridge University Press.

Kelemen, R. Daniel (2010). Globalizing European Union Environmental Policy. *Journal of European Public Policy*, 17(3), 335–49.

Kelemen, R. Daniel, & Vogel, David (2010). Trading Places: The Role of the United States and the European Union in International Environmental Politics. *Comparative Political Studies*, 43(4), 427–56.

Kellow, Aynsley (2000). Norms, Interests and Environment NGOs: The Limits of Cosmopolitanism. *Environmental Politics*, 9(3), 1–22.

Kennan, George F. (1970). To Prevent a World Wasteland: A Proposal. *Foreign Affairs*, 48(3), 401–13.

Keohane, Robert O. (1988). International Institutions: Two Approaches. *International Studies Quarterly*, 32(4), 379-396.

Keohane, Robert O., Haas, Peter M., & Levy, Marc A. (1993). The Effectiveness of International Environmental Institutions. In Peter M. Haas, Robert O. Keohane, & Marc A. Levy (Eds.), *Institutions for the Earth: Sources of Effective International Environmental Protection*. Cambridge, MA: MIT Press, pp. 3–24.

Keohane, Robert O., & Nye, Joseph S., Jr. (Eds.) (1971). *Transnational Relations and World Politics*. Cambridge, MA: Harvard University Press.

Keohane, Robert O., & Victor, David G. (2011). The Regime Complex for Climate Change. *Perspectives on Politics*, 9(01), 7–23.

Kettlewell, Ursula (1992). The Answer to Global Pollution – A Critical Examination of the Problems and Potential of the Polluter-Pays Principle. *Colorado Journal of International Environmental Law & Policy*, 3(2), 429–78.

Kingsbury, Benedict (2007). Global Environmental Governance as Administration: Implications for International Law. In Daniel Bodansky, Jutta Brunnée, & Ellen Hey (Eds.), *The Oxford Handbook of International Environmental Law*. Oxford: Oxford University Press, pp. 63–84.

Kinkela, David (2013). The Paradox of US Pesticide Policy during the Age of Ecology. In Erika Marie Bsumek, David Kinkela, & Mark Atwood Lawrence (Eds.), *Nation-States and the Global Environment: New Approaches to International Environmental History*. Oxford: Oxford University Press, pp. 115–34.

Kirton, John J., & Kokotsis, Ella (2016). *The Global Governance of Climate Change: G7, G20, and UN Leadership*. London: Routledge.

Kjellén, Bo (2008). *A New Diplomacy for Sustainable Development: The Challenge of Global Change*. London: Routledge.

Knox, John H. (2018). The Past, Present, and Future of Human Rights and the Environment. *Wake Forest Law Review*, 53, 649–65.

Knudsen, Tonny Brems (2018). Fundamental Institutions and International Organizations: Theorizing Continuity and Change. In Tonny Brems Knudsen & Cornelia Navari (Eds.), *International Organization in the*

Anarchical Society: The Institutional Structure of World Order. Basingstoke: Palgrave, pp. 23–50.

Kohler, Pia M. (2019). *Science Advice and Global Environmental Governance: Expert Institutions and the Implementation of International Environmental Treaties*. London: Anthem Press.

Kopra, Sanna (2018). *China and Great Power Responsibility for Climate Change*. London: Routledge.

(2019). China, Great Power Management, and Climate Change: Negotiating Great Power Climate Responsibility in the UN. In Tonny Brems Knudsen & Cornelia Navari (Eds.), *International Organization in the Anarchical Society: The Institutional Structure of World Order*. Basingstoke: Palgrave Macmillan, pp. 149–173.

KPMG (2017). The Road Ahead. The KPMG Survey of Corporate Responsibility Reporting 2017. Available at: https://assets.kpmg/content/dam/kpmg/xx/pdf/2017/10/kpmg-survey-of-corporate-responsibility-reporting-2017.pdf

Krisch, Nico (2014). The Decay of Consent: International Law in an Age of Global Public Goods. *American Journal of International Law*, 108(1), 1–40.

Kupper, Patrick (2003). Die '1970er Diagnose'. Grundsätzliche Überlegungen zu einem Wendepunkt der Umweltgeschichte. *Archiv für Sozialgeschichte*, 43, 325–48.

La Vina, Antonio G.M. (2002). A Mandate for a Biosafety Protocol. In Christoph Bail, Robert Falkner, & Helen Marquard (Eds.), *The Cartagena Protocol on Biosafety: Reconciling Trade in Biotechnology with Environment and Development?* London: Earthscan, pp. 34–43.

Lane, Ann (2007). Modernising the Management of British Diplomacy: towards a Foreign Office Policy on Policy-Making? *Cambridge Review of International Affairs*, 20(1), 179-193.

Lantis, Jeffrey S. (2017). Theories of International Norm Contestation: Structure and Outcomes. In *Oxford Research Encyclopedia of Politics*. Available at: https://oxfordre.com/politics/view/10.1093/acrefore/9780190228637.001.0001/acrefore-9780190228637-e-590

Larson, Eric Thomas (2005). Why Environmental Liability Regimes in the United States, the European Community, and Japan Have Grown Synonymous with the Polluter Pays Principle. *Vanderbilt Journal of Transnational Law*, 38, 541–75.

Lee, Yok-shiu F., & So, Alvin Y. (1999). Introduction. In Yok-shiu F. Lee & Alvin Y. So (Eds.) *Asia's Environmental Movements: Comparative Perspectives*. Armonk: M.E. Sharpe.

Legault, L.H.J. (1971). The Freedom of the Seas: A License to Pollute. *University of Toronto Law Journal*, 21, 211–21.

Lekan, T.M. (2004). *Imagining the Nation in Nature: Landscape Preservation and German Identity, 1885–1945*. Cambridge, MA: Harvard University Press.

Lenton, Timothy M., Rockström, Johan, Gaffney, Owen, Rahmstorf, Stefan, Richardson, Katherine, Steffen, Will, & Schellnhuber, Hans Joachim (2019). Climate Tipping Points – Too Risky to Bet Against. *Nature*, 575, 592–5.

Leonard, L. Larry (1941). Recent Negotiations Toward the International Regulation of Whaling. *American Journal of International Law*, 35(1), 90–113.

Levy, David L. (1997). Business and International Environmental Treaties: Ozone Depletion and Climate Change. *California Management Review*, 39 (3), 54–71.

Levy, David L., & Newell, Peter J. (Eds.) (2005). *The Business of Global Environmental Governance*. Cambridge, MA: MIT Press.

Li, Chien-hui (2000). A Union of Christianity, Humanity, and Philanthropy: The Christian Tradition and the Prevention of Cruelty to Animals in Nineteenth-Century England. *Society & Animals*, 8(3), 265–85.

Lightfoot, Simon, & Burchell, Jon (2005). The European Union and the World Summit on Sustainable Development: Normative Power Europe in Action? *Journal of Common Market Studies*, 43(1), 75–95.

Linklater, Andrew (2011). *The Problem of Harm in World Politics: Theoretical Investigations*. Cambridge: Cambridge University Press.

Linklater, Andrew, & Suganami, Hidemi (2006). *The English School of International Relations: A Contemporary Reassessment*. Cambridge: Cambridge University Press.

Linsenmaier, Thomas (2018). World Society as Collective Identity: World Society, International Society, and Inclusion/Exclusion from Europe. *International Politics*, 55(1), 91–107.

Lister, Jane, Poulsen, René Taudal, & Ponte, Stefano (2015). Orchestrating Transnational Environmental Governance in Maritime Shipping. *Global Environmental Change*, 34, 185–95.

Litfin, Karen T. (Ed.) (1998). *The Greening of Sovereignty in World Politics*. Cambridge, MA: MIT Press.

Little, Richard (2000). The English School's Contribution to the Study of International Relations. *European Journal of International Relations*, 6(3), 395–422.

(2009). History, Theory and Methodological Pluralism in the English School. In Cornelia Navari (Ed.), *Theorising International Society: English School Methods*. Basingstoke: Palgrave Macmillan, pp. 78–103.

Lomborg, Bjørn (2003). *The Skeptical Environmentalist: Measuring the Real State of the World*. Cambridge: Cambridge University Press.

Lovelock, James (2009). *The Vanishing Face of Gaia: A Final Warning*. New York: Basic Books.

M'Gonigle, R. Michael, & Zacher, Mark W. (1981). *Pollution, Politics, and International Law: Tankers at Sea*. Berkeley: University of California Press.

MacDonald, Mary (1998). *Agendas for Sustainability: Environment and Development into the Twenty-First Century*. London: Routledge.

Macekura, Stephen (2011). The Limits of the Global Community: The Nixon Administration and Global Environmental Politics. *Cold War History*, 11(4), 489–518.

(2015). *Of Limits and Growth: The Rise of Global Sustainable Development in the Twentieth Century*. Cambridge: Cambridge University Press.

MacKenzie, John M. (1988). *The Empire of Nature: Hunting, Conservation and British Imperialism*. Manchester: University of Manchester.

Maddock, Rowland T. (1994). Japan and Global Environmental Leadership. *Journal of Northeast Asian Studies*, 13(4), 37-48.

Malhotra, Saloni (2017). The International Crime That Could Have Been but Never Was: An English School Perspective on the Ecocide Law. *Amsterdam Law Forum*, 9(3), 49–70.

Manners, Ian (2008). The Normative Ethics of the European Union. *International Affairs*, 84(1), 45–60.

Manning, C.A.W. (1962). *The Nature of International Society*. London: Bell and Sons.

Manulak, Michael W. (2017). Developing World Environmental Cooperation: the Founex Seminar and the Stockholm Conference. In Wolfram Kaiser & Jan-Henrik Meyer (Eds.), *International Organizations and Environmental Protection: Conservation and Globalization in the Twentieth Century*. New York: Berghahn Books, pp. 103–127.

Marchant, Gary E., & Abbott, Kenneth W. (2012). International Harmonization of Nanotechnology Governance Through Soft Law Approaches. *Nanotechnology Law & Business*, 9, 393–410.

Markham, William T. (2008). *Environmental Organizations in Modern Germany: Hardy Survivors in the Twentieth Century and Beyond*. New York, Berghahn Books.

Marsh, George P. (1865). *Man and Nature; or Physical Geography as Modified by Human Action*. New York: Charles Scribner.

Martens, Kerstin (2001). Non-Governmental Organisations as Corporatist Mediator? An Analysis of NGOs in the UNESCO System. *Global Society*, 15(4), 387–404.

Martinez-Alier, Joan (2002). *The Environmentalism of the Poor: A Study of Ecological Conflicts and Valuation*. Cheltenham, Edward Elgar.

Marx, Axel (2019). Public-Private Partnerships for Sustainable Development: Exploring Their Design and Its Impact on Effectiveness. *Sustainability*, 11 (4), 1–9.

Matagne, Patrick (1998). The Politics of Conservation in France in the 19th Century. *Environment and History*, 4(3), 359–67.

Mather, A.S., & Fairbairn, J. (2000). From Floods to Reforestation: The Forest Transition in Switzerland. *Environment and History*, 6(4), 399–421.

Mathews, Jessica T. (1989). Redefining Security. *Foreign Affairs*, 68(2), 162–77.

May, James R., & Daly, Erin (2015). *Global Environmental Constitutionalism*. Cambridge: Cambridge University Press.

Mayall, James (1990). *Nationalism and International Society*. Cambridge: Cambridge University Press.

(2000). *World Politics: Progress and Its Limits*. Cambridge: Polity Press.

(2009). The Limits of Progress: Normative Reasoning in the English School. In Cornelia Navari (Ed.), *Theorising International Society: English School Methods*. Basingstoke: Palgrave Macmillan, pp. 209–26.

Mazower, Mark (2012). *Governing the World: The History of an Idea*. London: Allen Lane.

McCloskey, J. Michael (2005). *In the Thick of It: My Life in the Sierra Club*. Washington: Island Press.

McCormick, John (1989). *Reclaiming Paradise: The Global Environmental Movement*. Bloomington: Indiana University Press.

(2001). *Environmental Policy in the European Union*. Basingstoke: Palgrave.

McDaniels, Jeremy, & Robins, Nick (2018). Greening the Rules of the Game: How Sustainability Factors Are Being Incorporated into Financial Policy and Regulation. *Inquiry Working Paper* 18/01, UNEP.

McDonald, Matt (2012). The Failed Securitization of Climate Change in Australia. *Australian Journal of Political Science*, 47(4), 579–92.

(2018). Climate Change and Security: Towards Ecological Security? *International Theory*, 10(2), 153–80.

McGraw, Désirée M. (2002). The Story of the Biodiversity Convention: From Negotiation to Implementation. In Philippe G. Le Prestre (Ed.), *Governing Global Biodiversity: The Evolution and Implementation of the Convention on Biological Diversity*. Aldershot: Ashgate, pp. 7–38.

McKeil, Aaron (2018). A Silhouette of Utopia: English School and Constructivist Conceptions of a World Society. *International Politics*, 55(1), 41–56.

McKeown, Ryder (2009). Norm Regress: US Revisionism and the Slow Death of the Torture Norm. *International Relations*, 23(1), 5–25.

McNeill, John R. (2000). *Something New Under the Sun: An Environmental History of the Twentieth Century*. London, Allen Lane.

McNeill, John R., & Engelke, Peter (2016). *The Great Acceleration*. Cambridge, MA: Harvard University Press.

Meadowcroft, James (2005). From Welfare State to Ecostate. In John Barry & Robyn Eckersley (Eds.), *The State and the Global Ecological Crisis*. Cambridge, MA: MIT Press, pp. 3–23.

(2012). Greening the State. In Paul F. Steinberg & Stacy D. VanDeveer (Eds.), *Comparative Environmental Politics: Theory, Practice, and Prospects*. Cambridge, MA: MIT Press, pp. 63–87.

Meadows, Donella H., Meadows, Dennis L., Randers, Jorgen, & Behrens, William W. (1972). *The Limits to Growth: A Report for the Club of Rome's Project on the Predicament of Mankind*. New York: Universe Books.

Meckling, Jonas (2011). *Carbon Coalitions: Business, Climate Politics, and the Rise of Emissions Trading*. Cambridge, MA: MIT Press.

Mehta, Sailesh, & Merz, Prisca (2015). Ecocide – A New Crime Against Peace? *Environmental Law Review*, 17(1), 3–7.

Merchant, Carolyn (2010). George Bird Grinnell's Audubon Society: Bridging the Gender Divide in Conservation. *Environmental History*, 15(1), 3–30.

Mertes, Tom (Ed.) (2004). *A Movement of Movements: Is Another World Really Possible?* London: Verso.

Meyer, Jan-Henrik (2017). From Nature to Environment: International Organizations and Environmental Protection before Stockholm. In Wolfram Kaiser & Jan-Henrik Meyer (Eds.), *International Organizations and Environmental Protection: Conservation and Globalization in the Twentieth Century*. New York: Berghahn Books, pp. 31–73.

Meyer, John M. (1997). Gifford Pinchot, John Muir, and the Boundaries of Politics in American Thought. *Polity*, 30(2), 267–84.

Milicay, Fernanda Maria (2015). The Common Heritage of Mankind: 21st Century Challenges of a Revolutionary Concept. In Lilian del Castillo

(Ed.), *Law of the Sea, from Grotius to the International Tribunal for the Law of the Sea*. Leiden: Brill Nijhoff, pp. 272–95.

Milkoreit, Manjana (2019). The Paris Agreement on Climate Change – Made in USA? *Perspectives on Politics*, 17(4), 1–19.

Miller, Marian (1995). *The Third World in Global Environmental Politics*. Boulder, CO: Lynne Rienner.

Mitchell, Ronald B. (1994). Regime Design Matters: Intentional Oil Pollution and Treaty Compliance. *International Organization*, 48(03), 425–58.

(2010). *International Politics and the Environment*. London: Sage.

Mitchell, Ronald B. (2002–2020). International Environmental Agreements Database Project (Version 2020.1). Available at: http://iea.uoregon.edu, accessed 30 May 2020.

Mitchell, Ronald B., Andonova, Liliana B., Axelrod, Mark, Balsiger, Jörg, Bernauer, Thomas, Green, Jessica F., Hollway, James, Kim, Rakhyun E., & Morin, Jean-Frédéric (2020). What We Know (and Could Know) About International Environmental Agreements. *Global Environmental Politics*, 20 (1), 103–21.

Morin, Jean-Frédéric, Dür, Andreas, & Lechner, Lisa (2018). Mapping the Trade and Environment Nexus: Insights from a New Data Set. *Global Environmental Politics*, 18(1), 122–39.

Morin, Jean Frédéric, Pauwelyn, Joost, & Hollway, James (2017). The Trade Regime as a Complex Adaptive System: Exploration and Exploitation of Environmental Norms in Trade Agreements. *Journal of International Economic Law*, 20(2), 365–90.

Morphet, Sally (1996). NGOs and the Environment. In Peter Willets (Ed.), *'The Conscience of the World': The Influence of Non-Governmental Organisations in the UN System*. London: Hurst & Co, pp. 116–46.

Morris, Ian (2011). *Why the West Rules – For Now: The Patterns of History and What They Reveal about the Future*. London: Profile Books.

Murphy, Roger (2018). *Challenges from Within*. London: Routledge.

Myers, Norman (1989). Environment and Security. *Foreign Policy*, (74), 23–41.

Naess, Arne (1973). The Shallow and the Deep, Long-range Ecological Movement. *Inquiry*, 16(1), 95–100.

Naím, Moisés (2009). Minilateralism: The Magic Number to Get Real International Action. *Foreign Policy*, (173), 135–36.

Najam, Adil (2003). The Case Against a New International Environmental Organization. *Global Governance: A Review of Multilateralism and International Organizations*, 9(3), 367–84.

(2005). Developing Countries and Global Environmental Governance: From Contestation to Participation to Engagement. *International Environmental Agreements: Politics, Law and Economics*, 5(3), 303–21.

Nash, Roderick F. (1970). The American Invention of National Parks. *American Quarterly*, 22(3), 726–35.

(1989). *The Rights of Nature: A History of Environmental Ethics*. Madison: The University of Wisconsin Press.

(2001). *Wilderness and the American Mind*. 4th ed. New Haven: Yale University Press.

Navari, Cornelia (Ed.) (2009). *Theorising International Society: English School Methods*. Basingstoke: Palgrave Macmillan.
Navari, Cornelia (2014). English School Methodology. In Cornelia Navari & Daniel M. Green (Eds.), *Guide to the English School in International Studies*. Chichester: Wiley, pp. 205–21.
(2018). Two Roads to World Society: Meyer's 'World Polity' and Buzan's 'World Society'. *International Politics*, 55(1), 11–25.
Neumann, Roderick P. (1996). Dukes, Earls, and Ersatz Edens: Aristocratic Nature Preservationists in Colonial Africa. *Environment and Planning D: Society and Space*, 14(1), 79–98.
(2001). Africa's 'Last Wilderness': Reordering Space for Political and Economic Control in Colonial Tanzania. *Africa*, 71(4), 641–65.
Neuner, Fabian G. (2020). Public Opinion and the Legitimacy of Global Private Environmental Governance. *Global Environmental Politics*, 20(1), 60–81.
Newell, Peter (2001). Managing Multinationals: The Governance of Investment for the Environment. *Journal of International Development*, 13(7), 907–19.
(2002). A World Environment Organization: The Wrong Solution to the Wrong Problem. *The World Economy*, 25(5), 659–71.
(2020). *Global Green Politics*. Cambridge: Cambridge University Press.
Nicholls, Yvonne I. (Ed.) (1973). *Source Book: Emergence of Proposals for Recompensing Developing Countries for Maintaining Environmental Quality*. Morges: IUCN.
Nicholson, Max (1970). *The Environmental Revolution*. London: Hodder & Stoughton.
Nielsen, L. Daniel, & Tierney, J. Michael (2003). Delegation to Internal Organizations: Agency Theory and World Bank Environmental Reform. *International Organization*, 57(2), 241–76.
Niemann, Holger, & Schillinger, Henrik (2017). Contestation 'All the Way Down'? The Grammar of Contestation in Norm Research. *Review of International Studies*, 43(1), 29–49.
North American Conservation Conference (1909). Declaration of Principles. Canada, Sessional Paper No. 90, 8-9 Edward VII. Available at: https://archive.org/stream/1909v43i17p90_0607/1909v43i17p90_0607_djvu.txt
Nyman, Jonna, & Zeng, Jinghan (2016). Securitization in Chinese Climate and Energy Politics. *Wiley Interdisciplinary Reviews: Climate Change*, 7(2), 301–13.
Oberthür, Sebastian, & Lefeber, René (2010). Holding Countries to Account: The Kyoto Protocol's Compliance System Revisited after Four Years of Experience. *Climate Law*, 1, 133.
O'Connor, Martin (Ed.) (1994). *Is Capitalism Sustainable? Political Economy and the Politics of Ecology*. New York: The Guildford Press.
Oels, Angela (2012). From 'Securitization' of Climate Change to 'climatization' of the Security Field: Comparing Three Theoretical Perspectives. In Jürgen Scheffran (Ed.), *Climate Change, Human Security and Violent Conflict*. Berlin: Springer, pp. 185–205.
Okereke, Chukwumerije (2008). Equity Norms in Global Environmental Governance. *Global Environmental Politics*, 8(3), 25–50.

Onuf, Nicholas (2002). Institutions, Intentions and International Relations. *Review of International Studies*, 28(2), 211–28.

Ophuls, William (1977). *Ecology and the Politics of Scarcity: Prologue to a Political Theory of the Steady State*. San Francisco, CA: W.H. Freeman.

Oreskes, Naomi, & Conway, Erik M. (2011). *Merchants of Doubt: How a Handful of Scientists Obscured the Truth on Issues from Tobacco Smoke to Global Warming*. New York: Bloomsbury.

Overdevest, Christine, & Zeitlin, Jonathan (2014). Assembling an Experimentalist Regime: Transnational Governance Interactions in the Forest Sector. *Regulation & Governance*, 8(1), 22–48.

Padgett, Stephen (2003). Between Synthesis and Emulation: EU Policy Transfer in the Power Sector. *Journal of European Public Policy*, 10(2), 227–45.

Padua, Jose Augusto (2000). 'Annihilating Natural Productions': Nature's Economy, Colonial Crisis and the Origins of Brazilian Political Environmentalism (1786–1810). *Environment and History*, 6(3), 255–87.

Palmujoki, Eero (2013). Fragmentation and Diversification of Climate Change Governance in International Society. *International Relations*, 27(2), 180–201.

Panke, Diana, & Petersohn, Ulrich (2016). Norm Challenges and Norm Death: The Inexplicable? *Cooperation and Conflict*, 51(1), 3–19.

Park, Susan (2005). Norm Diffusion within International Organizations: A Case Study of the World Bank. *Journal of International Relations and Development*, 8(2), 111-41.

(2007). The World Bank Group: Championing Sustainable Development Norms? *Global Governance*, 13(4), 535-56.

(2013). Transnational Environmental Activism. In Robert Falkner (Ed.), *The Handbook of Global Climate and Environment Policy*. Cheltenham: John Wiley & Sons Ltd., pp. 268–85.

Parson, Edward (2003). *Protecting the Ozone Layer: Science and Strategy*. Oxford: Oxford University Press.

Paterson, Matthew (2005). Global Environmental Governance. In Alex J. Bellamy (Ed.), *International Society and Its Critics*. Oxford, Oxford University Press, pp. 163–77.

Pattberg, Philipp (2005). What Role for Private Rule-Making in Global Environmental Governance? Analyzing the Forest Stewardship Council (FSC). *International Environmental Agreements: Politics, Law and Economics*, 5(2), 175–89.

(2007). *Private Institutions and Global Governance: The New Politics of Environmental Sustainability*. Cheltenham: Edward Elgar.

(2017). The Emergence of Carbon Disclosure: Exploring the Role of Governance Entrepreneurs. *Environment and Planning C: Politics and Space*, 35(8), 1437–55.

Pattberg, Philipp, Biermann, Frank, Chan, Sander, & Mert, Ayşem (2012). Introduction: Partnerships for Sustainable Development. In Philipp Pattberg, Frank Biermann, Sander Chan, & Ayşem Mert (Eds.), *Public-private Partnerships for Sustainable Development: Emergence, Influence and Legitimacy*. Cheltenham: Edward Elgar, pp. 1–18.

Pattberg, Philipp, Widerberg, Oscar, & Kok, Marcel T.J. (2019). Towards a Global Biodiversity Action Agenda. *Global Policy*, 10(3), 385–390.

Payne, Rodger A. (2001). Persuasion, Frames and Norm Construction. *European Journal of International Relations*, 7(1), 37–61.

Pearson, Chris (2015). Environments, States and Societies at War. In Michael Geyer & Adam Tooze (Eds.), *The Cambridge History of the Second World War: Volume 3: Total War: Economy, Society and Culture* (Vol. 3). Cambridge: Cambridge University Press, pp. 220–44.

Pella, John Anthony (2013). Thinking Outside International Society: A Discussion of the Possibilities for English School Conceptions of World Society. *Millennium*, 42(1), 65–77.

Pepper, David (1993). *Eco-socialism: From Deep Ecology to Social Justice*. London: Routledge.

(1996). *Modern Environmentalism: An Introduction*. London: Routledge.

Peritore, N. Patrick (1999). *Third World Environmentalism: Case Studies from the Global South*. Gainesville: University Press of Florida.

Peters, Birgit (2017). Unpacking the Diversity of Procedural Environmental Rights: The European Convention on Human Rights and the Aarhus Convention. *Journal of Environmental Law*, 30(1), 1–27.

Pickering, Jonathan, McGee, Jeffrey S., Stephens, Tim, & Karlsson-Vinkhuyzen, Sylvia I. (2018). The Impact of the US Retreat from the Paris Agreement: Kyoto Revisited? *Climate Policy*, 18(7), 818–27.

Pincetl, Stephanie (1993). Some Origins of French Environmentalism: An Exploration. *Forest & Conservation History*, 37(2), 80–9.

Pinchot, Gifford (1910). *The Fight for Conservation*. New York: Doubleday, Page.

Ponte, Stefano (2019). *Business, Power and Sustainability in a World of Global Value Chains*. London: Zed Books Ltd.

Ponte, Stefano, & Cheyns, Emmanuelle (2013). Voluntary Standards, Expert Knowledge and the Governance of Sustainability Networks. *Global Networks*, 13(4), 459–77.

Ponte, Stefano, & Daugbjerg, Carsten (2015). Biofuel Sustainability and the Formation of Transnational Hybrid Governance. *Environmental Politics*, 24 (1), 96–114.

Popovic, Neil A.F. (1995). In Pursuit of Environmental Human Rights: Commentary on the Draft Declaration of Principles on Human Rights and the Environment. *Columbia Human Rights Law Review*, 27, 487–603.

Prendergast, David K., & William M. Adams (2003). Colonial Wildlife Conservation and the Origins of the Society for the Preservation of the Wild Fauna of the Empire (1903–1914). *Oryx*, 37(2), 251–60.

Pryde, Philip R. (1991). *Environmental Management in the Soviet Union*. Cambridge: Cambridge University Press.

Radkau, Joachim (2011). *Die Ära der Ökologie: Eine Weltgeschichte*. Munich: C.H. Beck.

Ralph, Jason (2007). *Defending the Society of States: Why America Opposes the International Criminal Court and Its Vision of World Society*. Oxford: Oxford University Press.

Rajamani, Lavanya (2012). The Changing Fortunes of Differential Treatment in the Evolution of International Environmental Law. *International Affairs*, 88 (3), 605–23.

(2016). Ambition and Differentiation in the 2015 Paris Agreement: Interpretative Possibilities and Underlying Politics. *International & Comparative Law Quarterly*, 65(2), 493–14.

Raustiala, Kal (1997). States, NGOs, and International Environmental Institutions. *International Studies Quarterly*, 41(4), 719–40.

(2002). The Architecture of International Cooperation: Transgovernmental Networks and the Future of International Law. *Virginia Journal of International Law*, 43, 1–92.

Reifschneider, Laura (2002). Global Industry Coalition. In Christoph Bail, Robert Falkner, & Helen Marquard (Eds.), *The Cartagena Protocol on Biosafety: Reconciling Trade in Biotechnology with Environment and Development?* London: Earthscan, pp. 273–77.

Reilly, Charles A. (1993). The Road from Rio. NGO Policy Makers and the Social Ecology of Development. *Grassroots Development: Journal of the Inter-American Foundation*, 17(1), 25–35.

Reinsberg, Bernhard, & Westerwinter, Oliver (2021). The Global Governance of International Development: Documenting the Rise of Multi-Stakeholder Partnerships and Identifying Underlying Theoretical Explanations. *The Review of International Organizations*, 16(1), 59–94.

Reus-Smit, Christian (1996). The Normative Structure of International Society. In Fen Osler Hampson & Judith Reppy (Eds.), *Earthly Goods: Environmental Change and Social Justice*. Ithaca: Cornell University Press, pp. 96–121.

(1997). The Constitutional Structure of International Society and the Nature of Fundamental Institutions. *International Organization*, 51(4), 555–89.

(1999). *The Moral Purpose of the State: Culture, Social Identity, and Institutional Rationality in International Relations*. Princeton: Princeton University Press.

(2002). Imagining Society: Constructivism and the English School. *British Journal of Politics and International Relations*, 4(3), 487–509.

(2009). Constructivism and the English School. In Cornelia Navari (Ed.), *Theorising International Society: English School Methods*. Basingstoke: Palgrave Macmillan, pp. 58–77.

(2018). *On Cultural Diversity: International Theory in a World of Difference*. Cambridge: Cambridge University Press.

Reus-Smit, Christian, & Tim Dunne (2017). The Globalization of International Society. In Tim Dunne & Christian Reus-Smit (Ed.), *The Globalization of International Society*. Oxford: Oxford University Press, pp. 18–40.

Ridley, Matt (1995). *Down to Earth: A Contrarian View of Environmental Problems*. London: Institute of Economic Affairs.

Rietig, Katharina (2014). 'Neutral' Experts? How Input of Scientific Expertise Matters in International Environmental Negotiations. *Policy Sciences*, 47(2), 141–60.

(2016). The Power of Strategy: Environmental NGO Influence in International Climate Negotiations. *Global Governance*, 22(2), 269–88.

Risse-Kappen, Thomas (Ed.) (1995). *Bringing Transnational Relations Back In: Non-State Actors, Domestic Structures and International Institutions*. Cambridge: Cambridge University Press.

Roberts, J. Timmons, Parks, Bradley C., & Vásquez, Alexis A. (2004). Who Ratifies Environmental Treaties and Why? Institutionalism, Structuralism and Participation by 192 Nations in 22 Treaties. *Global Environmental Politics*, 4(3), 22–64.

Robertson, Thomas (2012). *The Malthusian Moment: Global Population Growth and the Birth of American Environmentalism*. New Brunswick: Rutgers University Press.

Robinson, Nicholas A. (2018). *Environmental Law: Is an Obligation Erga Omnes Emerging*. Panel Discussion at the United Nations (New York), 4 June 2018. Available at: www.iucn.org/sites/dev/files/content/documents/2018/environ mental_law_is_an_obligation_erga_omnes_emerging_interamcthradvisoryo pinionjune2018.pdf

Roe, Paul (2004). Securitization and Minority Rights: Conditions of Desecuritization. *Security Dialogue*, 35(3), 279–94.

Roger, Charles, & Dauvergne, Peter (2016). The Rise of Transnational Governance as a Field of Study. *International Studies Review*, 18(3), 415–37.

Roger, Charles, Hale, Thomas, & Andonova, Liliana (2017). The Comparative Politics of Transnational Climate Governance. *International Interactions*, 43 (1), 1–25.

Roger, Charles B. (2020). *The Origins of Informality: Why the Legal Foundations of Global Governance Are Shifting, and Why It Matters*. Oxford: Oxford University Press.

Rootes, Christopher (2006). Facing South? British Environmental Movement Organisations and the Challenge of Globalisation. *Environmental Politics*, 15 (5), 768–86.

(2007). Nature Protection Organizations in England. In C.S.A. Van Koppen & William T. Markham (Eds.), *Protecting Nature: Organizations and Networks in Europe and the United States*. Cheltenham: Edward Elgar, pp. 34–62.

Rollins, William H. (1997). *A Greener Vision of Home: Cultural Politics and Environmental Reform in the German Heimatschutz Movement, 1904–1918*. Ann Arbor: University of Michigan Press.

(1999). Imperial Shades of Green: Conservation and Environmental Chauvinism in the German Colonial Project. *German Studies Review*, 22 (2), 187–213.

Ross, Corey (2015). Tropical Nature as Global Patrimoine: Imperialism and International Nature Protection in the Early Twentieth Century. *Past & Present*, 226(suppl_10), 214–39.

(2017). *Ecology and Power in the Age of Empire: Europe and the Transformation of the Tropical World*. Oxford: Oxford University Press.

Rothschild, N. Charles (1913). International Conference for the Global Protection of Nature, 27 November. Kew National Archives, FO 881/ 10351.

Ruggie, John Gerard (1982). International Regimes, Transactions, and Change: Embedded Liberalism in the Postwar Economic Order. *International Organization*, 36(2), 379–415.

(1998). *Constructing the World Polity: Essays on International Institutionalization*. London: Routledge.

Runge, C. Ford (2001). A Global Environment Organization (GEO) and the World Trading System. *Journal of World Trade*, 35(4), 399–426.

Sachs, Ignacy (1974). Environment and Styles of Development. *Economic and Political Weekly*, 9(21), 828–37.

Safranski, Rüdiger (2009). *Romantik: Eine deutsche Affäre*. Frankfurt/M.: Fischer.

Sale, Kirkpatrick (1985). *Dwellers in the Land: The Bioregional Vision*. San Francisco: Sierra Club Books.

Sand, Peter H. (2007). The Evolution of International Environmental Law. In Daniel Bodansky, Jutta Brunnée, & Ellen Hey (Eds.), *The Oxford Handbook of International Environmental Law*. Oxford: Oxford University Press, pp. 29–43.

Sands, Philippe, & Peel, Jacqueline (2012). *Principles of International Environmental Law*. Cambridge: Cambridge University Press.

Sandholtz, Wayne (2008). Dynamics of International Norm Change: Rules against Wartime Plunder. *European Journal of International Relations*, 14(1), 101–31.

Sarre, Philip (1995). Towards Global Environmental Values: Lessons from Western and Eastern Experience. *Environmental Values*, 4(2), 115–27.

Scarce, Rik (2016). *Eco-Warriers: Understanding the Radical Environmental Movement*. London: Routledge.

Schäfer, Mike S., Scheffran, Jürgen, & Penniket, Logan (2016). Securitization of Media Reporting on Climate Change? A Cross-National Analysis in Nine Countries. *Security Dialogue*, 47(1), 76–96.

Schaper, Marcus (2007). Leveraging Green Power: Environmental Rules for Project Finance. *Business and Politics*, 9(3), 1–27.

Schleifer, Philip (2013). Orchestrating Sustainability: The Case of European Union Biofuel Governance. *Regulation & Governance*, 7(4), 533–46.

(2017). Private Regulation and Global Economic Change: The Drivers of Sustainable Agriculture in Brazil. *Governance*, 30(4), 687–703.

Schlosberg, David (1999). *Environmental Justice and the New Pluralism: The Challenge of Difference for Environmentalism*. Oxford: Oxford University Press.

Schmelzer, Matthias (2012). The Crisis before the Crisis: The 'Problems of Modern Society' and the OECD, 1968–74. *European Review of History: Revue europeenne d'histoire*, 19(6), 999–1020.

Schofer, Evan, & Hironaka, Ann (2005). The Effects of World Society on Environmental Protection Outcomes. *Social Forces*, 84(1), 25–47.

Schouenborg, Laust (2011). A New Institutionalism? The English School as International Sociological Theory. *International Relations*, 25(1), 26–44.

Schreurs, Miranda A. (2004). Assessing Japan's Role as a Global Environmental Leader. *Policy and Society*, 23(1), 88–110.

Schreurs, Miranda (2013). Regionalism and Environmental Governance. In Robert Falkner (Ed.), *The Handbook of Global Climate and Environment Policy*. Cheltenham: Wiley-Blackwell, pp. 358–73.

Schroeder, Heike, & Lovell, Heather (2011). The Role of Non-Nation-State Actors and Side Events in the International Climate Negotiations. *Climate Policy*, 12(1), 23–37.

Schumacher, Ernst F. (1973). *Small Is Beautiful: A Study of Economics as if People Mattered*. London: Blond & Briggs.

Scott, James B. (1923). *Robert Bacon. Life and Letters*. New York: Doubleday, Page & Co.

Scott, Shirley V. (2012). The Securitization of Climate Change in World Politics: How Close Have We Come and Would Full Securitization Enhance the Efficacy of Global Climate Change Policy? *Review of European, Comparative & International Environmental Law*, 21(3), 220–30.

(2015). Implications of Climate Change for the UN Security Council: Mapping the Range of Potential Policy Responses. *International Affairs*, 91 (6), 1317–33.

(2018). The Attitude of the P5 Towards a Climate Change Role for the Council. In Shirley V. Scott & Charlotte Ku (Eds.), *Climate Change and the UN Security Council*. Cheltenham: Edward Elgar, pp. 209–28.

Scott, Shirley V, & Ku, Charlotte (Eds.) (2018). *Climate Change and the UN Security Council*. Cheltenham: Edward Elgar Publishing.

Scott, Shirley V., & Oriana, Lucia Meilin (2017). Resisting Japan's Promotion of a Norm of Sustainable Whaling. In Alan Bloomfield & Shirley V. Scott (Eds.), *Norm Antipreneurs and the Politics of Resistance to Global Normative Change*. London: Routledge, pp. 108–24.

Scruton, Roger (2012). *Green Philosophy: How to Think Seriously About the Planet*. London, Atlantic Books.

Selin, Henrik (2013). Global Chemicals Politics and Policy. In Robert Falkner (Ed.), *The Handbook of Global Climate and Environment Policy*. Cheltenham: John Wiley & Sons Ltd, pp. 107–23.

Selin, Henrik, & VanDeveer, Stacy D. (2006). Raising Global Standards: Hazardous Substances and E-waste Management in the European Union. *Environment*, 48(10), 6–18.

(2015). *European Union and Environmental Governance*. London: Routledge.

Shabecoff, Philip (1993). *A Fierce Green Fire: The American Environmental Movement*. New York: Hill and Wang.

Shapiro, Judith (2012). *China's Environmental Challenges*. Cambridge: Polity Press.

Shelton, Dinah (2009). Common Concern of Humanity. *Environmental Policy & Law*, 39(2), 83–6.

Shiva, Vandana (1993). The Greening of the Global Reach. In Jeremy Brecher, John Browne Childs, & Jill Cutler (Eds.), *Global Visions: Beyond the New World Order*. Boston, MA: South End Press, pp. 53–60.

Sierra Club (1892). Articles of Incorporation. 4 June. Available at: www.sierraclub.org/articles-incorporation

Sikkink, Kathryn (2014). Latin American Countries as Norm Protagonists of the Idea of International Human Rights. *Global Governance*, 20(3), 389–404.

Simlinger, Florentina, & Mayer, Benoit (2019). Legal Responses to Climate Change Induced Loss and Damage. In Reinhard Mechler, Laurens M. Bouwer, Thomas Schinko, & Swenja Surminski (Eds.), *Loss and Damage from Climate Change: Concepts, Methods and Policy Options*. Springer, pp. 179–203.

Simma, Bruno (1994). From Bilateralism to Community Interest in International Law. *Recueil des Cours*, 250, 217–384.

Simpson, Gerry (2004). *Great Powers and Outlaw States: Unequal Sovereigns in the International Legal Order*. Cambridge: Cambridge University Press.

Skodvin, Tora, & Andresen, Steinar (2006). Leadership Revisited. *Global Environmental Politics*, 6(3), 13–27.

Slaughter, Anne-Marie (2004). *A New World Order*. Princeton: Princeton University Press.

Sluga, Glenda (2010). UNESCO and the (One) World of Julian Huxley. *Journal of World History*, 21(3), 393–418.

Smith, Michael B. (2001). 'Silence, Miss Carson!' Science, Gender, and the Reception of Silent Spring. *Feminist Studies*, 27(3), 733–52.

Stec, Stephen (2010). Humanitarian Limits to Sovereignty: Common Concern and Common Heritage Approaches to Natural Resources and Environment. *International Community Law Review*, 12(3), 361–89.

Steinberg, Ted (2002). *Down to Earth: Nature's Role in American History*. Oxford: Oxford University Press.

Sterling, Claire (1970). The U.N. and World Pollution. *Washington Post*, 28 July.

Stern, Nicholas (2007). *The Economics of Climate Change: The Stern Review*. Cambridge: Cambridge University Press.

Stiles, Daniel (2004). The Ivory Trade and Elephant Conservation. *Environmental Conservation*, 31(4), 309–21.

Stivachtis, Yannis A. (2018). 'International Society' versus' 'World Society': Europe and the Greek War of Independence. *International Politics*, 55(1), 108–24.

Stivachtis, Yannis A., & McKeil, Aaron (2018). Conceptualizing World Society. *International Politics*, 55(1), 1–10.

Stoll, Mark (2015). *Inherit the Holy Mountain: Religion and the Rise of American Environmentalism*. Oxford: Oxford University Press.

Stone, Christopher D. (1972). Should Trees Have Standing? Toward Legal Rights for Natural Objects. *Southern California Law Review*, 45, 450–501.

Stone, Randall W. (2013). Informal Governance in International Organizations: Introduction to the Special Issue. *The Review of International Organizations*, 8 (2), 121–36.

Streck, Charlotte (2001). The Global Environment Facility – A Role Model for International Governance? *Global Environmental Politics*, 1(2), 71–94.

(2020). Filling in for Governments? The Role of the Private Actors in the International Climate Regime. *Journal for European Environmental & Planning Law*, 17(1), 5–28.

Stroikos, Dimitrios (2018). Engineering World Society? Scientists, Internationalism, and the Advent of the Space Age. *International Politics*, 55(1), 73–90.

Sundin, Bosse (2005). Nature as Heritage: The Swedish Case. *International Journal of Heritage Studies*, 11(1), 9–20.

Suzuki, Shogo (2009). *Civilization and Empire: China and Japan's Encounter with European International Society*. London: Routledge.

Suzuki, Shogo, Zhang, Yongjin, & Quirk, Joel (Eds.) (2014). *International Orders in the Early Modern World: Before the Rise of the West*. London: Routledge.

Szarka, Joseph (2002). *The Shaping of Environmental Policy in France*. New York: Berghahn Books.

Tal, Alon (Ed.) (2006). *Speaking of Earth: Environmental Speeches that Moved the World*. New Brunswick: Rutgers University Press.

Talberg, Anita, Christoff, Peter, Thomas, Sebastian, & Karoly, David (2018). Geoengineering Governance-by-Default: An Earth System Governance Perspective. *International Environmental Agreements: Politics, Law and Economics*, 18(2), 229–53.

Tams, Christian J., & Tzanakopoulos, Antonios (2010). Barcelona Traction at 40: the ICJ as an Agent of Legal Development. *Leiden Journal of International Law*, 23(4), 781–800.

Tanasescu, Mihnea (2013). The Rights of Nature in Ecuador: The Making of an Idea. *International Journal of Environmental Studies*, 70(6), 846–61.

Tarrow, Sidney (2005a). *The New Transnational Activism*. Cambridge: Cambridge University Press.

(2005b). The Dualities of Transnational Contention: 'Two Activist Solitudes' or a New World Altogether? *Mobilization: An International Quarterly*, 10(1), 53–72.

Thomas, Michael Durant (2015). Climate Securitization in the Australian Political–Military Establishment. *Global Change, Peace & Security*, 27(1), 97–118.

Thoreau, Henry D.T. (1968 [1854]). *Walden*. London: Everyman's Library.

Tilley, Helen (2011). *Africa as a Living Laboratory: Empire, Development, and the Problem of Scientific Knowledge, 1870–1950*. Chicago: University of Chicago Press.

Tinker, Catherine (1995). Responsibility for Biological Diversity Conservation Under International Law. *Vanderbilt Journal of Transnational Law*, 28, 777–821.

Tosun, Jale (2013). How the EU Handles Uncertain Risks: Understanding the Role of the Precautionary Principle. *Journal of European Public Policy*, 20(10), 1517–28.

Trachtman, Joel P. (2017). WTO Trade and Environment Jurisprudence: Avoiding Environmental Catastrophe. *Harvard International Law Journal*, 58, 273.

Trombetta, Maria Julia (2008). Environmental Security and Climate Change: Analysing the Discourse. *Cambridge Review of International Affairs*, 21(4), 585–602.

Tucker, Richard P. (2013). The International Environmental Movement and the Cold War. In Richard H. Immerman & Petra Goedde (Eds.), *The Oxford Handbook of the Cold War*. Oxford: Oxford University Press, pp. 565–83.

Tyrrell, Ian (2015). *Crisis of the Wasteful Nation: Empire and Conservation in Theodore Roosevelt's America*. Chicago: The University of Chicago Press.

Uekötter, Frank (2004). Wie neu sind die Neuen Sozialen Bewegungen? Revisionistische Bemerkungen vor dem Hintergrund der umwelthistorischen Forschung. *Mitteilungsblatt des Instituts für soziale Bewegungen*, (31), 115–38.

Ullman, Richard (1983). Redefining Security. *International Security*, 8(1), 129–53.

Underdal, Arild (2010). Complexity and Challenges of Long-Term Environmental Governance. *Global Environmental Change*, 20(3), 386–93.

United Nations (1972). United Nations Conference on the Human Environment. Stockholm. List of NGO Observers. A/CONF.48/INF. 6.

(1973). Report of the United Nations Conference on the Human Environment. Stockholm, 5–16 June 1972. A/CONF.48/14/Rev.1

(2011). Kyoto Protocol to the United Nations Framework Convention on Climate Change. Canada: Withdrawal. C.N.796.2011.TREATIES-1 (Depositary Notification).

United Nations Development Program (1994). *Human Development Report 1994: New Dimensions of Human Security*. New York: Oxford University Press.

United Nations Economic and Social Council (1994), Commission on Human Rights. Review of Further Development in Fields with which the Sub-Commission has been Concerned: Human Rights and the Environment, Final Report. E/CN.4/Sub.2/1994/9.

United Nations Environment Programme (2018). *The Emissions Gap Report 2018*. Nairobi: UNEP.

United Nations Framework Convention on Climate Change (2016), Conference of the Parties, Report of the Conference of the Parties on its twenty-first session, held in Paris from 30 November to 13 December 2015, FCCC/CP/2015/10/Add.1.

United Nations General Assembly (1968). Problems of the Human Environment, 1733rd plenary meeting, 3 December. A/RES/2398(XXIII).

(1989). Resolution: United Nations Conference on Environment and Development. A/RES/44/228.

(1992). Report of the United Nations Conference on Environment and Development. A/CONF.151/26 (Vol. I).

(2012). Resolution: Follow-up to paragraph 143 on human security of the 2005 World Summit Outcome. A/RES/66/290.

United Nations Security Council (2011). Statement by the President of the Security Council. S/PRST/2011/15.

United States Department of State (1971). Bureau of Intelligence and Research, Intelligence Note, RSGN-16, 12 August. Available at: https://2001-2009.state.gov/documents/organization/52438.pdf

(1997). *Environmental Diplomacy: The Environment and U.S. Foreign Policy*. Washington, DC.

Van Asselt, Harro (2016). The Role of Non-State Actors in Reviewing Ambition, Implementation, and Compliance under the Paris Agreement. *Climate Law*, 6(1-2), 91–108.

Van der Ven, Hamish, Bernstein, Steven, & Hoffmann, Matthew (2017). Valuing the Contributions of Nonstate and Subnational Actors to Climate Governance. *Global Environmental Politics*, 17(1), 1–20.

Vanhala, Lisa, & Hestbaek, Cecilie (2016). Framing Climate Change Loss and Damage in UNFCCC Negotiations. *Global Environmental Politics*, 16(4), 111–29.

Vidal, John, & Bowcott, Owen (2016). ICC Widens Remit to Include Environmental Destruction Cases. *The Guardian*, 15 September.

Victor, David G. (2006). Toward Effective International Cooperation on Climate Change: Numbers, Interests and Institutions. *Global Environmental Politics*, 6 (3), 90–103.

Vincent, R. John (1986). *Human Rights and International Relations*. Cambridge: Cambridge University Press.

Vogel, David (1995). *Trading Up. Consumer and Environmental Regulation in a Global Economy*. Cambridge, MA: Harvard University Press.

Vogler, John, & Bretherton, Charlotte (2006). The European Union as a Protagonist to the United States on Climate Change. *International Studies Perspectives*, 7(1), 1–22.

Vogler, John, & Stephan, Hannes (2007). The European Union in Global Environmental Governance: Leadership in the Making? International Environmental Agreements: Politics, *Law and Economics*, 7(4), 389–413.

Vormedal, Irja (2011). From Foe to Friend? Business, the Tipping Point and U.S. Climate Politics. *Business and Politics*, 13(3), 1–29.

Vukić, N. Markota, Vuković, Renata, & Calace, Donato (2018). Non-Financial Reporting as a New Trend in Sustainability Accounting. *Journal of Accounting and Management*, 7(2), 13–26.

Wade, Robert (1997). Greening the Bank: The Struggle over the Environment, 1970–1995. In Devesh Kapur, John P. Lewis, & Richard Webb (Eds.), *The World Bank: Its First Half Century* (Vol. 2). Washington, DC: Brookings Institution, pp. 611–734.

Waltz, Kenneth N. (1979). *Theory of International Politics*. New York: McGraw-Hill.

Wang, Cheng-Tong Lir, & Hosoki, Ralph Ittonen (2016). From Global to Local: Transnational Linkages, Global Influences, and Taiwan's Environmental NGOs. *Sociological Perspectives*, 59(3), 561–81.

Wæver, Ole (1995). Securitization and Desecuritization. In Ronnie D. Lipschutz (Ed.), *On Security*. New York: Columbia University Press, pp. 46–87.

Wapner, Paul (1996). *Environmental Activism and World Civic Politics*. Albany: State University of New York Press.

(1998). Reorienting State Sovereignty: Rights and Responsibilities in the Environmental Age. In Karen T. Litfin (Ed.), *The Greening of Sovereignty in World Politics*. Cambridge, MA: MIT Press, pp. 275–97.

(2002). Horizontal Politics: Transnational Environmental Activism and Global Cultural Change. *Global Environmental Politics*, 2(2), 37–62.

(2003). World Summit on Sustainable Development: Toward a Post-Jo'burg Environmentalism. *Global Environmental Politics*, 3(1), 1–10.

Ward, Barbara (1966). *Spaceship Earth*. New York: Columbia University Press.

Ward, Barbara, and René Dubos (1972). *Only One Earth: The Care and Maintenance of a Small Planet*. Harmondsworth: Penguin.

Warde, Paul, Robin, Libby, & Sörlin, Sverker (2018). *The Environment: A History of the Idea*. Baltimore, MD: Johns Hopkins University.

Warlenius, Rikard (2018). Decolonizing the Atmosphere: The Climate Justice Movement on Climate Debt. *The Journal of Environment & Development*, 27 (2), 131–55.

Watson, Adam (1992). *The Evolution of International Society: A Comparative Historical Analysis*. London: Routledge.

Weale, Albert (1992). *The New Politics of Pollution*. Manchester: Manchester University Press.

Weart, Spencer R. (2003). *The Discovery of Global Warming*. Cambridge, MA: Harvard University Press.
(2011). Global Warming: How Skepticism Became Denial. *Bulletin of the Atomic Scientists*, 67(1), 41–50.
(2012). *The Rise of Nuclear Fear*. Cambridge, MA: Harvard University Press.
Weaver, Duncan (2018). The Aarhus Convention and Process Cosmopolitanism. *International Environmental Agreements: Politics, Law and Economics*, 18(2), 199–213.
Weinert, Matthew S. (2011). Reframing the Pluralist–Solidarist Debate. *Millennium*, 40(1), 21–41.
(2018). Reading World Society Phenomenologically: An Illustration Drawing upon the Cultural Heritage of Humankind. *International Politics*, 55(1), 26–40.
Weiss, Edith B. (2011). The Evolution of International Environmental Law. *Japanese Yearbook of International Law*, 54, 1–27.
Weiss, Thomas G. (2016). *Global Governance: Why? What? Whither?* Cambridge: Polity Press.
Welsh, Jennifer M. (2019). Norm Robustness and the Responsibility to Protect. *Journal of Global Security Studies*, 4(1), 53–72.
Wendt, Alexander (1999). *Social Theory of International Politics*. Cambridge: Cambridge University Press.
(2003). Why a World State Is Inevitable. *European Journal of International Relations*, 9(4), 491–542.
Whalley, John, & Zissimos, Ben (2002). An Internationalisation-based World Environmental Organisation. *The World Economy*, 25(5), 619–42.
Wheeler, F.B. (1970). Letter to Mr Arculus. Science and Technology Department, 30 November. Folio 98. FCO 55/384. Kew National Archives.
Wheeler, Nicholas J. (2003). *Saving Strangers: Humanitarian Intervention in International Society*. Oxford: Oxford University Press.
Wheeler, Nicholas J. & Tim Dunne (1996). Hedley Bull's Pluralism of the Intellect and Solidarism of the Will. *International Affairs*, 72(1), 91–107.
White, Damian Finbar, Rudy, Alan P., & Wilbert, Chris (2007). Anti-environmentalism: Prometheans, Contrarians and Beyond. In Jules Pretty, Andrew S. Ball, Ted Benton, Julia Guivant, David R. Lee, David Orr, Max J. Pfeffer, & Hugh Ward (Eds.), *The Sage Handbook of Environment and Society*. London: Sage, pp. 124–41.
White, Lynn (1967). The Historical Roots of Our Ecologic Crisis. *Science*, 155 (3767), 1203–7.
White House, The (2014). Remarks by the President at U.N. Climate Change Summit. Office of the Press Secretary. 23 September. Available at: https://obamawhitehouse.archives.gov/the-press-office/2014/09/23/remarks-president-un-climate-change-summit
Wiener, Antje (2004). Contested Compliance: Interventions on the Normative Structure of World Politics. *European Journal of International Relations*, 10(2), 189–234.
(2008). *The Invisible Constitution of Politics: Contested Norms and International Encounters*. Cambridge: Cambridge University Press.

(2018). *Contestation and Constitution of Norms in Global International Relations*. Cambridge: Cambridge University Press.

Wiener, Antje, & Puetter, Uwe (2009). The Quality of Norms Is What Actors Make of It Critical-Constructivist Research on Norms. *Journal of International Law & International Relations*, 5, 1–16.

Wiener, Jonathan B. (2007). Precaution. In Daniel Bodansky, Jutta Brunnée, & Ellen Hey (Eds.), *The Oxford Handbook of International Environmental Law*. Oxford: Oxford University Press, pp. 597–612.

Wight, Martin (1977). *Systems of States*. Leicester: Leicester University Press.

(1991). *International Theory: The Three Traditions*. London: Leicester University Press.

Willetts, Peter (1996). From Stockholm to Rio and Beyond: The Impact of the Environmental Movement on the United Nations Consultative Arrangements for NGOs. *Review of International Studies*, 22(1), 57–80.

Williams, John (2005). Pluralism, Solidarism and the Emergence of World Society in English School Theory. *International Relations*, 19(1), 19–38.

(2011). Structure, Norms and Normative Theory in a Re-defined English School: Accepting Buzan's Challenge. *Review of International Studies*, 37 (03), 1235–53.

(2014). The International Society – World Society Distinction. In Cornelia Navari & Daniel M. Green (Eds.), *Guide to the English School in International Studies*. Chichester: Wiley & Sons, pp. 127–42.

(2015). *Ethics, Diversity, and World Politics: Saving Pluralism From Itself?* Oxford: Oxford University Press.

Williams, Michael C. (2003). Words, Images, Enemies: Securitization and International Politics. *International Studies Quarterly*, 47(4), 511–31.

Wilson, Jeffrey K. (2012). *The German forest: Nature, identity, and the contestation of a national symbol, 1871-1914* (Vol. 11). Toronto: University of Toronto Press.

Wilson, Peter (2012). The English School Meets the Chicago School: The Case for a Grounded Theory of International Institutions. *International Studies Review*, 14(4), 567–90.

Wing, John T. (2012). Keeping Spain Afloat: State Forestry and Imperial Defense in the Sixteenth Century. *Environmental History*, 17(1,: 116–45.

Wirth, John D. (1996). The Trail Smelter Dispute: Canadians and Americans Confront Transboundary Pollution, 1927–41. *Environmental History*, 1(2), 34–51.

Wissenburg, Marcel (2013). *Green Liberalism: The Free and the Green Society*. London: Routledge.

Wöbse, Anna-Katharina (2008). Oil on Troubled Waters? Environmental Diplomacy in the League of Nations. *Diplomatic History*, 32(4), 519–37.

(2011). 'The World after All Was One': The International Environmental Network of UNESCO and IUPN, 1945–1950. *Contemporary European History*, 20(3), 331–48.

(2012). *Weltnaturschutz: Umweltdiplomatie in Völkerbund und Vereinten Nationen 1920-1950*. Frankfurt/M.: Campus Verlag.

Wohl, Anthony S. (1983). *Endangered Lives: Public Health in Victorian Britain*. London: JM Dent and Sons Ltd.

Wolfrum, Rüdiger (1983). The Principle of the Common Heritage of Mankind. *Heidelberg Journal of International Law*, 43, 312–37.
World Commission on Environment and Development (1987). *Our Common Future*. Oxford: Oxford University Press.
Worster, Donald (1994). *Nature's Economy: A History of Ecological Ideas*. 2nd ed. Cambridge: Cambridge University Press.
Wright, Christopher (2012). Global Banks, the Environment, and Human Rights: The Impact of the Equator Principles on Lending Policies and Practices. *Global Environmental Politics*, 12(1), 56–77.
(2013). Global Finance and the Environment. In Robert Falkner (Ed.), *The Handbook of Global Climate and Environment Policy*. Cheltenham: John Wiley & Sons Ltd, pp. 428–45.
Wulf, Andrea (2015). *The Invention of Nature: The Adventures of Alexander von Humboldt, the Lost Hero of Science*. London: John Murray.
Wurzel, Rüdiger, & Connelly, James (Eds.) (2010). *The European Union as a Leader in International Climate Change Politics*. London: Routledge.
Wynn, Graeme (1979). Pioneers, Politicians and the Conservation of Forests in Early New Zealand. *Journal of Historical Geography*, 5(2), 171–88.
Zahar, Alexander (2015). *International Climate Change Law and State Compliance*. London: Routledge.
Zartman, I William, & Berman, Maureen R. (1982). *The Practical Negotiator*. New Haven: Yale University Press.
Zedan, Hamdallah (2002). The Road to the Biosafety Protocol. In Christoph Bail, Robert Falkner, & Helen Marquard (Eds.), *The Cartagena Protocol on Biosafety: Reconciling Trade in Biotechnology with Environment and Development?* London: Earthscan, pp. 23–33.
Zhang, Yongjin (2003). The 'English School' in China: A Travelogue of Ideas and their Diffusion. *European Journal of International Relations*, 9(1), 87–114.
(2016). China and Liberal Hierarchies in Global International Society: Power and Negotiation for Normative Change. *International Affairs*, 92(4), 795–816.
Zito, Anthony R. (2005). The European Union as an Environmental Leader in a Global Environment. *Globalizations*, 2(3), 363–75.

Index

152 *Cecelia Lynch*
Wrestling with God
Ethical Precarity in Christianity and International Relations
151 *Brent J. Steele*
Restraint in International Politics
150 *Emanuel Adler*
World Ordering
A Social Theory of Cognitive Evolution
149 *Brian C. Rathbun*
Reasoning of State
Realists and Romantics in International Relations
148 *Silviya Lechner and Mervyn Frost*
Practice Theory and International Relations
147 *Bentley Allan*
Scientific Cosmology and International Orders
146 *Peter J. Katzenstein and Lucia A. Seybert* (eds.)
Protean Power
Exploring the Uncertain and Unexpected in World Politics
145 *Catherine Lu*
Justice and Reconciliation in World Politics
144 *Ayşe Zarakol* (ed.)
Hierarchies in World Politics
143 *Lisbeth Zimmermann*
Global Norms with a Local Face
Rule-of-Law Promotion and Norm-Translation
142 *Alexandre* Debs and *Nuno P. Monteiro*
Nuclear Politics
The Strategic Causes of Proliferation
141 *Mathias Albert*
A theory of world politics
140 *Emma Hutchison*
Affective communities in world politics
Collective emotions after trauma
139 *Patricia Owens*
Economy of force
Counterinsurgency and the historical rise of the social
138 *Ronald R. Krebs*
Narrative and the making of US national security
137 *Andrew Phillips and J. C. Sharman*
International order in diversity
War, trade and rule in the Indian Ocean

136 *Ole Jacob Sending, Vincent Pouliot and Iver B. Neumann (eds.)*
Diplomacy and the making of world politics

135 *Barry Buzan and George Lawson*
The global transformation
History, modernity and the making of international relations

134 *Heather Elko McKibben*
State strategies in international bargaining
Play by the rules or change them?

133 *Janina Dill*
Legitimate targets?
Social construction, international law, and US bombing

132 *Nuno P. Monteiro*
Theory of unipolar politics

131 *Jonathan D. Caverley*
Democratic militarism
Voting, wealth, and war

130 *David Jason Karp*
Responsibility for human rights
Transnational corporations in imperfect states

129 *Friedrich Kratochwil*
The status of law in world society
Meditations on the role and rule of law

128 *Michael G. Findley, Daniel L. Nielson and J. C. Sharman*
Global shell games
Experiments in transnational relations, crime, and terrorism

127 *Jordan Branch*
The cartographic state
Maps, territory, and the origins of sovereignty

126 *Thomas Risse, Stephen C. Ropp and Kathryn Sikkink (eds.)*
The persistent power of human rights
From commitment to compliance

125 *K. M. Fierke*
Political self-sacrifice
Agency, body and emotion in international relations

124 *Stefano Guzzini*
The return of geopolitics in Europe?
Social mechanisms and foreign policy identity crises

123 *Bear F. Braumoeller*
The great powers and the international system
Systemic theory in empirical perspective

122 *Jonathan Joseph*
The social in the global
Social theory, governmentality and global politics

121 *Brian C. Rathbun*
Trust in international cooperation
International security institutions, domestic politics and American multilateralism

120 *A. Maurits van der Veen*
Ideas, interests and foreign aid

119 *Emanuel Adler and Pouliot Vincent (eds.)*
International practices

118 *Ayşe Zarakol*
After defeat
How the East learned to live with the West

117 *Andrew Phillips*
War, religion and empire
The transformation of international orders

116 *Joshua Busby*
Moral movements and foreign policy

115 *Séverine Autesserre*
The trouble with the Congo
Local violence and the failure of international peacebuilding

114 *Deborah D. Avant, Martha Finnemore and Susan K. Sell* (eds.)
Who governs the globe?

113 *Vincent Pouliot*
International security in practice
The politics of NATO-Russia diplomacy

112 *Columba Peoples*
Justifying ballistic missile defence
Technology, security and culture

111 *Paul Sharp*
Diplomatic theory of international relations

110 *John A. Vasquez*
The war puzzle revisited

109 *Rodney Bruce Hall*
Central banking as global governance
Constructing financial credibility

108 *Milja Kurki*
Causation in international relations
Reclaiming causal analysis

107 *Richard M. Price*
Moral limit and possibility in world politics

106 *Emma Haddad*
The refugee in international society
Between sovereigns

105 *Ken Booth*
Theory of world security

104 *Benjamin Miller*
States, nations and the great powers
The sources of regional war and peace

103 *Beate Jahn (ed.)*
Classical theory in international relations

102 *Andrew Linklater and Hidemi Suganami*
The English School of international relations
A contemporary reassessment

101 *Colin Wight*
Agents, structures and international relations
Politics as ontology

100 *Michael C. Williams*
The realist tradition and the limits of international relations

99 *Ivan Arreguín-Toft*
How the weak win wars
A theory of asymmetric conflict

98 *Michael Barnett and Raymond Duvall (eds.)*
Power in global governance

97 *Yale H.* Ferguson *and Richard W. Mansbach*
Remapping global politics
History's revenge and future shock

96 *Christian Reus-Smit (ed.)*
The politics of international law

95 *Barry Buzan*
From international to world society?
English School theory and the social structure of globalisation

94 *K. J. Holsti*
Taming the sovereigns
Institutional change in international politics

93 *Bruce Cronin*
Institutions for the common good
International protection regimes in international security

92 *Paul Keal*
European conquest and the rights of indigenous peoples
The moral backwardness of international society

91 *Barry* Buzan *and Ole Wæver*
Regions and powers
The structure of international security

90 *A. Claire Cutler*
Private power and global authority
Transnational merchant law in the global political economy

89 *Patrick M. Morgan*
Deterrence now

88 *Susan Sell*
Private power, public law
The globalization of intellectual property rights

87 *Nina Tannenwald*
The nuclear taboo
The United States and the non-use of nuclear weapons since 1945

86 *Linda Weiss*
States in the global economy
Bringing domestic institutions back in

85 *Rodney Bruce Hall and Thomas J. Biersteker (eds.)*
The emergence of private authority in global governance

84 *Heather Rae*
State identities and the homogenisation of peoples

83 *Maja Zehfuss*
Constructivism in international relations
The politics of reality

82 *Paul K. Ruth and Todd Allee*
The democratic peace and territorial conflict in the twentieth century

81 *Neta C. Crawford*
Argument and change in world politics
Ethics, decolonization and humanitarian intervention

80 *Douglas Lemke*
Regions of war and peace

79 *Richard Shapcott*
Justice, community and dialogue in international relations

78 *Phil Steinberg*
The social construction of the ocean

77 *Christine Sylvester*
Feminist international relations
An unfinished journey

76 *Kenneth A. Schultz*
Democracy and coercive diplomacy

75 *David Houghton*
US foreign policy and the Iran hostage crisis

74 *Cecilia Albin*
Justice and fairness in international negotiation

73 *Martin Shaw*
Theory of the global state
Globality as an unfinished revolution

72 *Frank C. Zagare and D. Marc Kilgour*
Perfect deterrence

71 *Robert O'Brien, Anne Marie Goetz, Jan Aart Scholte and Marc Williams*
Contesting global governance
Multilateral economic institutions and global social movements

70 *Roland Bleiker*
Popular dissent, human agency and global politics
69 *Bill McSweeney*
Security, identity and interests
A sociology of international relations
68 *Molly Cochran*
Normative theory in international relations
A pragmatic approach
67 *Alexander Wendt*
Social theory of international politics
66 *Thomas Risse, Stephen C. Ropp and Kathryn Sikkink (eds.)*
The power of human rights
International norms and domestic change
65 *Daniel W. Drezner*
The sanctions paradox
Economic statecraft and international relations
64 *Viva Ona Bartkus*
The dynamic of secession
63 *John A. Vasquez*
The power of power politics
From classical realism to neotraditionalism
62 *Emanuel Adler* and *Michael Barnett (eds.)*
Security communities
61 *Charles Jones*
E. H. Carr and international relations
A duty to lie
60 *Jeffrey W. Knopf*
Domestic society and international cooperation
The impact of protest on US arms control policy
59 *Nicholas Greenwood Onuf*
The republican legacy in international thought
58 *Daniel S. Geller and J. David Singer*
Nations at war
A scientific study of international conflict
57 *Randall D. Germain*
The international organization of credit
States and global finance in the world economy
56 *N. Piers Ludlow*
Dealing with Britain
The Six and the first UK application to the EEC
55 *Andreas Hasenclever, Peter Mayer and Volker Rittberger*
Theories of international regimes
54 *Miranda A. Schreurs and Elizabeth C. Economy (eds.)*
The internationalization of environmental protection

53 *James N. Rosenau*
Along the domestic-foreign frontier
Exploring governance in a turbulent world

52 *John M. Hobson*
The wealth of states
A comparative sociology of international economic and political change

51 *Kalevi J. Holsti*
The state, war, and the state of war

50 *Christopher Clapham*
Africa and the international system
The politics of state survival

49 *Susan Strange*
The retreat of the state
The diffusion of power in the world economy

48 *William I. Robinson*
Promoting polyarchy
Globalization, US intervention, and hegemony

47 *Roger Spegele*
Political realism in international theory

46 *Thomas J. Biersteker and Cynthia Weber (eds.)*
State sovereignty as social construct

45 *Mervyn Frost*
Ethics in international relations
A constitutive theory

44 *Mark W. Zacher with Brent A. Sutton*
Governing global networks
International regimes for transportation and communications

43 *Mark Neufeld*
The restructuring of international relations theory

42 *Thomas Risse-Kappen (ed.)*
Bringing transnational relations back in
Non-state actors, domestic structures and international institutions

41 *Hayward R. Alker*
Rediscoveries and reformulations
Humanistic methodologies for international studies

40 *Robert W. Cox with* Timothy J. Sinclair
Approaches to world order

39 *Jens Bartelson*
A genealogy of sovereignty

38 *Mark Rupert*
Producing hegemony
The politics of mass production and American global power

37 *Cynthia Weber*
Simulating sovereignty
Intervention, the state and symbolic exchange

36 *Gary Goertz*
Contexts of international politics
35 *James L. Richardson*
Crisis diplomacy
The Great Powers since the mid-nineteenth century
34 *Bradley S. Klein*
Strategic studies and world order
The global politics of deterrence
33 *T. V. Paul*
Asymmetric conflicts
War initiation by weaker powers
32 *Christine Sylvester*
Feminist theory and international relations in a postmodern era
31 *Peter J. Schraeder*
US foreign policy toward Africa
Incrementalism, crisis and change
30 *Graham Spinardi*
From Polaris to Trident
The development of US Fleet Ballistic Missile technology
29 *David A. Welch*
Justice and the genesis of war
28 *Russell J. Leng*
Interstate crisis behavior, 1816-1980
Realism versus reciprocity
27 *John A. Vasquez*
The war puzzle
26 *Stephen Gill (ed.)*
Gramsci, historical materialism and international relations
25 *Mike Bowker and Robin Brown (eds.)*
From cold war to collapse
Theory and world politics in the 1980s
24 *R. B. J. Walker*
Inside/outside
International relations as political theory
23 *Edward Reiss*
The strategic defense initiative
22 *Keith Krause*
Arms and the state
Patterns of military production and trade
21 *Roger Buckley*
US-Japan alliance diplomacy 1945-1990
20 *James N. Rosenau and Ernst-Otto Czempiel* (eds.)
Governance without government
Order and change in world politics
19 *Michael Nicholson*
Rationality and the analysis of international conflict

18 *John Stopford and Susan Strange*
Rival states, rival firms
Competition for world market shares

17 *Terry Nardin and David R. Mapel (eds.)*
Traditions of international ethics

16 *Charles F. Doran*
Systems in crisis
New imperatives of high politics at century's end

15 *Deon Geldenhuys*
Isolated states
A comparative analysis

14 *Kalevi J. Holsti*
Peace and war
Armed conflicts and international order 1648-1989

13 *Saki Dockrill*
Britain's policy for West German rearmament 1950-1955

12 *Robert H. Jackson*
Quasi-states
Sovereignty, international relations and the third world

11 *James Barber* and *John Barratt*
South Africa's foreign policy
The search for status and security 1945-1988

10 *James Mayall*
Nationalism and international society

9 *William Bloom*
Personal identity, national identity and international relations

8 *Zeev Maoz*
National choices and international processes

7 *Ian Clark*
The hierarchy of states
Reform and resistance in the international order

6 *Hidemi Suganami*
The domestic analogy and world order proposals

5 *Stephen Gill*
American hegemony and the Trilateral Commission

4 *Michael C. Pugh*
The ANZUS crisis, nuclear visiting and deterrence

3 *Michael Nicholson*
Formal theories in international relations

2 *Friedrich V. Kratochwil*
Rules, norms, and decisions
On the conditions of practical and legal reasoning in international relations and domestic affairs

1 *Myles L. C. Robertson*
Soviet policy towards Japan
An analysis of trends in the 1970s and 1980s

CPSIA information can be obtained
at www.ICGtesting.com
Printed in the USA
LVHW010534010921
696659LV00003B/226